Identifying and Regulating
CARCINOGENS

Office of Technology Assessment Task Force

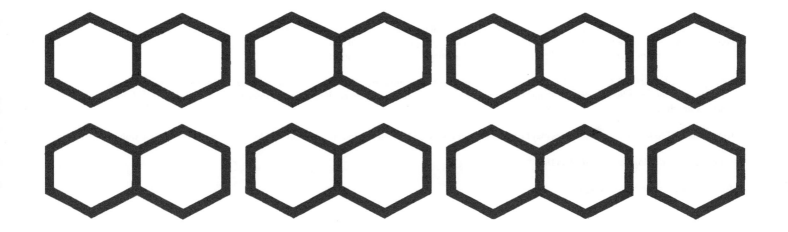

CRC Press
Taylor & Francis Group
Boca Raton London New York

CRC Press is an imprint of the
Taylor & Francis Group, an **informa** business

First published 1988 by CRC Press
Taylor & Francis Group
6000 Broken Sound Parkway NW, Suite 300
Boca Raton, FL 33487-2742

Reissued 2018 by CRC Press

Library of Congress Cataloging-in-Publication Data

Identifying and regulating carcinogens.

Bibliography: p.
1. Chemicals—Law and legislation—United States. 2. Carcinogens—Safety regulations—United States. 3. Chemicals—United States—Testing.   4. Carcinogenicity testing.   5. Health risk assessment—United States.

KF3958.I34   1988   344.73'0424   88-554
ISBN 0-87371-152-1   347.304424

A Library of Congress record exists under LC control number: 88000554

Publisher's Note
The publisher has gone to great lengths to ensure the quality of this reprint but points out that some imperfections in the original copies may be apparent.

Disclaimer
The publisher has made every effort to trace copyright holders and welcomes correspondence from those they have been unable to contact.

ISBN 13: 978-1-315-89424-9 (hbk)
ISBN 13: 978-1-351-07334-9 (ebk)

Visit the Taylor & Francis Web site at http://www.taylorandfrancis.com and the
CRC Press Web site at http://www.crcpress.com

Foreword

Interest in reducing and preventing exposures to carcinogenic chemicals has been expressed by Congress in statutes authorizing the regulation of food additives, drugs, consumer products, occupational exposures, air and water pollutants, drinking water contaminants, pesticides, toxic chemicals, and hazardous wastes.

This study was requested by the House Government Operations Committee and its Subcommittee on Intergovernmental Relations and Human Resources. In it, OTA describes policies issued by Federal agencies concerning the identification, assessment, and regulation of carcinogenic chemicals; the chemicals that have been regulated because of carcinogenic risk; the Federal Government's carcinogenicity testing program; and the results of OTA's analysis of the extent of agency action on chemicals determined to be carcinogenic.

Identifying and Regulating Carcinogens covers 16 different Federal agencies, programs, and activities: the Occupational Safety and Health Administration, the Mine Safety and Health Administration, the National Institute for Occupational Safety and Health, the Consumer Product Safety Commission, and the Food and Drug Administration, including its activities on food and color additives, human drugs, and animal drugs. For the Environmental Protection Agency (EPA), this work discusses the Carcinogen Assessment Group, as well as EPA standard-setting on carcinogens under the Clean Air Act, Clean Water Act, Safe Drinking Water Act, Toxic Substances Control Act, the Federal Insecticide, Fungicide, and Rodenticide Act, Resource Conservation and Recovery Act, and the Comprehensive Environmental Response, Compensation, and Liability Act. Finally, the activity of the National Toxicology Program is described.

A large number of Federal agency personnel and others assisted in providing information for this work. We thank them for their assistance. Key OTA staff involved with this book were Karl Kronebusch, Sylvia Tognetti, and Neil Holtzman. Carl Cranor worked on this project while serving as a Congressional Fellow of the American Philosophical Association.

Office of Technology Assessment

The Office of Technology Assessment (OTA) was created in 1972 as an analytical arm of Congress. OTA's basic function is to help legislative policymakers anticipate and plan for the consequences of technological changes and to examine the many ways, expected and unexpected, in which technology affects people's lives. The assessment of technology calls for exploration of the physical, biological, economic, social, and political impacts that can result from applications of scientific knowledge. OTA provides Congress with independent and timely information about the potential effects—both beneficial and harmful—of technological applications.

Requests for studies are made by chairmen of standing committees of the House of Representatives or Senate; by the Technology Assessment Board, the governing board of OTA; or by the Director of OTA in consultation with the Board.

The Technology Assessment Board is composed of five members of the House, five members of the Senate, and the OTA Director, who is a non-voting member.

OTA has studies under way in nine program areas: energy and materials; industry, technology, and employment; international security and commerce; biological applications; food and renewable resources; health; communication and information technologies; oceans and environment; and science, education, and transportation.

Contents

Preface

This volume, *Identifying and Regulating Carcinogens*, is an excellent resource describing the various Federal Agency programs, guidelines, and requirements regarding potential exposure to carcinogens in food and water, the workplace, the environment, drugs, and other consumer products. It covers 17 different programs, activities, and Federal Agencies.

The reader is provided insight into the historic background and rationale behind current Federal policy relating to public health and potential carcinogen hazards. The necessary differentiation among hazard identification, risk characterization, and risk management is clearly described. *Identifying and Regulating Carcinogens* provides a single, concise focal point for the scientific, technical, and lay reader seeking to understand the current Federal position on carcinogen identification, testing, assessment, and regulatory requirements.

Chapter 1
Introduction and Summary

CONTENTS

Figures

Introduction and Summary

INTRODUCTION

Over the years, laws have been enacted to protect the health of Americans, with particular emphasis on protection against cancer. By and large, these laws provide for reducing or eliminating exposures to external chemical carcinogens with which people come into contact—in the food supply; in drinking water; in pharmaceutical drugs and other consumer products; in work environments; in ambient air, water, and soil. Most cases of cancer, however, are not caused by these types of carcinogenic exposures.

Instead, according to the best interpretation of the evidence currently available, most result from "lifestyle" factors, of which the details are only slowly becoming clear. One—tobacco smoking—stands out the clearest of all, and alone is the cause of more than one-third of all deaths from cancer each year in the United States. The more poorly defined lifestyle factors include such items as overall dietary balance and aspects of sexual behavior; others, slightly better defined, include exposure to sunlight (see OTA 1981 for a fuller discussion of causes of cancer). In addition to lifestyle factors, viruses are potentially great, but currently unquantifiable, contributors to the overall cancer burden. Nevertheless, those carcinogenic chemicals that *can* be identified specifically and can be controlled are important for those very reasons: they are avoidable. And often, unlike cigarette smoking, exposure to them is involuntary. Furthermore, the potential for introducing new, potent, carcinogens is very real.

For the laws addressing chemical carcinogens to be effective, there must be means of identifying substances that have caused, or would cause, human beings to get cancer. Once the substances have been identified, regulatory decisions can be made about whether and how to control exposures. Both the process for finding out which substances already in the human environment are causing cancer in the population (through epidemiologic studies) and the process for predicting carcinogenicity in humans *before* people are exposed (by testing in the laboratory and in experimental animals) are imperfect, and interpretation of the results of such studies is contentious. While efforts to develop improved methods for identifying carcinogens continue, current and past regulatory decisions have, of necessity, embodied many untested and some untestable assumptions.

This OTA background paper responds to a request from the House Committee on Government Operations and its Subcommittee on Intergovernmental Relations and Human Resources to examine Federal activity in testing chemicals for carcinogenicity and the use of test results by regulatory agencies.

In this background paper, OTA addresses the following specific questions:

- What policies for regulating carcinogens have Federal agencies adopted? What guidance do these policies provide about identifying, assessing, and regulating chemical carcinogens? What kind of evidence, human or animal, do the agencies require to identify a chemical qualitatively as carcinogenic? How do the agencies intend to conduct quantitative risk assessments?
- What chemicals have actually been regulated? What evidence provided the basis for these regulations? How long does the regulatory process take?
- How is Federal carcinogenicity testing organized? How are chemicals chosen for such testing? After the chemicals are tested, are the chemicals that test positive regulated? Have agencies regulated the chemicals listed in the Federal Government's *Annual Report on Carcinogens*?

Chapter 2 of this background paper compares the formal Federal policies for identifying and assessing the risks from carcinogenic chemicals. Chapter 3 lists the carcinogenic chemicals that have been regulated by each Federal regulatory agency. Federal agencies with the greatest roles in regulating chemical carcinogens are the Food and Drug Administration (FDA) for foods, cosmetics, and human and animal drugs; the Occupational Safety and Health Administration (OSHA) for worker exposure in most industries; the Mine Safety and Health Administration (MSHA) for worker exposure in mines; the Consumer Product Safety Commission (CPSC) for consumer products; and the Environmental Protection Agency (EPA). EPA is charged with regulating air pollutants under the Clean Air Act (CAA); water pollutants under Clean Water Act (CWA); drinking water contaminants under the Safe Drinking Water Act (SDWA); pesticides under the Federal Insecticide, Fungicide, and Rodenticide Act (FIFRA); toxic chemicals under the Toxic Substances Control Act (TCSA); and hazardous wastes under the Resource Conservation and Recovery Act (RCRA) and the Comprehensive Environmental Response, Compensation, and Liability Act (CERCLA).

Chapter 4 describes the National Toxicology Program (NTP), the home of the Federal testing program, and its carcinogenicity testing. Chapter 5 examines the regulatory responses to positive results from Federal carcinogenicity bioassays and to the chemicals listed in the *Annual Report*

on Carcinogens. Appendix A describes the Federal statutes that have been most important in regulating carcinogenic chemicals.

The scope of this background paper is limited to "chemicals" that have been tested, listed, or regulated by the Federal Government for carcinogenicity. The term "chemical" is used broadly here to encompass substances, mixtures, groups of substances, and exposures. This background paper does not examine the regulation of radiation sources licensed by the Nuclear Regulatory Commission, electronic radiation (including, for example, x-ray machines, which are regulated by FDA), ultraviolet radiation, alcohol, and tobacco. Depending on statutory mandate, Federal regulatory decisions can be based on such factors as control technologies and costs, in addition to risks. Agency procedures for developing information on these factors will not be discussed in this background paper. Moreover, while very important, other related efforts not covered here are those of industry and the private sector to test chemicals for carcinogenicity and implement voluntary controls to reduce exposures to carcinogens.

SUMMARY

Agency Policies

Over the last decade, several Federal agencies have issued guidelines and policies detailing how they intend to identify, evaluate, and regulate carcinogens. These guidelines encompass the design of animal carcinogenicity bioassays, the interpretation of data from human and animal studies, and the assumptions that should or will be made when assessing human risk from such studies.

The assumptions in these documents represent scientific views and policy judgments about carcinogen assessment. Some assumptions are made because, though appropriate data might be obtained with current techniques, the data are simply not available in a particular case. Other more general assumptions take the place of experimental evidence that may be developed with further research. Finally, some assumptions are employed because of ethical considerations and the inher-

ent limits of experimental methods. The use of assumptions, the frequent absence of data, the potential economic implications of government regulation, and underlying political disputes about the desirability of regulation, combine to make the assessment of carcinogenicity and the development of corresponding regulations subjects of intense debates.

It is now common to distinguish between risk assessment and risk management: risk assessment characterizes the adverse health effects of human exposures to environmental hazards; risk management is the choosing of regulatory options. Both risk assessment and risk management incorporate policy choices and reflect the values of the risk assessors and managers. Some agencies have attempted to establish separate staffs for the two tasks, but this separation does not eliminate the need to make policy choices about the assumptions used in risk assessments.

The values and policy preferences of decision-makers, risk assessors, and representatives of industry, labor unions, environmental organizations, and public interest groups often differ. Scientists disagree about the nature of scientific evidence. These differences explain some of the past controversies over the regulation of specific carcinogenic chemicals and the development of agency policies.

In 1983, a committee of the National Research Council recommended the development of uniform guidelines for conducting risk assessments. The committee described several advantages and disadvantages of such guidelines. They have the advantages of promoting quality control, consistency, predictability, public understanding, administrative efficiency, and improvements in methods. In addition, guidelines serve an important role within agencies in training new staff in agency practices. The potential disadvantages of such guidelines include oversimplification, inappropriate mixing of scientific knowledge with risk assessment policy, misallocation of agency resources to the task of developing guidelines, and insensitivity to scientific developments. Some have hoped that policies for assessing and regulating carcinogens would speed regulatory activity. Others have tried to use such policies to change the direction of risk assessment and regulation.

While much effort has been devoted to developing guidelines and policies for carcinogen assessment and regulation, it is not clear how much effect they have actually had. They do provide points of reference for discussions on particular regulatory issues. Nevertheless, while there are important disagreements among regulatory agencies, industries, and other groups on general issues, many disagreements concern interpretations of evidence in particular cases. Everyone may agree, for example, that animal data can be used to identify potential human carcinogens, yet they may disagree about the applicability of results from particular animal experiments in assessing particular chemicals, especially commercially important ones. Adoption of general guidelines cannot resolve these specific disputes.

Agency policies and guidelines have varied considerably in their flexibility, formality, and comprehensiveness. They have also evolved, generally becoming more complex and detailed.

This background paper considers two distinct, but related, types of guidelines: agency requirements for animal carcinogenicity studies; and agency policies on identifying, assessing, and regulating carcinogens.

Required Animal Testing

FDA and EPA have required industry to conduct carcinogenicity testing of food and color additives, animal drugs, animals, human drugs, pesticides, and toxic substances.

FDA requires carcinogenicity testing for a proposed food additive only if it falls into certain chemical categories and its expected concentration in food exceeds specified levels. For an animal drug, testing may be required depending on the expected extent of its use in animals, the levels of drug residues, and the potential toxicity of the drug as determined from chemical structure, short-term tests, and other data. FDA requires carcinogenicity testing for new human drugs that are expected to have chronic or widespread use, although this requirement has not been applied to drugs marketed prior to 1968. For some of these older drugs, which are used widely today, FDA has requested studies from NTP rather than from drug manufacturers.

EPA may require animal carcinogenicity studies of pesticides when they generate some toxicologic concern, when they will be used on food, or when their use will result in significant human exposure. Considerable delays have occurred in requiring test data on pesticides marketed prior to 1972. Under TSCA, EPA may require testing for new chemicals or for existing chemicals.

Guidelines for Testing Protocols

OTA compared the bioassay study designs for suspected carcinogens that are specified by several Federal agencies. FDA and EPA have issued guidelines for the design of toxicologic studies, including those of carcinogenicity. FDA has relied on nonregulatory guidelines, such as its "Red Book," for studies required of new food and color additives. For human drugs, a joint workshop sponsored by FDA and the Pharmaceutical Manufacturers' Association (PMA) discussed the design of studies. Although FDA decided not to issue guidelines under its own name, the guidelines were published by PMA. EPA issued as regulations separate testing guidelines for pesticides and for toxic substances. The National Cancer Institute (NCI) and NTP test guidelines were also considered in this OTA comparison.

Federal agency guidelines are generally consistent about major features of the study design. They specify testing in two animal species, which in practice are usually rats and mice. The two oldest guidelines (those of NCI and PMA) require at least two dose groups in addition to a control group. All other guidelines suggest the use of three dose groups and a control group. The guidelines agree that to maximize the sensitivity of a study in detecting carcinogenic effects, the highest dose in the study must be set as high as possible without shortening the animals' lives because of non-carcinogenic toxic effects.

Risk Assessment Policies

OTA also compared Federal agencies' policies on identifying and assessing carcinogens. These policies were issued under a variety of circumstances and are organized in different ways. In some cases, the policies are relatively informal statements of current scientific understanding about how carcinogens might be identified. In other cases, they constitute formally adopted regulations, specifying how an agency will identify carcinogens and limiting the kinds of arguments and evidence to be considered in specific regulatory proceedings. In between these two extremes, some documents outline an agency's standard procedures and discuss problematic areas of interpretation, including the inference assumptions that the agency will use.

Several agency policies have taken a regulatory form, for example, OSHA's 1980 policy. OSHA intended to collect evidence and testimony on "generic" issues in carcinogen identification and

regulation, make decisions on these issues, and then rely on these decisions and presumptions in future proceedings. The policy might be termed a "presumption-rebuttal" approach, providing strong presumptions and limited room for rebuttal. The framers of this policy hoped it would limit debate in subsequent regulatory proceedings and thereby speed carcinogen regulation. That hope has not been realized. Two carcinogens with occupational exposures, ethylene oxide and asbestos, have been regulated since the publication of OSHA's policy.

CPSC attempted to adopt carcinogen assessment guidelines in 1978. CPSC was sued, and the guidelines were struck down by a reviewing court. Subsequent to this decision, CPSC formally withdrew its policy.

FDA has been working on a regulatory definition of allowable animal drug residues in human food since 1973. This definition specifies how sensitive an analytic technique must be, hence the definition is called "sensitivity of method" (SOM). It was first proposed in 1973, made final in 1977, challenged in court and sent back to FDA, reproposed in 1979, then proposed for a third time in 1985. The final rule has still not been issued.

Other agency policies provide guidelines for conducting risk assessments. EPA's 1976 "interim" guidelines and its 1986 carcinogen risk assessment guidelines are examples of this approach, which discusses scientific issues, sets forth flexible assumptions, and specifies an analysis based on the weight of the evidence.

A 1979 Interagency Regulatory Liaison Group (IRLG) policy and a 1985 guideline issued by the White House Office of Science and Technology Policy (OSTP) both discussed current knowledge of carcinogenesis and related risk assessment techniques. These documents are important because they represent the results of extensive discussions among scientists from many agencies. One goal of these discussions was to develop a consensus among the agencies on these issues.

Not all agency programs have adopted policies on risk assessment. For example, FDA does not have a formal risk assessment policy on food and color additives. However, FDA's Center for Food

Safety and Applied Nutrition has established a formal committee for considering evidence on the carcinogenicity of food and color additives. FDA has published a policy on regulating additives with carcinogenic impurities and has developed a policy incorporating a de minimis approach to regulating the safety of food and color additives. For evaluating the safety of human drugs, FDA requires different kinds of tests depending on the expected duration of human use of the drug, but it has never issued guidelines for evaluating or assessing animal carcinogenicity test results. It does provide guidance, however, on preparing data for statistical analysis.

Carcinogen Assessment Policies

A National Research Council committee divided risk assessment into four distinct parts: hazard identification, dose-response assessment, exposure assessment, and risk characterization (137). Hazard identification is the qualitative identification of a substance as a human or animal carcinogen. In dose-response assessment, the relationship between the level of exposure or the dose and the incidence of disease is described. The two most important aspects of the second step are extrapolating from information on incidence at high doses to predict incidence at lower doses and, in the case of risk assessments based on animal data, converting animal doses into equivalent human doses. Exposure assessment estimates the frequency, duration, and intensity of human exposures to the agent in question. Finally, risk characterization relies on information from both dose-response and exposure assessments to estimate the expected risk, as well as to explain the nature of the risk and any uncertainties in assessing it.

Figure 1-1 illustrates these steps, which eventually lead to information useful for risk management decisions. Each step involves some uncertainty, owing either to inadequate data on the particular agent or to uncertainty about its mechanisms of toxicity.

Hazard Identification

In many situations of regulatory interest, there are few toxicity data of any sort. When data are available, the agencies value epidemiologic studies as the most conclusive evidence for human carcinogenicity, presume that substances found to be carcinogenic in animals in long-term bioassays present carcinogenic hazards to humans, and use short-term test results as supportive information. Analyses of structure-activity relationships (analyses based on the structural similarity of a substance to other known carcinogens) are used mostly when there are no other data (e.g., to identify new chemicals that should have additional testing prior to large-scale manufacture).

All Federal policies accept the use of animal data in predicting human effects. While it is not known with certainty that all animal carcinogens are also human carcinogens, most well-studied human carcinogens show some evidence of carcinogenicity in animals.

While agencies accept animal data, determining exactly what evidence demonstrates that a substance is an animal carcinogen is more complex. Generally, the agencies accept data derived from use of the maximum tolerated dose, and then use the increased incidence of malignant or benign tumors to demonstrate carcinogenicity. Policies usually state that positive results in animals outweigh negative epidemiologic results, and that positive results in one species outweigh negative results in another.

Dose-Response Assessment

Prior to 1970, there was considerable doubt about the utility of quantitative assessments. During the 1970s and 1980s, the agencies began using these assessments for carcinogens. In 1973, FDA specified the use of quantitative risk assessment in the proposed SOM for evaluating animal drugs. In 1978 and 1979, FDA conducted risk assessments for the environmental contaminants aflatoxins and polychlorinated biphenyls (PCBs). FDA first used risk assessment to determine the risk of carcinogenic impurities of color additives in 1982 and of food and color additives themselves in 1985 and 1986. The first EPA risk assessment, in 1975, concerned vinyl chloride. In 1976, EPA established its Carcinogen Assessment Group and published its "interim" guidelines on risk assessment. CPSC's first use of risk assessment came with its evalua-

Figure 1-1.—Elements of Risk Assessment and Risk Management

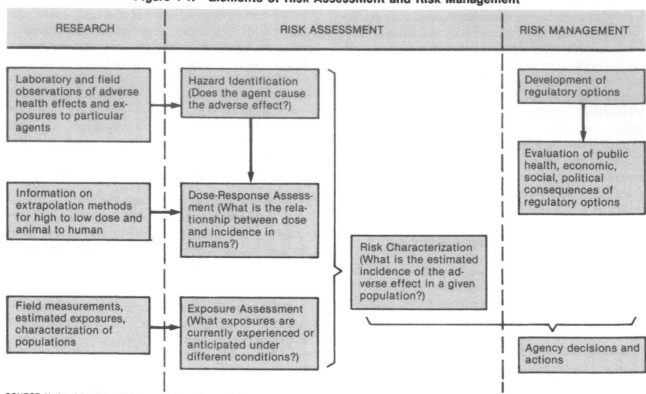

SOURCE: National Academy of Sciences/National Research Council, *Risk Assessment in the Federal Government: Managing the Process* (Washington, DC: National Academy Press, 1983).

tion of tris(2,3-dibromopropyl)phosphate (Tris) in 1977. While OSHA had first prepared a quantitative risk assessment in 1976 for worker exposure to coke oven emissions, it resisted calls for increased use of these assessments until the Supreme Court's 1980 decision on the benzene standard. Today, although there are still many uncertainties associated with quantitative risk assessment, all of these agencies use it.

The agencies all assume that human risk estimates can be derived from animal data, that carcinogenic chemicals do not have no-effects thresholds, and that risk estimates should be based on results from the most sensitive animal species. All the agencies use mathematical models that assume low-dose linearity for extrapolating from the doses tested in the animal experiment to the doses of regulatory interest, although they differ on the mathematical technique to use, whether the focus should be on the "upper confidence limit" or the "maximum likelihood estimate," and the

method of converting animal doses into human doses. The general approach is to develop risk estimates with assumptions designed to err on the side of safety. The agency policies do not distinguish among chemicals thought to have different mechanisms of action (e.g., between "initiators" and "promoters"). The agencies are only beginning to explore the use of pharmacokinetic modeling techniques, and thus have not discussed these in detail in their policies.

Exposure Assessment

Agency policies give much less detailed guidance on how human exposures to specific chemicals should be estimated. While EPA has issued exposure guidelines, the predominant approach in those guidelines and in actual agency practice is to make evaluations case by case. The lack of detailed guidelines does not diminish the great importance of considering exposure in estimating human risk.

Risk Characterization

Several policies discuss risk characterization, mentioning alternative ways to describe estimated risk and various sources of uncertainty. Some policies also specify a method of classifying carcinogens, for example, by the weight of evidence for carcinogenicity. Considering the weight of evidence, that is, using all available information on a chemical's effects, has received more attention in recent policies.

Federal Assessment and Regulation of Carcinogens

Federal statutes authorize agencies to set exposure standards, residue limits, tolerances, and emissions standards for carcinogenic chemicals found in air, water, food, and the workplace. Some statutes authorize or require the outright banning of carcinogenic substances or products containing them; in other cases, agencies may set rules for a product's use.

Under this authority, a number of carcinogens have been regulated, although the agencies have not acted on all of the exposures known to present carcinogenic risk. While some time is required to prepare the analyses necessary for regulatory action and to respond to public comment, there have also been lengthy delays between knowing the outcome of human epidemiologic studies or animal bioassays and publishing proposed regulations, and delays between the publication of proposed and final rules. Regulations on carcinogens have frequently been challenged in court by industry, labor unions, environmental organizations, or other groups. In some cases the courts have ruled that the agencies exceeded their authority, although in other cases the courts have compelled the agencies to act.

Many chemical exposure limits set by the government or recommended by private individuals and organizations were established primarily to protect people from noncarcinogenic toxicities—effects that manifest themselves at the time of exposure or shortly thereafter. But cancer is an insidious disease. People can be exposed to carcinogens at levels that do not cause any immediately apparent adverse effects. These exposures, however, can crucially injure individual cells, leading to cancer many years later. Thus, regulatory standards to protect the public from carcinogen exposures will need to be set at levels much lower than those designed to protect against acute toxicities.

In general, a standard that reduces exposures based on concern for one health effect will do so for all health effects associated with that chemical. But a standard based on noncarcinogenic toxicities may not reduce exposures sufficiently to protect against cancer. Significantly, many Federal standards regulating carcinogenic chemicals were set originally to protect against noncarcinogenic toxicities and have not been updated to take account of carcinogenic effects.

OSHA

Congress passed the Occupational Safety and Health Act in 1970. In 1971, OSHA adopted a large number of startup standards, setting exposure limits on about 400 specific chemicals. These exposure standards consisted largely of the 1968 recommendations of the American Conference of Governmental Industrial Hygienists (ACGIH) and had been developed primarily to protect workers from noncarcinogenic toxicities. While the ACGIH recommendations are updated annually, OSHA standards are not.

From 1972 to 1986, OSHA issued health standards covering 22 carcinogens, many of which had been regulated by the 1971 standards. Most of these carcinogen standards have aroused controversy. Of 9 final actions on carcinogens regulated individually (including 2 on asbestos), 7 resulted in court challenges. In OSHA's regulation of a group of 14 carcinogens, the final standards for 2 chemicals were challenged. Permanent standards for 2 chemicals were struck down as a result of such challenges.

National Institute for Occupational Safety and Health (NIOSH)

One role of NIOSH is to identify substances that pose potential health problems and recommend exposure levels to OSHA. However, OSHA has not responded to many NIOSH recommendations. Since 1971, NIOSH recommendations have addressed 71 different chemicals or processes that they determined to be carcinogenic. OSHA has issued health standards for 21 of the 71 chemicals or processes. Two of these OSHA standards were struck down by the courts. Thus, 19 of the 71 NIOSH recommendations on carcinogens have actually been addressed by OSHA regulations. Of the 50 chemicals or processes that are not the subjects of a final OSHA standard based on carcinogenicity, many are still regulated under the 1971 startup standards. OSHA has proposed regulations for four, but is actively working on a final standard for only one. No OSHA proposals have been issued for the remaining 46 chemicals or processes.

OSHA has criticized the quality of early NIOSH criteria documents, yet OSHA's failure to respond with standards highlights OSHA's regulatory difficulties. Increasingly, OSHA's regulatory agenda is being set by outside groups, in the form of petitions, court orders, congressional directives, and EPA referrals, including those on seven chemical carcinogens that EPA formally or informally referred under TSCA. OSHA has proposed a standard for one of these referred substances.

MSHA

MSHA regulation covers coal mines and metal and nonmetal mines. Regulation of toxic exposures in mines consists largely of reference to the 1972 and 1973 recommendations of ACGIH, depending on the type of mine. The ACGIH recommendations are updated annually, while MSHA has changed few of its standards.

In the late 1970s, MSHA regulated asbestos exposures for surface mines (using the exposure limit OSHA issued in 1972) and the chemicals OSHA included in its "14-carcinogens standard." MSHA has also proposed revised standards for underground exposure to radon daughters.

OSHA set a stricter standard for asbestos in 1986, but MSHA has not followed suit. Moreover, MSHA's current asbestos standard does not apply to exposures in underground coal mines. The increased use of diesel engines in underground coal mines has exposed workers to fumes. While MSHA has standards for such exposures in metal and nonmetal mines, these standards were not based on carcinogenicity. MSHA is developing a proposed standard for diesel exposures in coal mines.

FDA Actions on Food and Color Additives

Since congressional enactment in 1958 of the Delaney clause, which prohibits the use of food additives determined to cause cancer, FDA has identified over 60 relevant carcinogenic chemicals. They include direct food additives, indirect food additives (chemicals that might migrate from packaging material or manufacturing processes into foods or beverages), color additives, cosmetic ingredients, contaminants or potential contaminants of food or color additives, and environmental or unavoidable contaminants of food.

The regulation of food additives received much public attention when FDA banned cyclamates and proposed to ban saccharin. FDA has actually banned seven direct food additives. Its proposed ban of saccharin was barred by congressional action.

The review of provisionally approved color additives, begun in 1962 under the Color Additive Amendments of 1960, has been lengthy. It has taken until now to obtain required toxicity data and make regulatory decisions about many of the substances on the list. FDA has banned a total of 10 color additives, while a number of other color additives were withdrawn from the market by their sponsors who sometimes chose not to conduct the FDA-required testing.

In the last few years FDA policy on regulating food and color additives has also changed. Prior to 1982, FDA banned several color additives because they were shown to be carcinogenic or contaminated with a carcinogen. Since 1982, FDA has permanently listed several color additives even though they contain known carcinogens. The new policy states that, if a color additive itself does not cause cancer in humans or animals, but a contaminant of the additive does, FDA will regulate this color additive based on the general safety provisions of the act. Under this policy, the carcinogenic impurities are not considered to trigger the requirements of the Delaney clause. FDA will estimate potential risk using quantitative risk assessment techniques and if the risk of the impurities is estimated to be low, FDA will permit the use of the color additive.

In 1985 and 1986, FDA took action to allow use of food and color additives that were themselves carcinogenic, basing its action on quantitative risk assessment. In 1985, FDA proposed to allow the continued use of methylene chloride for decaffeinating coffee by limiting the allowable residue, rather than to ban the chemical's use entirely. Several color additives were identified by FDA as carcinogenic in 1982 and 1983 based on the results of animal bioassays. After performing risk assessment calculations, FDA announced in 1986 that it was permanently listing these additives because their estimated carcinogenic risks were low. FDA believes such actions are legally permissible under the interpretation that the Food, Drug, and Cosmetic Act allows FDA to ignore de minimis risks, despite the seemingly absolute language of the Delaney clause. In February 1987, FDA argued further that because the estimated risk in humans was low, the color additives in question would not be considered, for purposes of the Delaney clause, to be animal carcinogens either.

Indirect food additives are generally packaging material—various plastics and adhesives used to hold foods and liquids—and materials that contaminate foods in the manufacturing process. FDA has banned two indirect food additives. Other indirect additives containing carcinogenic impurities have been regulated by prescribing conditions for "safe use."

In the mid-1970s, FDA prohibited the use of bottles made from polymers of acrylonitrile and vinyl chloride, because these chemicals might leach into liquids. FDA's position was rejected by the courts. In the 1980s, FDA issued a rule to allow acrylonitrile copolymer bottles and proposed to allow polyvinyl chloride bottles, arguing that new manufacturing technology can ensure minimal leaching from these bottles.

FDA can set regulatory tolerances or action levels for environmental or unavoidable contaminants. It has set tolerances for PCB contamination of fish and action levels for aflatoxins, dimethylnitrosamines (in malt beverages), and N-nitrosamines (in baby bottle nipples).

FDA Actions on Animal Drugs

FDA has identified 14 chemicals associated with animal drugs that might leave carcinogenic residues in animal tissues. Such residues had been subject to the Delaney clause, but in 1962 Congress amended the Food, Drug, and Cosmetics Act to permit the use of carcinogenic drugs in animals, providing carcinogenic residues cannot be detected in meat or milk using FDA-approved methods. FDA has banned diethylstilbestrol (DES) from use in animals and has required residue studies on six other substances. FDA has proposed to withdraw approval for seven. One animal drug was withdrawn by the sponsor and there is no reported action for several others. As mentioned above, FDA has been working for 14 years on regulatory guidelines specifying the SOM for determining the presence of harmful animal drug residues.

FDA Actions on Human Drugs

In regulating carcinogens in human drugs, FDA has issued rules on six substances or groups of substances. Two were removed from the market, one was voluntarily recalled, and cautionary labeling was required on three. When a drug is determined to be carcinogenic, the drug's labeling for physicians is usually updated informally. Many, but not all, carcinogenic drugs on the market are, in fact, anticancer drugs. Treatment in these cases involves balancing the risk of future cancer against the benefit of treating a diagnosed cancer today.

CPSC

Since its creation in 1970, CPSC has evaluated and attempted to regulate or begun to regulate eight chemicals (or groups of chemicals) for carcinogenicity. CPSC regulations have often been overruled by the courts. although in the case of Tris-treated children's pajamas, CPSC developed an alternative strategy to remove the product from the market. In 1981, CPSC issued a rule regulating hazardous urea-formaldehyde foam insulation (UFFI), a rule that was also struck down by the courts.

In other cases, use of chemicals in consumer products stopped, even though regulation was not final or had been overturned in court. In some cases, CPSC has been able to negotiate voluntary actions by manufacturers, such as the 1979 voluntary recall of hairdryers containing asbestos shields.

EPA Actions Under the Clean Air Act

Since the 1970 enactment of the Clean Air Act, EPA has, often under legal pressure, listed seven carcinogens and issued hazardous air pollutant emission standards on six, although one of these actions was based on noncarcinogenic toxicity.

Although the Clean Air Act provides EPA one year to issue regulations after a substance is listed, this deadline was met only in the case of vinyl chloride. EPA has taken an average of almost 4½ years from the date of listing to final action for the six carcinogens on which it has issued final rules. During the time between the listing and regulation of benzene, one major industrial source of benzene had changed its process and eliminated release of the chemical.

EPA has created a new type of action in addition to listing: an "intent to list" decision. According to EPA, the intent to list a substance as a hazardous pollutant does not legally bind the agency as does a listing decision. EPA has indicated the intent to list for 10 substances, but none as yet has been listed and therefore none regulated.

EPA Actions Under the Clean Water Act

Important amendments to the Clean Water Act were enacted in 1972, 1977, 1981, and 1987. From 1972 to 1975, EPA issued toxic effluent standards for six categories of pollutants, under court order. In a consent decree, EPA agreed to regulate toxic pollutants by industry and by specifying the technology to be used. EPA agreed to issue effluent limitations for 65 categories of toxic substances, including 29 judged to be carcinogenic according to the water quality criteria documents that were also developed under this decree.

EPA has focused on 126 chemicals within these 65 classes of pollutants, but not all of these chemicals are regulated for every industry. In addition,

EPA has not established effluent limitations for toxic pollutants from the organic chemicals industry, and current regulation of the pesticides industry does not limit the discharges of most toxic pollutants in that industry. EPA had issued new regulations for the pesticides industry, but they were challenged in court and are now being reconsidered by EPA. Again, this regulatory activity has taken considerable time (from the 1976 consent decree until today), has involved the courts on a number of occasions, and is not yet finished. Further, while the list of 126 chemicals was chosen based on known toxicity and probable presence in water, and represented the best efforts of the participants at the time, more recent data reveal that many of the chemicals most commonly found in industrial discharges are not on this list.

EPA has also prepared nonbinding water quality criteria documents for States to use in developing water quality standards and requirements for specific discharge permits. However, only 7 of the 29 water quality criteria set for carcinogens have been adopted by one or more States. For only one of these substances (arsenic) have more than one-fourth of the States issued a water quality standard, although in some States that have not taken legislative action, individual discharge permits impose limitations based on the water quality criteria.

EPA Actions Under the Safe Drinking Water Act

In 1975, EPA issued the "interim" drinking water standards still used today for several inorganic and organic chemicals and for microbial contaminants. These standards were based on the 1962 recommendations of the U.S. Public Health Service for noncarcinogenic toxicities. EPA also issued regulations for radionuclides in 1976 and for total trihalomethanes in 1979, two groups of substances presenting carcinogenic hazards.

Following the congressionally mandated reports on drinking water by the National Academy of Sciences (the first of six volumes was published in 1977), EPA was required to publish proposed recommended maximum contaminant levels (RMCLs) and then to issue maximum contaminant levels (MCLs) for particular chemicals found in drinking water. The MCLs are to be set as close to the RMCLs as is feasible. After considering a 1978 proposed regulation to set generic standards for treating surface water supplies, EPA decided to continue focusing on individual substances.

In 1982 and 1983, EPA published two Advanced Notices of Proposed Rule-making (ANPRMs) listing 83 chemicals of concern. In 1983 and 1985, it proposed RMCLs for inorganic substances, volatile organic compounds, and synthetic organic compounds. EPA issued final RMCLs for eight volatile organic compounds in November 1985. It has not yet issued final RMCLs for the inorganic substances and the synthetic organic compounds, and has not proposed RMCLs for radionuclides. To date, EPA has issued final MCLs for nine chemicals, five of which are judged to have sufficient evidence for carcinogenicity, and one to have limited evidence.

Congress was concerned that drinking water standards were not being set quickly enough, so in the 1986 reauthorization of the act, it set deadlines for EPA to regulate the 83 chemicals that had been identified as candidates for regulation in 1982 and 1983. These 83 substances include 51 in the process of being regulated. In addition, 52 health advisories have been issued by EPA. Many of these provide information on potential carcinogens in drinking water. (See Note, page 22.)

EPA Actions Under FIFRA

To prevent unreasonable adverse effects on health and the environment, FIFRA authorizes EPA to screen pesticides before they enter the market and to regulate through reregistration the pesticides that were already on the market in 1972. In both cases, EPA may require manufacturers to conduct toxicity tests, including long-term bioassays for carcinogenicity.

FIFRA was substantially rewritten in 1972. At that time there were about 50,000 pesticide products and 600 active ingredients previously registered by the Federal Government that needed reregistration under the new law. The reregistration process has taken longer than originally anticipated. It was to have been completed by 1976, but in 1975 Congress extended the deadline to 1977, and in 1978 Congress dropped the deadline

completely because of the large number of substances not yet reregistered. This task will occupy EPA for many years.

For a number of active ingredients subject to reregistration, EPA has lacked sufficient information to judge their carcinogenic effects. EPA is taking steps to obtain this information. Still, as of March 31, 1986, it had identified at least 81 carcinogenic active pesticide ingredients. Of these, 18 have been canceled or restricted, Daminozide (Alar) is still undergoing review, and 15 have been voluntarily canceled. However, cancellations often cover only some uses. Other uses of the pesticide continue, although EPA may set additional requirements, for example, requiring workers to wear protective clothing. Special Reviews (SRs) for the substances EPA canceled or restricted required from 13 to 88 months, taking an average of about 44 months.

Another 18 chemicals have also been subjects of SRs. The SRs have been completed for 10 carcinogens, and these chemicals have not been canceled based on EPA judgments weighing risks and benefits. For the remaining 8 chemicals, SRs are not yet complete. Finally, EPA has identified 29 carcinogens, but has not started SR or cancellation proceedings for any of these.

Thus, EPA has identified 47 carcinogenic active pesticide ingredients that have not been canceled. For 13 of these EPA has determined that low exposure, low risk, or the weight of evidence for carcinogenicity suggest no action need be taken.

In addition to considering active ingredients, EPA has indicated that about 55 inert ingredients are of "high concern," with 28 of these showing carcinogenic effects. In 1987, EPA announced for the first time that it was taking steps to address some of the hazards of these ingredients.

EPA Actions Under TSCA

EPA actions under TSCA cover both new and existing chemicals. For new chemicals, the principal focus is the premanufacture review process. If after review of the manufacturer's premanufacture notice (PMN), EPA decides that there is cause for concern, it can request or require that additional toxicity testing be done, that certain controls be used when working with the chemical, and that the manufacturer notify EPA before beginning a significant new use of the chemical.

From mid-1979, when the PMN program began, until September 1986, EPA received 7,356 valid PMNs. Of these, 80 percent or 5,671 required no further action, according to EPA. Of the remaining chemicals, 523 were subject to some kind of action; an unknown number of these raised concerns about carcinogenicity. About half the time, EPA attention led to the manufacturer's informally and voluntarily agreeing to testing, control actions, or withdrawal of the PMN. For the remaining cases, EPA took more formal action, although often with the manufacturer's consent.

The lack of information in the PMNs is a potential problem. In 1983, OTA found that about half the submitted PMNs reported no toxicity information and "only 17 percent of PMNs have any test information about the likelihood of the substance's causing cancer, birth defects or mutations." Because many PMNs do not provide any toxicity test data, EPA uses information on chemical structure-activity relationships to attempt to predict the hazards that a substance may present.

For existing chemicals, EPA can require toxicity and environmental effects testing, designate the chemical for accelerated review, or require manufacturers to report on production and uses, provide EPA with any studies they have conducted, or report significant new uses. EPA can also issue regulations restricting or banning the production of a chemical or limiting its uses.

TSCA established an Interagency Testing Committee (ITC) to make recommendations on needed testing for toxicity and environmental effects. In the early years of the program, EPA's responses to the ITC recommendations provoked concern, both because of EPA delays in deciding whether to test and because of the particular administrative arrangements chosen for obtaining test data. In addition to the ITC recommendations, EPA could select other chemicals for testing. So far, this has not occurred often, although this may be changing.

A rule issued under section 8(a) of TSCA requires manufacturers to provide information about the production and uses of a chemical, while a rule adopted under section 8(d) requires that manufacturers submit to EPA unpublished health and safety studies. EPA has issued 8(a) and 8(d) rules for all the substances recommended by ITC, but until recently for few additional chemicals. EPA has recently received data from manufacturers as part of its effort to update its inventory on all chemicals in commerce.

Sufficient toxicity information is available on some existing chemicals to show they are carcinogenic. For these chemicals, the issues are determining whether the risks of cancer are "unreasonable" and what actions may be needed to reduce or eliminate such risks. EPA's Office of Toxic Substances, which is in charge of the TSCA program, has identified 38 chemicals or chemical classes as carcinogenic and has prepared risk assessments for 21 of these.

But beyond the development of risk assessments and the gathering of other information, regulatory actions on existing chemicals have been limited. Four chemicals have been designated for an accelerated review under section 4(f) (4,4'-methylenedianiline, 1,3-butadiene, formaldehyde, methylene chloride). Consideration of the regulation of occupational exposures to these chemicals has been referred formally or informally to OSHA since TSCA provides for referrals if EPA believes another agency may be able to address a hazard. Under TSCA authority, EPA began proposing Significant New Use Rules (SNURs) for existing chemicals considered to be carcinogenic. However, actions on carcinogens began in 1984, nearly 7 years after TSCA's enactment. For carcinogenic chemicals, EPA has now proposed six SNURs on eight existing chemicals and has issued four.

Section 6 of TSCA provides wide-ranging authority to limit production and uses of chemicals, including the authority to ban a substance. EPA has proposed section 6 action on PCBs, asbestos, chlorofluorocarbons, and metalworking fluids. PCBs were banned by Congress in TSCA itself; EPA regulations cover implementing that ban and arranging for disposal of PCBs. EPA has also banned propellant uses of chlorofluorocarbons, but it has not yet taken action on the most important uses of this group of chemicals, which are used in refrigeration and air-conditioning. Finally, EPA has issued rules on identification of asbestos in schools and proposed rules to require removal in certain cases. EPA has also regulated asbestos exposures for certain workers not covered by the OSHA asbestos standard, although it has not taken final action on a major proposal to limit and eventually ban asbestos use. The proposal on metalworking fluids is also not yet final.

EPA Actions Under RCRA

RCRA regulates the generators, transporters, storers, and disposers of hazardous wastes. EPA's lists of hazardous wastes cover 361 commercial chemicals and 85 industrial waste processes. When possible, EPA has emphasized waste streams from commercial processes rather than specific hazardous substances, to relieve waste generators of testing burdens and uncertainties in "relating a waste containing many substances to a list of specific substances." EPA has also issued a list of toxic chemicals as Appendix VIII of its RCRA standards. Wastes containing chemicals on this list may be deemed hazardous wastes.

EPA has made limited changes in its list of RCRA hazardous wastes. For example, since 1980 EPA has added five wastes to the RCRA list. In the 1984 RCRA amendments Congress employed "hammers"—congressionally enacted prohibitions against disposal of certain groups of chemicals unless EPA has acted to specify treatment techniques for those wastes. In addition, Congress mandated that EPA review, over a 3-year period, the entire RCRA list of hazardous wastes.

EPA Actions Under CERCLA

Commonly known as Superfund, CERCLA was enacted in 1980. CERCLA requires EPA to identify reportable quantities for hazardous substances and set requirements for notification of environmental releases.

Congress specifically included in the definition of hazardous substances those chemicals already regulated under several environmental statutes. In addition, Congress set reportable quantities for these substances at 1 pound (except for reportable quantities specified under the Clean Water Act) until EPA could set more appropriate reportable quantities. In May 1983, EPA published its initial list of hazardous substances. Since 1983, 19 substances have been added to the CERCLA list yielding a total of 717 substances. Most of the regulatory activity on the CERCLA list has been in modifying the reportable quantities. In 1987, EPA proposed modified reportable quantities for CERCLA carcinogens. Of the CERCLA hazardous substances, 191 have been identified by EPA as "potential carcinogens" or as substances "having carcinogenic potential."

EPA's Carcinogen Assessment Group (CAG)

As mentioned above, CAG was established in 1976 to centralize the conduct of carcinogen risk assessments at EPA. Major CAG assessments are thorough reviews of the carcinogenic risks of particular chemicals, including both qualitative evaluation of the weight of evidence for carcinogenicity and quantitative dose-response estimates. To date, CAG has prepared full assessments on 57 chemicals.

Office of Management and Budget (OMB)

Although not a regulatory agency, OMB has become an important actor in developing Federal regulations through their review of proposed regulations under Executive order 12291 and the Paperwork Reduction Act. This review has led to delays in proposing and issuing standards on carcinogens. OMB has also publicly questioned some of the regulatory agencies' assumptions in conducting risk assessments. The methods OMB used in commenting on a proposed OSHA formaldehyde standard ran counter to some of the assumptions typically used by the regulatory agencies and incorporated in agency policies on identifying and assessing carcinogens.

Type of Evidence: Human or Animal Data

Agencies use the hazard data available at the time of their action, most generally, data from human or animal studies. OTA has attempted to characterize the evidence that agencies have used in regulating carcinogens.

FDA has relied mostly on animal evidence in evaluating food additives, color additives, human drugs, and animal drugs.

CPSC has used both human and animal evidence, although in its action on Tris and attempted regulation of formaldehyde, it relied upon animal evidence only.

Of the 57 chemicals covered by CAG health assessments, 40 have been assessed based on "sufficient" animal evidence. Nine more were supported by sufficient human evidence and all but one of these were also supported by sufficient animal evidence. EPA judged the remaining 8 chemicals to have inadequate human evidence and limited animal evidence.

Most cancellations and restrictions of pesticides have been based on the results of carcinogenicity tests in at least two animal species. Nearly all TSCA hazard identifications and risk assessments are based on animal data.

There is some evidence of the carcinogenicity of the 35 chemicals proposed for regulation under the Safe Drinking Water Act, but EPA believes that the evidence for the carcinogenicity of 8 of these in drinking water has not been established and thus is basing RMCLs for these chemicals on noncarcinogenic effects. EPA's classification of the other 27 drinking water contaminants as carcinogens relied mostly on animal evidence.

The original RCRA list of hazardous wastes and CERCLA list of hazardous waste reportable quantities were developed largely without specific concern for carcinogenicity, although the original regulations on which these lists were based may have had this concern. Recently proposed adjustments in the CERCLA list of reportable quantities classify 191 chemicals as potential carcinogens: 14 based on sufficient human evidence, 110 on sufficient animal evidence, and 20 on limited animal evidence. Most (40) of the remaining chemicals were classified based on a parent element (e.g., inorganic compounds of arsenic were classified based on the carcinogenicity of arsenic), although for 7 chemicals EPA had no evidence of carcinogenicity.

OSHA and EPA's Clean Air Program have based regulation on human data most of the time, though there are indications this may be changing.

Of OSHA's eight regulations on individual carcinogens, seven were based on at least some evidence of human carcinogenicity. The other carcinogen, 1,2-dibromo-3-chloropropane (DBCP), was regulated primarily because it caused infertility in men. The evidence of its carcinogenicity consists of animal data. Regulation of three carcinogens under the "14-carcinogen standard" was based on human evidence, that of nine on animal evidence. The remaining two substances were regulated because of their chemical relationship to other carcinogens. Most OSHA regulations of carcinogens based only on animal evidence occurred with the regulation of the 14 carcinogens in 1974. Standards since then have been based mostly on human data, although for OSHA's 1984 regulation of ethylene oxide the primary evidence for its carcinogenicity is animal data. The primary evidence for several chemicals now being considered for regulation, including formaldehyde and methylene chloride, is animal evidence.

For the five substances regulated primarily as carcinogens under the Clean Air Act, EPA has relied on human evidence of carcinogenicity. EPA's intent-to-list decisions for eight of ten substances have relied on animal bioassays for evidence of carcinogenicity; the other two substances show both animal and human evidence of carcinogenicity.

The National Toxicology Program (NTP)

Since 1961, the Federal Government has been developing a testing program for determining the carcinogenicity of chemicals, first at NCI, and since 1978, at NTP. The program encompasses long-term animal studies and other tests to determine carcinogenic activity. NTP is probably the largest such testing program in the world, and is thus important in advancing knowledge of carcinogenic chemicals.

Early testing at NCI focused primarily on understanding the etiology and biological mechanisms of cancer. In the late 1960s, the Federal Government expanded carcinogenicity testing. Today, NTP bioassays and other tests provide important information for developing risk assessments and issuing regulations.

NTP was created to coordinate the toxicity testing of the then Department of Health, Education, and Welfare and to provide a mechanism for regulatory agencies (and others) to request bioassays on chemicals of regulatory interest. The NTP budget consists of contributions from several different agencies in the Department of Health and Human Services (FDA/National Center for Toxicological Research (NCTR), CDC/NIOSH, and the National Institute of Environmental Health Sciences (NIEHS)), although the lion's share of funds derive from NIEHS. The Director of NIEHS is also the Director of NTP. Activities of the contributing agencies are coordinated by the NTP Steering Committee, which consists of the heads of these agencies and the NTP Director. Formal authority to approve and monitor the general plan of NTP activities is vested in an Executive Committee that consists of the heads of the four major health and environmental regulatory agencies (CPSC, EPA, FDA, and OSHA), the heads of four research agencies (National Institutes of Health (NIH), NCI, NIEHS, and NIOSH) and the Assistant Secretary for Health of the Department of Health and Human Services (DHHS). This structure allows both the regulatory agencies and research agencies a voice in planning and operating NTP.

The nomination of chemicals for NTP testing is invited from any source, including the regulatory and research agencies. NTP's established procedures to evaluate nominations include review by the interagency Chemical Evaluation Committee, solicitation of public comments, review by NTP's Board of Scientific Counselors, and final decision by the NTP Executive Committee.

After selection, a protocol is prepared and testing begins. Testing consists of various preliminary studies, a long-term dosing regimen (which by itself takes 2 years), sacrifice, and pathologic examination, including microscope studies of tissues and tumor diagnoses. NTP has established procedures for ensuring the quality of these diagnoses, which are crucial to determining the final bioassay results. The resulting data are analyzed and the draft technical report is submitted to a peer review committee. Peer reviewers have the training and experience appropriate to judge the quality of the bioassay and to interpret bioassay results. NTP has chosen to include on its peer review committees people of different perspectives, including academics and representatives of industry, environmental organizations, and labor unions.

The number of chemicals tested depends primarily on the resources available. The NTP budget increased approximately 40 percent between 1979 and 1981. From fiscal year 1981 to 1987 the total NTP budget rose from $70.5 to $77.9 million, which, after adjustment for inflation, represents a small decline. Budget reductions necessitated by the Gramm-Rudman-Hollings Act have affected NTP. Recently, NCTR discontinued long-term NTP animal tests on one antihistamine and continued two other tests only because NIEHS agreed to pay 75 percent of the costs to complete the 2-year exposure phase. NTP has now agreed to fund completion of these two studies. Given current resources, more chemicals are nominated than can be tested.

The entire process—nomination, selection, preliminary testing, chronic testing, necropsy, data analysis, review, and publication—is a long one. OTA examined the process for a group of chemicals reviewed by NTP's Chemical Evaluation Committee in fiscal year 1981 and 1982. None of these chemicals has passed through the entire testing process. Of the 30 chemicals approved for testing in those 2 years, 4 have reached the stage of chronic testing.

The time from nomination to selection is more than 2 years for most chemicals. Some shortening of this period should be possible. But much of the remaining time required (between selection and beginning chronic exposures) is difficult to shorten because it is used to develop information important for the design, conduct, and interpretation of the bioassay.

The nomination process raises at least two issues. First, nominations and selections are important because they may set the regulatory agenda for the following decade. Today, several agencies are working on regulations for such chemicals as methylene chloride, 1,3-butadiene, 4,4'-methylene dianiline, and benzene, which NTP tests showed to be carcinogenic. These test results and the resulting regulatory action proceed in part from selection decisions of a number of years ago.

Second, NTP's recent decisions on testing the benzodiazepines (which include Valium and Librium) raise the issue of who should pay for carcinogenicity testing—government manufacturers, drug sponsors, pesticide registrants, or others. There are advantages to testing through common protocols and in the Federal Government's program. There is also reason to argue that the manufacturers and sponsors of chemicals have a responsibility to pay for the toxicity tests of their products.

Regulatory Responses to NCI/NTP Test Results and the *Annual Report*

NCI/NTP Bioassay Results

As of June 1987, the NCI/NTP bioassay program has completed testing of 308 chemicals in a total of 327 studies. Chemicals are typically tested in both sexes of rats and mice, for a total of four "experiments." At the end of the study, the results of each experiment are classified as clear evidence, some evidence, equivocal evidence, or no evidence for carcinogenicity, or as an inadequate test.

OTA has analyzed the regulatory uses of the NCI and NTP test results subject to peer review and audit approval by September 1986. These results represent 284 chemicals studied in 295 tests. For the analysis, "clear evidence" and "some evidence" for carcinogenicity were grouped as "positive" results. The chemicals tested were grouped based on the number of the four experiments for each that showed positive results. Of the 284 chemicals, 36 yielded four positive results, 25

three positives, 51 two positives, and 32 one positive result, for a total of 144 chemicals testing positive in at least one experiment.

OTA did not incorporate any additional data on the affected animal tumor sites, on whether both high and low doses (or all three doses in a three-dose experiment) produced a response, or on chemicals' estimated potencies. The grouping of substances for this analysis is also based only on the results of NCI/NTP testing. OTA has not used the bioassay results of others or the results of human epidemiologic studies.

Annual Report on Carcinogens

In 1978, Congress mandated that the DHHS publish an annual report listing all known carcinogenic substances and substances reasonably thought to be carcinogenic to which a significant number of people in the United States are exposed. Furthermore, the report is to describe regulatory actions on these substances, and estimate how much those actions have reduced risk. The legislation's first sponsors thought this discussion would help focus on chemical exposures that still present risks, and thus on areas for regulatory activity.

The substances discussed in the report are chosen by an interagency committee, including representatives of CPSC, EPA, FDA, NCI, NIEHS, NIOSH, the National Library of Medicine, and OSHA. The committee bases its decisions on the previous *Annual Report*, lists of chemicals judged to be supported by sufficient evidence for carcinogenicity by the International Agency for Research on Cancer (IARC), and animal testing results from NTP and other peer-reviewed studies. They publish the list of possible additions for comments and then make their final selections. The latest *Annual Report*, the fourth, lists a total of 148 substances, groups of substances, and exposures. For this analysis, OTA eliminated double-counted chemicals in this list for a total of 145 chemicals.

OTA Analysis

OTA examined regulatory responses to three groups of chemicals: all NCI/NTP-tested chemicals with at least one positive experiment, the

NCI/NTP chemicals with three or four positive experiments, and the chemicals listed in the fourth *Annual Report on Carcinogens*. While OTA analyzed the three separately, in fact there is some overlap of the three lists. All the chemicals testing positive in three or four experiments of course also tested positive in at least one experiment. In addition, many of the chemicals with three or four positive results have been listed in the *Annual Report*.

OTA focused on the chemicals of potential regulatory interest for each agency or program: the chemicals found in specific environmental media, such as air or drinking water, occupational settings, consumer products, pesticides, food, and drugs. Information on exposures is, unfortunately, often simply unavailable. Quantitative information is particularly difficult to obtain. So OTA relied on information on estimated production levels, estimated number of workers exposed, and qualitative data on the presence of particular chemicals in given situations. Even using this information on regulatory jurisdictions, OTA found apparent gaps in regulatory coverage. Figure 1-2 summarizes OTA's analysis of agency actions and nonactions on chemicals in their jurisdictions.

The impact of these regulatory gaps on human health depends on factors not analyzed by OTA, including the extent and magnitude of exposures, the potency of the chemicals, and other potentially synergistic or antagonistic exposures and risk factors. Many agency analyses conducted to develop information prior to regulation on information hazards, risks, control technologies, costs, and other factors—have not been included in the actions discussed here.

Regulation of Chemicals Tested by NCI/NTP

While a number of regulatory actions appear to have been based directly on positive NCI/NTP test results, there also appear to be substantial gaps in regulatory activity. In the NCI/NTP bioassay program, 144 chemicals tested positive in at least one experiment. Considering each agency and program individually reveals that no agency has regulated more than a third of the chemicals with positive test results. More typically, an individual agency will have acted out of concern for carcinogenicity on 5 to 30 of the 144 chemicals.

FDA has taken action on 17 of the 48 positive NCI/NTP chemicals associated with food additives, color additives, or cosmetics. The balance have been evaluated, but have not been subject to further action. FDA has acted on 4 of the 5 positive NCI/NTP chemicals associated with animal drugs, and 6 of the 12 positive NCI/NTP chemicals that are human drugs bear labeling that warns of carcinogenicity. OSHA has set exposure standards for 29 of the 53 positive NCI/NTP chemicals that are of interest in the workplace, although 27 of these 29 are regulated by standards based on concern for noncarcinogenic toxicity, which were adopted by OSHA in 1971. NIOSH has provided OSHA with recommendations on 31 of the 62 positive NCI/NTP chemicals in its OTA-defined jurisdiction. Regulatory action or voluntary exposure reductions have occurred for 8 of the 14 positive NCI/NTP chemicals in CPSC's jurisdiction. EPA has listed under the Clean Air Act 2 of 12 positive NCI/NTP chemicals within the act's jurisdiction. Water quality criteria have been prepared for 14 of the 27 positive NCI/NTP chemicals in the jurisdiction of the Clean Water Act. Of the 14 positive NCI/NTP chemicals in the jurisdiction of the Safe Drinking Water Act, 12 have been addressed by some regulatory attention, although for many of these, the regulatory process is not yet finished. EPA has developed information on 53 of the 144 positive NCI/NTP chemicals in the TSCA's jurisdiction. For 5 of the 144 chemicals, EPA has issued SNURs, begun accelerated reviews, or taken action under section 6 of the act. Under FIFRA, there have been EPA-ordered or voluntary cancellations for 13 of the 22 positive NCI/NTP chemicals used as active pesticide ingredients. Of the 144 positive NCI/NTP chemicals, 41 have been included in RCRA's list of hazardous wastes or its Appendix VIII list, while 47 of the 144 positive NCI/NTP chemicals are listed under CERCLA. CAG has prepared health assessments for 22 of the 144 positive NCI/NTP chemicals. No actions have occurred for 43 of the 144 positive NCI/NTP chemicals.

Figure 1-2.—Agency Actions on *Annual Report* and Positive NCI/NTP Chemicals[a]

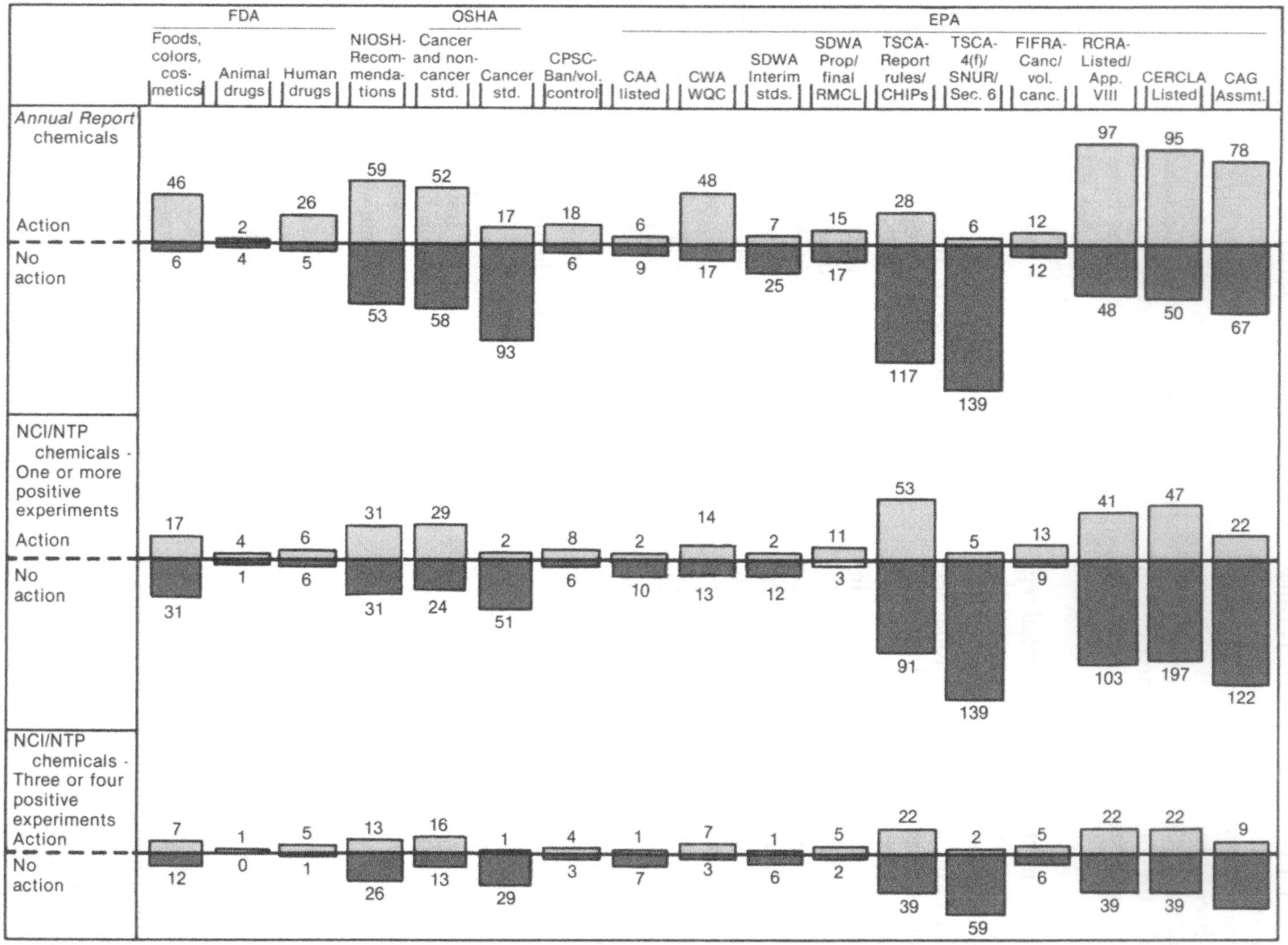

[a]For each agency or program, OTA included only chemicals in the OTA-defined jurisdiction for that agency or program. Agency decisions that regulation is not necessary or appropriate were included in the no action groups. Because of overlap between the three lists of chemicals, it is not appropriate to add them together. All actions through July 1987 are represented in this figure.

Key to acronyms: CAA—Clean Air Act; CAG—Carcinogen Assessment Group; CERCLA—Comprehensive Environmental Response, Compensation, and Liability Act; CHIPs—Chemical Hazard Information Profiles; CPSC—Consumer Product Safety Commission; EPA—Environmental Protection Agency; FDA—Food and Drug Administration; FIFRA—Federal Insecticide, Fungicide, and Rodenticide Act; NCI—National Cancer Institute; NIOSH—National Institute for Occupational Safety and Health; NTP—National Toxicology Program; OSHA—Occupational Safety and Health Administration; RCRA—Resource Conservation and Recovery Act; RMCL—recommended maximum contaminant level; SDWA—Safe Drinking Water Act; SNUR—Significant New Use Rule; TSCA—Toxic Substance Control Act; WQC—Water Quality Criteria.

SOURCE: Office of Technology Assessment, 1987.

Limiting attention to those chemicals with three or four positive experiments reveals that agencies and programs have each acted on 1 to 22 of the 61 NCI/NTP chemicals with these results. Chemicals with three or four positive experiments will generate greater concern because in these cases there are positive results from both rats and mice. FDA has taken some regulatory action on 7 of the 19 chemicals with three or four positive experiments associated with food or color additives or cosmetics. The one animal drug with three or four positive results has been revoked while 5 of the 6 chemicals with three or four positive experiments have been removed from human drugs or have been labeled for carcinogenicity. OSHA has regulated 16 of the 30 chemicals with three or four positive experiments that are in its jurisdiction. One of these standards is based on carcinogenicity. NIOSH has made recommendations on 13 of the 39 chemicals in its jurisdiction with three or four positive results. In CPSC's jurisdiction, 4 of 7 chemicals have been subject to regulatory or voluntary action. Under the Clean Air Act, EPA has listed one of eight chemicals with three or four positive results. Water quality criteria have been issued for 7 of 10 chemicals in the Clean Water Act jurisdiction, and some regulatory action has occurred for 6 of the 7 chemicals under the jurisdiction of the Safe Drinking Water Act. Information has been developed under TSCA for 22 of the 61 chemicals with three or four positive experiments and SNURs, accelerated reviews, and section 6 actions have addressed 2 of the 61. EPA-ordered and voluntary cancellations have occurred for 5 of the 11 active pesticide ingredients with three or four positive experiments. RCRA lists include 22 of the 61 chemicals with three or four positive experiments, and the CERCLA list covers 22 of the 61. CAG assessments address 9 of the 61. No actions have addressed 23 of the 61 chemicals with three or four positive experiments.

Regulation of Chemicals Listed in the *Annual Report on Carcinogens*

All the *Annual Report* chemicals have been addressed by at least one agency, although a large number of these chemicals have not been acted on by all the agencies and programs that might have an interest in them. Except for chemicals on the lists adopted under RCRA and CERCLA, no agency has regulated as many as half the chemicals included in the *Annual Report*. Generally, agencies have acted on 5 to 60 of these 145 *Annual Report* chemicals.

FDA has acted on 46 of the 52 *Annual Report* chemicals in its jurisdiction for food and color additives and cosmetics, and on 2 of the 6 *Annual Report* chemicals used as animal drugs. Of the 31 *Annual Report* chemicals with human drug uses, 26 have been removed from the market or have carcinogenicity warning labels. OSHA has exposure standards for 52 of 110 *Annual Report* chemicals in its jurisdiction; 17 of these standards are based on carcinogenicity. All *Annual Report* chemicals are covered by OSHA's hazard communication standard. NIOSH has made recommendations on 59 of the 112 *Annual Report* chemicals in its jurisdiction. Voluntary and regulatory actions have been taken on 18 of the 23 *Annual Report* chemicals in CPSC's jurisdiction. EPA listings under the Clean Air Act address 6 of 15 *Annual Report* chemicals in the act's jurisdiction. For 48 of 65 *Annual Report* chemicals in the jurisdiction of the Clean Water Act, water quality criteria have been prepared. Interim standards under the Safe Drinking Water Act, and the current RMCL/MCL process address 21 of 32 *Annual Report* chemicals within the act's jurisdiction. EPA has developed information on 28 of the 145 *Annual Report* chemicals in the TSCA jurisdiction and issued SNURs, started accelerated reviews, or section 6 actions on 6 of the 145. EPA-ordered and voluntary cancellations have affected 12 of the 24 *Annual Report* chemicals used as active ingredients in pesticides. The RCRA lists address 97, and the CERCLA lists 95 of the 145 *Annual Report* chemicals. CAG assessments cover 78 of the 145.

Comments on the OTA Analysis

In comments on a draft of this background paper, officials of Federal regulatory agencies emphasized their belief that they have acted appropriately in regulating the chemicals tested by NCI/NTP and the chemicals in the *Annual Report*. They pointed out that statutes require they assess the risks and benefits of using chemicals, and the technical feasibility and costs of regulatory action. Because of these considerations, as well as their judgments about the weight of evidence for carcinogenicity, in some cases they have decided not to regulate substances. In other cases, the chemicals are being considered as subjects of regulatory action.

Future Improvements

Today the hope for a more complete understanding of cancer causation rests on research into biochemical markers, pharmacokinetics, and molecular mechanisms. Nevertheless, science cannot now answer all the questions that are raised in this field. Even in the face of such uncertainty, however, it is important to take action to protect public health.

Ever since the development of carcinogenicity bioassays, there has been skepticism about the reliability of animal results for estimating human risk. The Federal agencies have usually assumed the usefulness of animal test results. However, regulated industries have often disputed these results in particular cases and express concern that society not impose unnecessary regulations. These disputes are not likely to go away.

To force regulatory action, Congress has legislated a variety of statutory mechanisms. The most common of these have been statutory deadlines, which have sometimes led to regulatory action, but are also frequently missed by the agencies. In the 1984 RCRA amendments, Congress included "hammers"—statutory provisions that go into effect if EPA misses particular deadlines. Congress has also mandated requirements, such as TSCA's ban of PCBs, and agency adoption or consideration of designated lists of chemicals. In one case (that of saccharin regulation), Congress prohibited an agency from acting. A final congressional mechanism is requiring agencies to consider or respond to recommendations of another agency or organization. For example, OSHA must consider the recommendations of NIOSH, EPA must respond to nominations of chemicals by the ITC, and, in the original Safe Drinking Water Act, EPA was to respond to National Academy of Sciences recommendations.

In light of the regulatory gaps revealed by OTA's analysis of agency responses to positive NCI/NTP bioassay results and the list of chemicals in the *Annual Report on Carcinogens*, Congress may wish to consider a statutory requirement mandating that agencies regulate these chemicals or at least publicly respond to these sources of information, even if, for various reasons, they choose not to regulate. On the other hand, such a requirement might make developing the *Annual Report* or selecting chemicals for NTP testing more difficult. In addition, regulatory action may not always be necessary and, if taken, may impose costs on regulated industries. Finally, in light of the importance exposures play in determining the need for regulation, it might be appropriate to develop additional information on the extent of human exposures to these chemicals.

PUBLISHER'S NOTE:

In 1984, recognizing the large number of drinking water additives (water treatment chemicals, pipes, coatings, etc.) to be evaluated for regulatory purposes, EPA requested proposals for the development of a drinking water additives program which would eventually replace EPA's Additives Advisory Program under the Safe Drinking Water Act. A consortium, led by the National Sanitation Foundation (NSF), responded and was awarded a cooperative agreement in 1985 to develop voluntary third-party consensus standards and a product certification program for all direct and indirect drinking water additives. Other consortium members include the American Water Works Association (AWWA), the American Water Works Association Research Foundation (AWWARF), the Association of State Drinking Water Administrators (ASDWA), and the Conference of State Health and Environmental Managers (COSHEM). The standards developed represent the coordinated efforts of manufacturers, regulators, water utilities, product users, and other interested parties. Both carcinogens and non-carcinogens, and EPA-regulated and non-regulated contaminants are addressed by the NSF standards.

Chapter 2

Policies for Testing, Assessing, and Regulating Carcinogens

CONTENTS

Policies for Testing, Assessing, and Regulating Carcinogens

INTRODUCTION

Over the last dozen years, health, safety, and environmental regulatory agencies have issued guidelines and policies on how they intend to identify, evaluate, and regulate carcinogens. Some guidelines and requirements address the design of toxicity tests in animals. Other policies describe the kinds of evidence, human or animal, that the agencies will use to identify and evaluate carcinogens. In these policies, agencies have given considerable attention to methods for predicting the nature and extent of possible human health risks based on human and animal data.

Some of the important issues in assessing potentially carcinogenic chemicals turn on the interpretation of test data, others on the use of assumptions (or "inference options"). These assumptions are derived from theories about cancer causation and decisions about appropriate public policy. OTA has identified four important kinds of assumptions:

1. assumptions used when data are not available in a particular case;
2. assumptions potentially testable, but not yet tested;[1]
3. assumptions that probably cannot be tested because of experimental limitations; and
4. assumptions that cannot be tested because of ethical considerations.

The lack of data and use of risk assessment assumptions, especially in conjunction with underlying political disputes about the desirability of government regulation, make this area of research the subject of lively debates.

This chapter will describe and compare the Federal agency policies that attempt to resolve certain issues in identifying carcinogens and assessing human risks. These policies include the guidelines on the design of animal bioassays for carcinogenicity, the guidelines governing the regulatory use of human epidemiologic data, animal toxicology tests and other information on toxicity, and the procedures for combining all this information in risk assessments.

The study of carcinogenesis is advancing rapidly. In this chapter, OTA has not attempted to summarize current scientific understanding, but only to describe and compare Federal agency policies on testing, assessing, and regulating carcinogenic chemicals. In addition to following procedures described in this chapter, agencies must also, prior to regulatory action, meet certain other statutory requirements. Depending on the statute, these may involve determining that the estimated risk is unreasonable or significant, that exposure reduction is technologically achievable, that the costs of control are economically achievable or proportionate to the benefits anticipated, and that the relevant statute authorizes regulatory activity for that hazard. These additional steps are not discussed in this chapter.

Types of Evidence

Four kinds of evidence may be used for qualitatively identifying carcinogens: epidemiologic studies, long-term animal bioassays, short-term tests, and structure-activity relationships. (See ref. 217 for a more detailed discussion of methods for identifying carcinogens.)

Epidemiologic studies collect information about human exposures and diseases. Reports of individual cases or clusters of cases are very often used to generate hypotheses for later study. In fact, many of the chemicals now determined to be human carcinogens were first identified in case reports by astute physicians. Larger epidemiologic studies are divided into descriptive, or correla-

---

[1] One important area of research is testing such assumptions and developing new experimental methods. Such work is taking place at the National Institute of Environmental Health Sciences and National Center for Toxicological Research.

tional, studies and analytic studies. Descriptive epidemiologic studies correlate risk factors (including exposures) and diseases or causes of death in populations. They are useful in generating hypotheses for further study and in providing clues about potential hazards. Analytic epidemiologic studies use comparison populations. In cohort studies, the comparison is made between a group exposed to the agent of interest and a group that is not exposed. For case-control studies, the comparison is made between people with a given disease and those without the disease.

Long-term animal bioassays are laboratory studies in which animals are exposed to a suspected hazard (for about 2 years in the case of rodents). The animals are examined for the presence of tumors and other signs of disease throughout the study. At the end of the study, the surviving animals are sacrificed. Tissues from these animals and from those that died during the study are given gross and microscopic examinations and tumors are diagnosed. The incidence of tumors in exposed and control groups is then compared.

Short-term tests examine genetic changes in laboratory cultures of cells, or in humans or other animals, or in lower organisms. These tests take relatively little time to perform. Short-term tests can be completed in days, weeks, or a few months, rather than requiring the several years needed to complete a bioassay in rodents.

Structure-activity relationships (SARs) in this context refer to associations between chemical structures and carcinogenicity. In a sense, judgments about them are "paper chemistry," because predictions are made about the carcinogenicity of substances based on previously observed associations between structure and toxicity, but without additional toxicity testing. The predictive value of using SARs is highest for chemicals within a class of closely related chemicals for which extensive carcinogenicity testing has already been conducted. Predictions based on SARs are less certain for classes of chemicals less extensively tested.

Many of the Federal carcinogen guidelines discuss the different roles to be played by the different kinds of evidence, as is discussed below. All of these policies value positive epidemiologic studies as the most conclusive evidence for hu-

man carcinogenicity, they generally presume that substances carcinogenic for animals in long-term bioassays should be treated as carcinogenic for humans, and they treat short-term test results as supporting information.

In practice, regulatory activity may be initiated based on positive human or long-term animal data. In most cases, if the only evidence consists of short-term test results, agencies will not initiate regulatory action to reduce exposures, although such test results might be the basis for requiring further animal testing. SARs are used mostly when no other data are available, for example, to identify new chemicals for which further testing is warranted prior to large-scale manufacture.

The relative ranking of these types of evidence is often an academic issue because for many types of chemicals, there are often few toxicity data of any sort, whether from human epidemiology, long-term animal bioassays, or short-term tests (138). In these situations, Federal agencies may be hampered in their efforts to protect public health.

Risk Assessment and Risk Management

It is common now to distinguish between risk assessment and risk management (111). This language was adopted in the report of a Committee on the Institutional Means for the Assessment of Risk to Public Health convened by the National Research Council of the National Academy of Sciences (NAS) (137). This committee described risk assessment as the process of characterizing the adverse health effects of human exposures to environmental hazards. Risk assessment relies on information from epidemiologic, clinical, toxicologic, and environmental research. Risk management, on the other hand, is the process of evaluating and choosing among regulatory options, based on information on economic, social, political, and engineering factors, as well as information on risk.

Some of the agency policies described below also outlined distinctions between risk assessment and risk management, predating the NAS report.

The Environmental Protection Agency (EPA) (1976) (293) describes two decisions: whether a substance poses a cancer risk and what regulatory action, if any, should be taken to reduce risk. The National Cancer Advisory Board (NCAB) (1977) (348), the oldest of the policy documents considered here in detail, argues that scientists play a major role in evaluating benefits and risks by providing and interpreting data, but "the final decision . . . must be made by society at large through informed governmental regulatory and legislative groups." Thus, the division, real or perceived, between "scientific data and interpretation" and "political decisions" has been noted for some time.

Risk assessment determines the qualitative nature of the risk posed by particular exposures to chemical or physical agents and quantifies the dimensions of that risk. The term "risk" has been used in many ways. OTA uses "risk" to mean the combined effects of the intrinsic hazard presented by the agent in question and the degree of exposure. Thus, an inherently very toxic agent may pose little risk when exposure levels are very low. Conversely, an agent of low intrinsic toxicity may be an important public health problem because a large number of people are exposed at fairly high levels.

Qualitative and Quantitative Risk Assessments

One distinction, frequently made in discussing policies on carcinogen regulation, is between the qualitative determination of a hazard and the quantitative evaluation of risk. The qualitative determination is a "yes" or "no" answer to the question: Does substance X cause cancer? These decisions may be difficult and may even include some quantitative analysis. For example, statistical techniques are used to determine whether an exposed group of people or animals have a significantly higher than expected incidence of tumors. In addition, qualitative determination depends on some interpretation, such as views on whether animal carcinogens are presumed to be human carcinogens, or whether benign tumors in animals indicate a hazard for humans.

Quantitative risk assessment starts with the qualitative determination that a substance does cause cancer and then goes on to ask: To what extent does exposure to a particular agent cause tumors? The answer involves four separate analytic exercises: developing a mathematical description of the dose-response relationship, extrapolating from animal data to human effects, developing information on human exposure levels, and using all this information to estimate individual risks and the number of expected cases in the human population.

Instead of "qualitative" and "quantitative" risk assessment, the NAS Committee on Risk Assessment used the terms hazard identification, dose-response assessment, exposure assessment, and risk characterization (137). These terms more clearly describe the separate analytic steps in a risk assessment, although the older terms will also be used in this background paper.

Hazard identification determines whether exposure to an agent increases the incidence of an adverse condition, for example, cancer in test animals. Dose-response assessment describes the relationship between the level of exposure or the dose and the incidence of disease. The two most important aspects of this step are extrapolating from information on incidence at high doses to predict incidence at lower doses and, in the case of risk assessments based on animal data, converting animal doses into equivalent human doses. Exposure assessment estimates the frequency, duration, and intensity of human exposures to the agent in question. Finally, risk characterization uses information from both dose-response and exposure assessments to estimate the expected incidence of the adverse health effect.

Inference Guidelines and Policies

All the steps described above involve uncertainties, some owing to the lack of data on particular agents, some to lack of knowledge concerning the causes and mechanisms of toxicity. Where the science is uncertain, inferences must be made. The NAS Committee used the term "components" to refer to the various points in the process where the risk assessor must choose among "scientifically plausible options." For example, one component would be the number of animal studies needed to be sure that the substance in question is truly a carcinogen. Some people are willing to act based on a single study in a single species, others want confirmation in a second species.

An "inference guideline" consists of assumptions that must be made to estimate human risk. The NAS Committee defined "risk assessment policy" as "the analytic choices that must be made in the course of a risk assessment. Such choices are based on both scientific and policy considerations" (137).

An agency might also adopt a risk management policy for choosing among regulatory options. In the committee's view, risk management policies should not be allowed to control risk assessment policy. While risk assessment and risk management are commonly distinguished, both are based on policy choices.

In addition to risk assessment guidelines and risk management policies, agencies have developed guidelines for conducting and evaluating animal toxicity tests. To some extent, these testing guidelines overlap with risk assessment guidelines. For example, both might specify whether benign tumors are to be considered with malignant tumors when evaluating the results of animal tests.

Agency policies and guidelines have varied in the degree of formality and in the basic approach they take toward evaluating evidence for risk assessments. Some policies, notably the cancer policy issued by the Occupational Safety and Health Administration (OSHA) and the sensitivity of method (SOM) guidelines proposed by the Food and Drug Administration (FDA), are intended to be binding regulations and were subject to notice-and-comment rulemaking. Other guidelines have been developed more informally by agency staff, printed, and made available to the public.

Rushefsky has classified agency carcinogen policies into three types: presumption-rebuttal, weight-of-the-evidence, and leave-it-to-the-scientists (180). The OSHA policy (276) represents the presumption-rebuttal approach. This policy approach uses the regulatory process, establishes "presumptions" and sets stringent conditions on

when and how these presumptions may be "rebutted." Other policies, particularly the latest policies of the White House Office of Science and Technology Policy (OSTP) (351) and the carcinogen risk assessment guidelines of EPA (284) take a weight-of-the-evidence approach, in which all relevant data are used. A weight-of-the-evidence approach is more flexible, and in implementation by the agencies, is more open to considering negative, as well as positive, data on carcinogenicity. The OSHA policy, on the other hand, restricted the circumstances in which negative data could be considered. In the third approach, leave-it-to-the-scientists, a separate body for conducting risk assessments is established. This represents the clearest separation of risk assessment from risk management. According to Rushefsky, only one agency policy, a paper prepared by OSTP staff in 1979 (23), adopts this approach, although other proposals for creating centralized science panels for developing or reviewing risk assessments are of this type (for a discussion of these proposals see ref. 217).

Interest groups have differed in their preferences for the different approaches. Industry groups have often supported various proposals to centralize risk assessments, while labor, public interest, and environmental organizations have opposed such proposals. Industry groups have also strongly endorsed the weight-of-the-evidence approach. Because of the importance labor, public interest, and environmental organizations place on the potential for harm to health, these groups want regulatory agencies to act on limited positive evidence and often when industry thinks the weight of the evidence does not support action.

Policies also vary in length, amount of detail, and complexity. Some, like EPA's "interim" guidelines of 1976 are only a few pages long, while, for example, the explanation of OSHA's policy occupies nearly 300 pages in the *Federal Register*.

Utility of Policies

The NAS committee cited above recommended that agencies adopt uniform risk assessment policies. Such guidelines have the advantages of promoting quality control, consistency, predictability, public understanding, administrative efficiency, and improvements in risk assessment methods. Potential disadvantages include over-simplification, inappropriate mixing of scientific knowledge with risk assessment policy, misallocation of agency resources to guideline development, and the freezing of science (137). An important use of guidelines within the agencies is in training junior staff in agency practices and procedures.

Some have hoped that risk assessment might be conducted as a neutral, nonpartisan, scientific enterprise. However, inference choices are necessary, and these, although often based on scientific understanding, are not empirically tested. Some hypotheses are extremely difficult to test experimentally; for instance, determining the doses that cause an increase in cancer risk of 1 percent would demand the use of 1,600 laboratory animals.[2] Others raise ethical issues, for example, evaluating the predictive value of animal test data by exposing human subjects to suspect carcinogens to follow them prospectively. Political and social values may also be reflected.

Policies may also reflect agency judgments on the acceptability of errors. From a regulatory perspective, two risks must be balanced:

> The first is the risk of taking precautionary action for a safe chemical (a regulatory false positive). The second is the risk of not controlling an unsafe chemical . . . (a regulatory false negative) (154).

The appropriate evaluation of an agency policy would not then be seen in whether the agency correctly identified every carcinogen and every non-carcinogen and placed them into the correct categories. Evaluation should be based on the overall success of the policy in improving public health. An important part of this is considering the costs of delaying public health protection (174). Some, however, argue that agency efforts to adopt "conservative" assumptions for developing risk assess-

---

[2]With a 95-percent confidence limit ranging from 0.5 percent to 1.5 percent (64).

ments are misguided, leading to substantial over-estimates of actual risks and distorting agency priorities (149,178).

In addition, agency guidelines are not, by themselves, sufficient to surmount two regulatory hurdles: the different perspectives of various interested parties and the importance of case-by-case interpretation.

In regulatory proceedings, the opinions of industry, labor, environmental groups, public interest organizations, and government are often substantially different. These groups place different values on the harm caused by unnecessarily regulating a chemical that later turns out to be safe and the harm caused by not regulating a chemical that turns out to be harmful. In a survey, Frances Lynn found evidence that there are links between political values, place of employment, and scientific beliefs. For example, industry scientists in Lynn's sample were less willing to accept animal data on carcinogenicity and more likely to believe in the existence of no-effects thresholds for carcinogens (see the discussion below) than were government scientists (112). In describing the history of Federal "cancer policies," Rushefsky points to the importance of political values in explaining some of the features of risk assessment policies (179,180). Another source of different perspectives is the various disciplines participants are trained in and the various scientific paradigms they work under (86).

Even when the agencies have policies, there will always be issues of interpretation in particular cases, especially for "flexible" policies. For example, if a policy establishes five categories, questions will arise on which category applies to a particular chemical. Even accepting the value of animal data in general, much regulatory debate on carcinogens centers on whether a particular animal study is reliable and on whether the data apply in particular cases. Arguments on particular cases are not likely to disappear, especially for commercially important chemicals.

HISTORY OF AGENCY POLICIES
The Food and Drug Administration

FDA was the first agency to set guidelines for toxicity assessment. FDA has responsibility for regulating the safety of foods, drugs, cosmetics, and medical devices. The 1958 Food Additives Amendment to the Food, Drug, and Cosmetic Act includes the Delaney clause, which proscribes the intentional use of food and color additives determined to be carcinogenic in either humans or animals. This clause does not apply to all food ingredients because some were considered to be "generally recognized as safe" or had been federally sanctioned prior to the 1958 amendment. Nevertheless, the general FDA policy (until recently) has been to ban food and color additives whenever they were determined to be carcinogenic. FDA has not explicitly specified any guidelines on interpreting carcinogenicity data.

FDA has specified the protocols for developing the animal data necessary to evaluate the safety of food and color additives. FDA first published toxicity testing guidelines, consisting of a series of papers by staff scientists, in a 1955 journal article (109). A revised version was published as a book in 1959 (267).

In 1970, an FDA advisory committee on protocols for safety evaluation prepared a report on designing experiments and on using animal data. FDA also made recommendations, specifying the use of at least two species, the maximum tolerated dose, and a two-generation bioassay design, in which exposure begins prior to conception and continues throughout the lifetime of the offspring. Reflecting the state of the science then, the committee concluded that "at the present time there is not enough information available to provide a basis for recommending any rapid [i.e., shortterm] test for carcinogenicity" (247). In 1982, FDA updated its guidelines on conducting animal toxicity tests (in the FDA "Red Book") (248).

As described later in this chapter, in the 1970s FDA began using quantitative risk assessments for certain environmental contaminants found in food. In the 1980s, FDA began applying these techniques to food and color additives, both when color additives are contaminated with small amounts of carcinogenic impurities and when the additive itself is determined to be carcinogenic. (FDA procedures for using such risk assessments are discussed in ch. 3.)

For animal drug residues in human food, the "DES proviso," part of the drug amendments of 1962, prohibits carcinogenic drug residues that can be detected by analytic methods approved by FDA. For years FDA has been working on a reg-

ulatory definition of what these approved methods would entail. The general label for these regulatory requirements is "sensitivity of method" or "SOM." SOM procedures were first proposed in 1973, finalized in 1977, challenged in court, withdrawn in 1978, and reproposed in 1979.

The 1973 proposal suggested use of a modified Mantel-Bryan procedure for extrapolating from effects at high doses to those at low doses (see the discussion later in this chapter on extrapolation models) and a risk cutoff of 1 in 100 million. This number means that exposures at the permissible limit would be associated with an upper bound estimate of 1 in 100 million people exposed.[3] In 1977, FDA issued a final rule keeping the Mantel-Bryan procedure but changing the risk cutoff to 1 in 1 million. The reproposal in 1979, kept this cutoff figure, but adopted linear extrapolation.

Subsequently, the responsibility for these regulations was transferred from FDA's Center for Food Safety to its Center for Veterinary Medicine. The guidelines were then reproposed in October 1985 (246). An approved analytic technique is defined as one that could detect residue concentrations as low as the level associated with an upperbound human risk estimate of 1 cancer for every 1 million persons exposed. Of course, this technique requires a risk assessment to estimate what residue levels correspond to this particular risk level.

In 1968 and 1973, FDA published guidelines on required toxicity information for investigating and marketing new human drugs (67). These guidelines specified an 18-month rat study and a 12-month study in dogs or monkeys, which was intended to cover both chronic toxicity and carcinogenicity. A 12-month rat study and a mouse carcinogenicity study could be substituted for the 18-month rat study (67).

In the 1970s, FDA and the Pharmaceutical Manufacturers' Association (PMA) convened a workshop to discuss toxicity testing for drugs, in-

---

[3]Agencies have not always distinguished clearly between risk estimates based on all cases of cancer and those based only on cancer deaths. Depending on the tumor site, the two estimates can differ (124). In this case, the FDA proposal referred only to "a minimal probability of risk to an individual (e.g., 1/100,000,000) . . ." (246).

cluding the length of carcinogenicity studies. While the workshop had been convened with the expectation that new guidelines would be issued, FDA decided not to update its own guidelines at that time. PMA, however, published guidelines in 1977 that reflected the workshop's consensus in requiring longer duration studies in two species (67). For carcinogenicity study designs for human drugs, FDA staff also refer to the "Red Book" guidelines for toxicity testing of food and color additives, the documents published by OSTP, and the report of the National Toxicology Program (NTP) Ad Hoc Panel on study design (258). No new formal guidelines for testing drugs have been issued, although FDA staff state that they are being developed (249).[4]

Nevertheless, in reviewing new drug applications, there is general understanding between FDA and industry about the evidence needed to obtain approval. The kinds of tests needed depend on the stage of clinical investigation and approval process, and the expected duration of human use of the drug (e.g., several days, up to 2 weeks, up to 3 months, 6 months to unlimited use). (For a summary, see ref. 218.)

For drugs expected to be continuously administered for 6 months or more, an application to conduct a Phase I or Phase II clinical investigation must include the results of 3-month animal toxicity studies conducted in two species. To initiate a Phase III trial, there must be information from two species given the drug for 6 months or more as part of ongoing studies of chronic toxicity and carcinogenicity. A New Drug Application (for a drug intended for chronic or repeated use in the general population) must now include the results of 18- to 24-month chronic studies in two rodent species (usually rats and mice) and a 12-month chronic study in a rodent species and a nonrodent species (e.g., dogs or monkeys).

FDA evaluates the evidence in the New Drug Application for therapeutic efficacy and potential risks of the drug. If FDA judges that the risks outweigh the benefits, the drug is not approved for marketing. If the benefits are thought to out-

weigh the risks, the drug is approved, but the labeling for the drug will discuss potential hazards, including any animal evidence for carcinogenicity (66). For any particular drug, the final decision depends on how "persuasive" or "alarming" the tumorigenic finding is, expected use of the drug, and the nature of alternative therapies (69).

The Environmental Protection Agency

EPA began developing carcinogen assessment guidelines during regulatory proceedings on the suspension and cancellation of several pesticides. In legal briefs written at the end of those proceedings, EPA attorneys summarized the expert testimony that the agency had received on evaluating carcinogenicity. These summaries were referred to as "cancer principles." (See box 2-A.)

Partly in response to criticism of these cancer principles, EPA established a permanent organizational unit, the Carcinogen Assessment Group (CAG), within EPA and developed a new set of guidelines (9,122,137). In May 1976, EPA published "interim" guidelines for assessing the health risks and economic impacts of suspected carcinogens (3,293). The text and explanation of these guidelines occupied less than four pages in the *Federal Register*.

In November 1977, the Environmental Defense Fund petitioned EPA to establish a policy on classifying and regulating carcinogenic air pollutants. In October 1979, EPA published its proposed airborne carcinogen policy. This policy has never been issued in final form, although agency staff indicate that they follow the outlines of this policy (103).

EPA issued water quality criteria documents under the Clean Water Act in response to a court order to assess the hazards and risks posed by a large group of substances. (See ch. 3 for details on the development of this list.) In March 1979, EPA made available a methodology for assessing human risk (methods for assessing other aspects of water quality, e.g., the hazard to aquatic life forms, had been prepared earlier). In November 1980, EPA announced the availability of the water quality criteria documents and published sum-

---

[4]FDA has provided guidance for statistical analysis of data for studies of human drugs, but this will not be discussed here.

Box 2-A.—Development of "Cancer Principles" at the Environmental Protection Agency

The substance of EPA "Cancer Principles" originated in the work of a group of scientists assembled by National Cancer Institute (NCI) scientist Umberto Saffiotti. In 1970, the group prepared a report to the Surgeon General, "Evaluation of Chemical Carcinogens." This report responded to another report prepared by the Food Protection Committee of the National Research Council of the National Academy of Sciences. This committee had suggested that regulators might allow potential carcinogens to be added to foods at "toxicologically insignificant levels." The committee also suggested that some substances might be considered safe without undergoing testing, if they had "been in commerical production for a substantial period" and that a "no carcinogenesis level" might be shown for an animal species, although there were no generally accepted ways of translating this threshold level to humans (122).

The 1970 report (250) by the Surgeon General's ad hoc committee represents one of the first "guidelines" for evaluating potential carcinogens:

Any substance which is shown conclusively to cause tumors in animals should be considered carcinogenic and therefore a potential cancer hazard for man . . .

No level of exposure to a chemical carcinogen should be considered toxicologically insignificant for man. For carcinogenic agents a "safe level for man" cannot be established by application of our present knowledge. The concept of "socially acceptable risk" represents a more realistic notion . . .

No chemical substance should be assumed safe for human consumption without proper negative lifetime biological assays of adequate size. The minimum requirements for carcinogenesis bioassays should provide for: adequate numbers of animals of at least two species and both sexes with adequate controls, subjected for their lifetimes to the administration of a suitable dose range, including the highest tolerated dose, of the test materials by routes of administration that include those by which man is exposed . . .

Evidence of negative results, under the conditions of the test used, should be considered superseded by positive findings in other tests . . .

The implication of potential carcinogenicity should be drawn from both tests resulting in the induction of benign tumors and those resulting in tumors which are more obviously malignant. . . .

The principle of zero tolerance for carcinogenic exposures should be retained in all areas of legislation presently covered by it and should be extended to cover other exposures as well. Only in the cases where contamination of an environmental source by a carcinogen has been proven to be unavoidable should exception be made to the principle of zero tolerance. Exceptions should be made only after the most extraordinary justification, including extensive documentation of chemical and biological analyses and a specific statement of the estimated risk for man, are presented. All efforts should be made to reduce the level of contamination to the minimum. Periodic review of the degree of contamination and the estimated risk should be made mandatory.

No substance developed primarily for uses involving exposure to man should be allowed for wide-spread human intake without having been . . . tested for carcinogenicity and found negative. . . . Any substance developed for use not primarily involving exposure in man but nevertheless resulting in such exposure, if found to be carcinogenic, should be either prevented from entering the environment or, if it already exists in the environment, progressively eliminated . . .

A unified approach to the assessment and prevention of carcinogenesis risks should be developed in the federal legislation; it should deal with all sources of human exposure to carcinogenic hazards . . .

An ad hoc committee of experts should be charged with the task of recommending methods for extrapolating dose-response bioassay data to the low response region . . .

At the EPA hearings on canceling registration of DDT, Saffiotti included parts of the ad hoc committee report in his testimony.[1] In the brief, which summarized the evidence in the DDT cancellation decision (315), EPA attorneys listed seven "general principles" for determining carcinogenic hazards that were drawn from the ad hoc committee report:

1. Any substance shown conclusively to produce tumors in animals should be deemed potentially carcinogenic in man, except when the effect is caused by physical induction, or where the route of administration is grossly inappropriate in terms of human exposure.
2. Carcinogenic data on man is acceptable only when it presents critically evaluated results of adequately conducted epidemiological studies.

---

[1]The NCI ad hoc committee report was also used by OSHA in justifying its "14-carcinogen standard."

3. No level of exposure to a chemical carcinogen should be considered toxicologically insignificant for man.
4. Carcinogenic bioassays should include two species of animals of both sexes, with adequate control animals, subject to lifetime administration of suitable doses, including highest tolerated doses, by routes of administration including those by which man is exposed.
5. Negative results should be considered superseded by positive results, which should be deemed definitive, unless new evidence conclusively proves that the positive results were not causally related to exposure.
6. An implication of potential carcinogenicity should be drawn both from tests which induce benign tumors and those resulting in tumors more obviously malignant.
7. The principle of zero tolerance is valid and should be expanded.

In a subsequent proceeding concerning the pesticides aldrin and dieldrin, the EPA brief listed nine "cancer principles":

1. A carcinogen is any agent which increases tumor induction in man or animals.
2. Well-established criteria exist for distinguishing between benign and malignant tumors; however, even the induction of benign tumors is sufficient to characterize a chemical as a carcinogen.
3. The majority of human cancers are caused by avoidable exposure to carcinogens.
4. While chemicals can be carcinogenic agents, only a small percentage actually are.
5. Carcinogenesis is characterized by its irreversibility and long latency period following the initial exposure to the carcinogenic agent.
6. There is great variation in individual susceptibility to carcinogens.
7. The concept of a "threshold" exposure level for a carcinogenic agent has no practical significance because there is no valid method for establishing such a level.
8. A carcinogenic agent may be identified through analysis of tumor induction results with laboratory animals exposed to the agent, or on a post hoc basis by properly conducted epidemiological studies.
9. Any substance which produces tumors in animals must be considered a carcinogenic hazard to man if the results were achieved according to the established parameters of a valid carcinogenesis test (quoted in 122).

In its notice proposing to suspend registration of the insecticides chlordane and heptachlor, EPA set forth principles very similar to these nine statements. Organizations, particularly from industry, and individuals outside EPA expressed concern about the principles' substantive content, and EPA staff scientists became concerned that these scientific principles had been formulated by EPA attorneys.

Later, Saffiotti prepared a draft summarizing 17 principles of carcinogenesis that had been used in previous proceedings. EPA attorneys attempted to have these principles included as "officially noticed facts" in the proceedings concerning the pesticide Mirex. Apparently a storm of protest followed, after which the 17 principles were reduced to "three basic facts":

1. There is presently no scientific basis concluding that there is a "no effect" level for chemical carcinogens.
2. Experimental data derived from mouse and rat studies can be used to evaluate whether there is a cancer risk to man.
3. All tumorigens must be regarded as potential carcinogens. For purposes of evaluating carcinogenicity hazard, no distinction should be made between the induction of tumors diagnosed as benign and the induction of tumors diagnosed as malignant (quoted in 122).

In April 1976, the Administrator of EPA decided that, while these proposed "facts" represented the best available evidence and were valid for supporting regulatory action, he wasn't prepared to designate them as "officially noticed facts" (122).

In 1975, while the effort to transform the principles into "officially noticed facts" was pending, an EPA scientist asked NCI to review EPA's cancer principles. This question was referred to a Subcommittee on Environmental Carcinogenesis of the National Cancer Advisory Board, which was asked in September 1975 "to develop general criteria for use in the assessment of whether specific environmental agents constitute a carcinogenic hazard in humans." EPA later withdrew its request for this effort, but the subcommittee, chaired by Phillipe Shubik, met in November 1975. The subcommittee finished a document in June 1976 that covered issues related to the identification of carcinogens (348). The report cautions that evidence of hazards must be evaluated case by case and that "criteria appropriate for one agency may not necessarily apply to another." In other respects, the conclusions of this report were similar to those in the other lists of "principles."

maries of the documents in the *Federal Register*. The policy described later in this chapter is found in an appendix to that announcement. The major change from the 1979 methodology to that described in 1980 was EPA's adoption of the linearized multistage model for extrapolating from high to low doses. The appendix describing carcinogen risk assessment was prepared by the staff of CAG. This publication was the most extensive explanation of their procedures available at that time.

In 1984, EPA published a proposed revision of its carcinogen assessment guidelines (309). EPA's purpose was "to promote quality and consistency of carcinogen risk assessments within the EPA and to inform those outside the EPA about its approach to carcinogen risk assessment." The guidelines were to "provide general directions for analyzing and organizing available data" and were not intended to alter risk management policies established under the various statutes administered by EPA. Also in November 1984, EPA published proposed guidelines for exposure assessment (310) and for mutagenicity and developmental toxicants risk assessments (311,313). Shortly thereafter, it published proposed guidelines for risk assessments of chemical mixtures (312). After making revisions and waiting for the Office of Management and Budget (OMB) to complete its review, EPA published the final version of these guidelines in September 1986 (284,285,286,287).

These last guidelines on carcinogen risk assessment consist of 10 pages in the *Federal Register*, describing the "general framework" to be used in assessing carcinogenic risk and "some salient principles to be used in evaluating the quality of data and in formulating judgments concerning the nature and magnitude of the cancer hazard from suspect carcinogens" (284). This policy outlines the various steps of risk assessment: hazard identification, dose-response assessment, exposure assessment, and risk characterization. Finally, the policy presents a "weight-of-the-evidence" classification system, with five basic categories. A chemical will be classified based on the nature (sufficient, limited, inadequate, etc.) of the evidence from human and animal studies. This classification has acquired important regulatory implications because EPA's Office of Drinking Water uses it to set rec-

ommended limits in drinking water and EPA's Office of Emergency Response uses it as part of a ranking system for adjusting reportable quantities of hazardous substances covered by the Comprehensive Environmental Response, Compensation, and Liability Act (CERCLA, commonly known as Superfund).

The Consumer Product Safety Commission

The Consumer Product Safety Commission (CPSC) published carcinogen assessment guidelines in 1978 and made them effective immediately (229). At the same time, CPSC provisionally classified perchloroethylene as a suspect carcinogen using the policy. Dow Chemical Company sued, claiming that even such a provisional classification harmed Dow. The court held that CPSC could not use the cancer policy in this manner until it was adopted in rulemaking procedures (45).

Subsequently, CPSC formally withdrew its cancer policy from the rulemaking process and decided to use the guidelines adopted by the Interagency Regulatory Liaison Group (IRLG), and more recently the guidelines issued by OSTP. (Even though CPSC's policy was withdrawn, its contents are still interesting in light of other Federal agency policies, and for this reason, the policy is discussed further below.)

The Occupational Safety and Health Administration

OSHA published a proposed regulation governing identification and regulation of carcinogens on January 20, 1977. OSHA held hearings, accumulated an extensive record, and published a final regulation in 1980. One important purpose of the policy was to improve the efficiency of the standards-setting process. OSHA officials argued that the slowness in setting standards was partly related to the many discussions, arguments, and lawsuits involved in every regulatory proceeding on carcinogens (365). The proposed policy generated considerable controversy about OSHA's identification and regulation of carcinogens, and dispute about the fraction of cancer incidence in the United States that can be attributed to occupational exposures. (See refs. 217 and 159 for discussions of this second issue.)

OSHA published its carcinogen policy as a binding regulation. Its intent was to collect evidence and testimony on "generic" issues in carcinogen identification and regulation, make decisions on these issues, and then rely on these decisions in future proceedings. (The policy uses a "presumption-rebuttal" approach (180).) The framers of this policy hoped that its use would speed the regulation of carcinogens by limiting debate about generic issues in the regulatory proceedings on individual carcinogens.

In contrast to other agencies' adoption of quantitative risk assessments for setting standards, the OSHA cancer policy stated that quantitative risk assessments would be used only to set priorities. Originally, some of the provisions of OSHA's carcinogen policy concerning risk management stated that once OSHA determined a substance to be a carcinogen, it would then set an exposure standard based only on feasibility. In 1981, OSHA amended its carcinogen policy to conform to the Supreme Court decision on OSHA's benzene standard, which provided that OSHA could only regulate exposures posing a "significant risk" to the health of workers and only if the regulation would significantly reduce the risk. The amendment allowed OSHA to consider the significance of estimated risk and feasibility in setting health standards for carcinogens. Regarding the specifics of risk assessment, OSHA did not change its judgments on the science of identifying carcinogens, although it did indicate that certain types of evidence and arguments that it had originally hoped to exclude from specific proceedings might be relevant to determining whether there was a significant risk. OSHA policy was also amended to reflect this conclusion (274). In 1982, OSHA suspended parts of the policy that required publication of lists of candidate carcinogens (after one list had been published in 1980) and requested public comment on more general issues concerning the substance of its policy (275). As yet, no changes have been made in the policy based on those comments. The policy was legally challenged by industry groups shortly after it was published in 1980, although the case was never argued and the suits have been dismissed.

Other Agencies and Interagency Efforts

OSTP and several interagency committees have worked on carcinogen assessment guidelines and regulatory policies. The Carter Administration established the IRLG, which initially had representatives from EPA, CPSC, OSHA, and FDA. Later, the Food Safety and Quality Service of the Department of Agriculture (USDA) joined. An IRLG working group, consisting of scientists from the IRLG agencies, the National Cancer Institute (NCI), and the National Institute of Environmental Health Sciences (NIEHS), published "Scientific Bases for Identification of Potential Carcinogens and Estimation of Risks" in July 1979. The report is noteworthy because it represents the first joint attempt of regulatory agencies to develop a consistent approach to identify carcinogens. However, some differences of opinion remained, especially concerning the desirability of quantitative risk assessment (137).

Another interagency group in the Carter Administration, the Regulatory Council, also prepared a document on carcinogen regulation. The Regulatory Council's conclusions on the science of identifying carcinogens relied heavily on the IRLG document, which was published as an appendix to the Council's document (354).

In 1979, several staff members of OSTP prepared a document to "stimulate development of a uniform decision-making framework to assure consistent Federal action regarding the identification, characterization, and control of potential human carcinogens." Making a distinction between "scientific data collection and analysis" and "regulatory decision-making," it examined only the former. This document was relatively short, consisting of short discussions of particular areas and giving the authors' recommendations for improvements in Federal decisionmaking. In particular, they suggested the coordination of Federal risk assessment activities under the aegis of NTP (23).

In 1981, the new Reagan Administration articulated a strong opposition to most government regulation. Even before the inauguration, David Stockman, the first Director of OMB under the Reagan Administration, had published a list of regulations he thought were undesirable. Within 2 months of taking office, Reagan created a task force on regulatory relief, chaired by the Vice President, and issued an Executive order providing for OMB review of agency regulatory proposals and final rules. The order stated that agencies could regulate only when the benefits of regulation exceeded its cost, except when this was prohibited by law. Administration officials asked affected businesses to inform them about regulations the businesses wanted changed. In 1981 the Reagan Administration also dissolved IRLG and the Regulatory Council.

In 1982, several events suggested the beginnings of a decidedly different approach to assessing carcinogenicity. In that year, EPA decided not to designate formaldehyde for priority review under the Toxic Substances Control Act (TSCA). In a memo to EPA Administrator Anne Burford, John Todhunter, EPA's Assistant Administrator for Pesticides and Toxic Substances, concluded that while formaldehyde appeared to be carcinogenic in rats, the carcinogenic potential of formaldehyde also seemed to "vary significantly with species and route." Moreover, although in certain exposure situations, formaldehyde could pose a human risk, the available epidemiologic information "supports the notion that any human problems . . . may be of low incidence or undetectable." Quantitative risk estimates fell into a range that Todhunter considered low. For these reasons, he did not think formaldehyde should be subject to an accelerated review under TSCA (155,196).

A second EPA document appearing in 1982 was a draft of guidelines for assessing carcinogenicity, specifically for developing water quality criteria. The draft described a weight-of-the-evidence stratification scheme, modeled after the scheme of the International Agency for Research on Cancer (IARC),[5] and suggested that regulatory distinctions be made between carcinogens that act by causing gene mutations and those that act by different mechanisms. For the latter, the draft suggested development of water quality standards using the "no observable effect level" (NOEL) (180,279). In 1983 it was also revealed that Rita

Lavelle had written a memo urging that trichloroethylene (TCE) be reevaluated and that EPA develop a "threshold model risk assessment for nongenotoxic chemicals such as TCE" (117).

A third draft document represented an administrationwide effort to revise agency practices on carcinogenicity risk assessment. In 1982, as part of the Reagan Administration's efforts to reduce the burden of government regulations and to develop a "scientifically sound basis for identifying and characterizing potential human carcinogens," OSTP convened an interagency committee to update the information contained in the 1979 IRLG document. The committee developed a "rough first draft" statement on "the current state of the science" (105).

The draft, which criticized many of the existing procedures used by regulatory agencies including the use of high-dose testing in animals and linear non-threshold extrapolation models, suggested distinctions based on mechanisms of action (e.g., between epigenetic and genotoxic agents) and the greater use of pharmacokinetic information in risk assessments (180).

The draft was circulated among a number of scientists and generated considerable controversy. Criticism especially focused on a chapter by John Todhunter, which suggested distinctions based on mechanisms. Congressional committees held hearings on this and other aspects of the Administration's regulatory policies in 1982 and 1983. In the wake of several revelations not directly related to the ongoing effort to develop "science principles" for carcinogenesis, most of the top EPA officials left office (362).

Review of the scientific basis for carcinogen risk assessment continued under Ronald Hart, the Director of FDA's National Center for Toxicological Research (NCTR). Drawing on scientists from NIEHS, NCI, NCTR, OSHA, CPSC, FDA, EPA, USDA, and OSTP, another draft was prepared and published for public comment and was generally received favorably (116).

A final version, "Chemical Carcinogens: A Review of the Science and Its Associated Principles, February 1985," was published by OSTP in March 1985 (351). It was republished in the journal *Envi-*

---

[5]EPA later adopted such a scheme in its 1986 policy, as discussed below.

ronmental Health Perspectives in 1986 under the authorship of the U.S. Interagency Staff Group on Carcinogens, reflecting the contributions of staff from all the agencies involved. This document is an extensive summary of the state of various scientific fields underlying risk assessment:

mechanisms of carcinogenesis, short-term tests, long-term bioassays, epidemiology, and exposure assessment. The document concluded with a discussion of the assumptions used in the process of risk assessment and included a series of summary principles.

GUIDELINES FOR CONDUCTING CARCINOGENICITY TESTING

Toxicity testing and interpreting test results are important features of several Federal regulatory and research efforts aimed at preventing exposures to carcinogens. In some circumstances, toxicity testing may include long-term bioassays to determine directly whether substances cause cancer in animals. Several laws that provide for carcinogen regulation allow Federal agencies to order regulated industries to conduct toxicity tests. The Federal Government's own carcinogenicity bioassay program was once housed at NCI, but is now coordinated by NTP. (See ch. 4.)

Required Carcinogenicity Testing

FDA requires carcinogenicity testing for some substances that are proposed as new, direct food or color additives. Decisions on whether a substance must be tested are made using a complex scheme based on the chemical structure of the substance (e.g., its chemical relation to known carcinogens) and the expected concentration of the substance in food. These two factors are used to classify the substance into one of three "concern levels." Only for the highest level are lifetime carcinogenicity bioassays required. For the other two levels, FDA requires that the substance be tested in a battery of short-term tests. The results of the short-term tests may alter the concern level of the substance and thus lead FDA to require a long-term bioassay.

For animal drugs that may leave potentially harmful residues in food, FDA uses three kinds of information to decide whether to require carcinogenicity testing:

- the potential toxicity of the drug, which is evaluated based on chemical structure and short-term and subchronic tests;
- the estimated level of use in food-producing animals; and

- the amount of drug residue expected to be consumed by a person during a single exposure.

For human drugs, FDA requires carcinogenicity testing for any new drugs expected to be used for chronic or repeated use, although these requirements are not found in any written guidelines or regulations. These requirements apply to new drugs and not to drugs that were approved prior to 1968 when FDA began to require chronic tests.

Under the Federal Insecticide, Fungicide, and Rodenticide Act (FIFRA), EPA may require animal carcinogenicity studies for registering new pesticides and reregistering existing pesticides. A carcinogenicity bioassay is required for these substances in three specific circumstances:

1. when the active ingredients, metabolites, degradation products or impurities are structurally related to recognized carcinogens, cause mutations in short-term tests, or produce a worrisome effect in subchronic studies;
2. when use of the pesticide will require that EPA or FDA issue a food tolerance limit or food additive regulation; and
3. when use of the pesticide will result in significant human exposure (e.g., in fabric treatments, insect repellants, and indoor pesticides).

Under TSCA, EPA may require testing either for new chemicals entering the market or for existing chemicals. In the latter case, EPA must conclude, first, that the chemical may present an unreasonable risk to health or the environment or that it may or will enter the environment in large quantities or present significant or substantial human exposures; and, second, that testing is necessary to provide more information about the chemical. (See Note, page 74.)

Analysis of Test Designs

In this section, OTA compares the carcinogenicity bioassay study designs required for food and color additives (245), for chronically used human drugs (161), for pesticides (332), and for tests ordered under TSCA (318). For animal drug carcinogenicity studies, FDA has not adopted separate guidelines but instead refers drug sponsors to the NCI test guidelines (251) and to the 1971 and 1982 versions of the guidelines for food additive testing (245,247). NCI developed a standard design for federally funded tests and published it in 1976 (251). NTP conducts most of its tests through the use of contract laboratories and, through its contractual "statement of work," sets the design of these studies (256). In addition to study designs used by NCI and NTP, the comparison covers the recommendations of a group of scientists whose findings were published by IARC (59) and of a panel of outside scientists convened by NTP (258). This comparison covers only written requirements and suggestions. OTA has not attempted to determine how well the conducted bioassays comply with these guidelines.

Not all regulatory agencies have specified guidelines for test design. OSHA, for example, rejected the specification of test protocols and data analysis in favor of reliance on informed scientific judgment. In part, as OSHA pointed out, this reflects its own regulatory purposes—OSHA uses whatever test data are available and does not require toxicity testing. FDA and EPA, on the other hand, can require industry to test.

General Provisions for Test Designs

Today there is relatively little controversy about the general design of carcinogenicity studies, and guidelines are relatively consistent in their requirements for study design. The basic study design uses two different animal species. Because of the relatively low cost and long experience using rats and mice, these two species are usually used. The animals must be free of disease and quarantined, then they are randomly assigned to different groups.

Exposure routinely begins by the time the animals are 6 weeks old and the study ends usually after 2 years of exposure. Exposure is preferably through the route that most closely imitates human exposure. For example, food additives should be tested by adding the suspect additive to the animals' feed, while airborne toxic substances should be tested by mixing the substance into the air the animals breathe.

Animals are randomly assigned to two or three treatment or exposure groups and a control group, which is not exposed at all. Care must be taken to ensure that the exposed animals and the control animals live under the same conditions, except for the exposure to the suspect substance.

Animals that die during the study are examined for signs of toxicity and for tumors. At the end of the study, all the surviving animals are killed and necropsy is performed. The various guidelines specify a complete examination for visible lesions and tumors, and list the organs that are to be prepared for microscopic examination, although they differ on the extent of microscopic examination required. After tumors have been diagnosed, statistical analyses are used to determine whether the exposed groups had a higher incidence of tumors than the control group.

Issues in Test Design

Table 2-1 presents several main issues in the design of carcinogenicity bioassays and the way they are handled by the various guidelines. As outlined in OTA's 1981 report, *Technologies for Determining Cancer Risks From the Environment*, the principal issues in study design are the following:

- *study plan*, including the selection of animal species, the number of animals for each dose level, and the dose levels themselves;
- *dosing regimen*, including the age at which to begin exposure, when to terminate exposure, and whether there should be an observation period between the end of exposure and the sacrifice of the animals;
- *pathology*, including the nature of the autopsy examination of the animals and the extent of microscopic examination; and
- *personnel qualifications*.

In addition, some of the guidelines discuss survival criteria, such as the number of animals that must survive to have a valid positive or negative study.

Some differences in terminology exist, but the guidelines are generally consistent about the important issues in study design. All the guidelines require that two different species be tested. NCI and NTP guidelines specify the strains of rats and mice to be used by testing programs. EPA's guidelines and PMA's drug testing guidelines specify the rat and the mouse as test animals, while the FDA "Red Book," IARC scientists, and the NTP Ad Hoc Panel suggest considering hamsters. The Ad Hoc Panel further encourages the search for other species for carcinogenicity testing.

All the guidelines specify that testing shall be done in both males and females, and all but 1 set of guidelines specify the size of each test group as 50 animals per dose. The NTP Statement of Work specifies 60 animals per dose, including 10 animals scheduled for interim sacrifice between the 12th and 18th month of the study. Several other guidelines mention that the number of animals should be increased if the researchers want to conduct an interim sacrifice, although no other guidelines require interim sacrifice. The number of animals needed for the chronic phase of testing depends on the number of doses.

NCI and PMA guidelines require at least two dose groups in addition to the control group.[6] All the other guidelines suggest the use of three dose groups and the unexposed control group. With 50 male rats, 50 female rats, 50 male mice and 50 female mice for each exposure level, a study using 2 exposure levels and controls uses 600 animals, a study with 3 exposure levels and controls requires 800.

All the guidelines provide that the highest dose level should be based on information gathered in a subchronic toxicity study (usually lasting about 90 days). However, slightly different terminology is used to refer to this dose level. The most common term in the toxicologic literature is "maximum tolerated dose" or MTD. Most of the guidelines refer instead to the "high dose level" or "high dose," perhaps to avoid the controversy that "maximum tolerated dose" has engendered. (As discussed below in the section on agency risk assessment policies, the reason for high-dose testing is to enable a study to best detect a carcinogenic response.)

In general terms, the high dose should be as high as possible without shortening the animals' lives from noncarcinogenic toxic effects. FIFRA, TSCA, and "Red Book" guidelines specify that the dose should be minimally toxic without substantially altering the normal lifespan of the animal. The NTP Statement of Work and NTP Ad Hoc Panel documents also state that the high dose should not affect the animals normal lifespan from effects other than carcinogenicity. NCI guidelines give more detail: the MTD should neither alter the lifespan (other than from carcinogenicity), clinical signs of toxicity, or pathological lesions (other than neoplasms) that shorten the animals' lives nor should it lead to more than a 10-percent decrement in weight gain in experimental animals relative to controls. PMA guidelines specify that the highest dose should be "slightly below toxic dose," without providing any further guidance. Important to the success of a bioassay is the professional judgment of the researchers conducting the study and analyzing the data from prechronic studies. To some extent, setting the highest dose requires an educated guess.

The low doses are often defined as fractions of the highest dose. For example, NCI guidelines set the second dose as one-half or one-fourth of the MTD. (This formula is also given by IARC scientists for studies aimed at only a qualitative determination of a substance's carcinogenicity.) PMA guidelines specify that the second dose should be greater than or equal to the expected equivalent human dose, but less than or equal to half the high dose. The FDA "Red Book" and TSCA guidelines specify the lowest of three doses to have "no indication of toxicity" and, generally, to be 10 percent of highest dose (FDA) or not less than 10 percent of the highest dose (TSCA). IARC scientists suggest, for studies gathering quantitative information, that the doses be scaled by factors of 3, 5, or 10. The NTP Ad Hoc Panel and TSCA guidelines mention that researchers should

---

[6]Today, however, virtually all drug carcinogenicity studies are conducted with three dose groups (249).

Table 2-1.—Test Design Issues

	NCI (251)	NTP-statement of work (256)	NTP Ad Hoc Comm. (258)	IARC (59)	FDA-food & color additives (248)	FDA/PMA-drugs (161)	TSCA (318)	FIFRA (332)
Animal species	2; NCI used primarily B6C3F1 mice & Fischer 344 or Osborne-Mendel rats	2; B6C3F1 mice & Fischer 344 rats (unless NTP specifies otherwise)	Recognizes rats, mice, hamsters as most popular; encourages search for other species	2; choose among rats, mice, hamsters	2; rodents: rats, mice, hamsters	2; rats & mice (pick strains w/low background incidence)	2; rats & mice	2; rats & mice preferred
Number at each dosage	50 male, 50 female	60 male, 60 female (allows for interim sacrifice of 10 animals)	50 male, 50 female; for special studies, number of groups & distribution may be altered	50 male, 50 female	50 male, 50 female (increase for interim sacrifice)	50 male, 50 female	50 male, 50 female (increase to allow for interim sacrifice)	50 male, 50 female (increase to allow for interim sacrifice)
Dosages	At least 2 plus control; MTD & MTD/2 or MTD/4; MTD defined as "highest dose . . . that can be predicted not to alter the animal's longevity from affects other than carcinogenicity," i.e., no more than 10% weight loss, no mortality or clinical signs of toxicity that shorten animal's life; desirable to have positive control group	3 plus control; high dose is "predicted not to alter normal longevity of the animals from effects other than carcinogenicity"	3 plus control; MTD identified in prechronic studies as dose which will not impair normal longevity from effects other than induction of tumors; use metabolic/pharmacokinetic studies to select lower doses; route should be same as human exposures; if using gavage, conduct pharmacokinetic studies to back up; explore alternatives to vegetable oil gavage	2 plus control for qualitative studies, 3 or more for studies to be used for quantitative assessment; select high dose "as one that produces some toxicity but *not* appreciable cell death or organ dysfunction, toxicity that impairs lifespan (other than tumors) or more than 10% decrement in weight gain compared to controls; lower doses; qualitative: MTD/2 to MTD/4, quantitative: scale by factors of 3,5,10	3 plus control; high dose should elicit minimal toxicity w/o substantially altering lifespan (other than effects related to tumors); lowest dose should induce no signs of toxicity (generally 10% of high dose); intermediate dose should be approx midway, depending on pharmacokinetics	2 or more plus control; high dose—slightly below toxic dose; low dose—greater than or equal to human dose level & less than or equal to one-half of high dose	3 plus control; high-dose level—minimally toxic w/o substantially altering normal lifespan; lowest dose—should not interfere w/ normal growth or show any other signs of toxicity, should not be less than 10% of high dose; intermediate—in between, depending on toxicokinetic properties, if known	3 plus control; highest dose level should be sufficiently high to elicit signs of minimal toxicity w/o substantially altering the normal lifespan
Required number surviving/Termination criteria	Terminate when survival reaches 10% within group	May terminate when cumulative mortality jeopardizes ability to draw conclusions on carcinogenicity; contractor must consult with NTP	—	Not satisfactory if mortality exceeds 50% before week 104 for rats, week 96 for mice; end study when mortality in control or low-dose groups equals 75%	Survival must be at least 50% at 24 months (rats) & 18 months (mice); no more than 10% lost due to autolysis, cannibalism, or management problems; may terminate "under special circumstances" if there are only 10 survivors in any group after 24 months (rats) or 18 months (mice), but minimum survival criteria must be met	Terminate if mortality reduces control group to less than 40% of original number of animals per sex	For valid negative study, survival must be greater than 50% in all dose groups at 24 months (rats) or 18 months (mice) & less than 10% loss due to autolysis, cannibalism, or management problems	Survival must be at least 50% at 18 months (rats) or 15 months (mice) & at least 25% at 24 months (rats) or 18 months (mice)

continued on next page

Start of dosing	Weanlings if possible, no older than 6 weeks	6-7 weeks old	Consider starting exposure in utero in certain circumstances	—	Begin exposure to parents prior to mating, exposure in utero & for life of offspring for non-nutritive additives & certain other substances	As soon as possible after weaning, ideally before 6 weeks not later than 8 weeks	As soon as possible after weaning, ideally before 6 weeks, not after 8 weeks
Duration of dosing	"Greater part" of the animals' lifespan—24 months	103 weeks (plus up to 2 more weeks to schedule necropsies)	NTP should do studies to determine optimal endpoint	See termination criteria above; in any case, no longer than 130 weeks for rats, 120 weeks for mice	At least 104 weeks, up to 130 weeks	24-30 months for rats, 8-24 months for mice	At least 24 months for rats, at least 18 months for mice
Observation period	May be desirable to hold animals for additional 3-6 months	1 week	Recommends against ending exposure prior to sacrifice unless there is concern about exposure to technicians	Prefers no observation period; if desired, treat to 104 weeks for rats 96 weeks, for mice	—	—	24-30 months for rats, 18-24 months for mice
Dosing frequency	7 days/week for food/water exposure; otherwise based on human exposures	—	—	—	7 days per week, use oral exposure route	Continuous, 7 days per week is preferable; 5 days per week is acceptable	Ideally 7 days per week; 5 days per week acceptable; oral route preferred provided substance is absorbed in GI tract
Organs & tissues examined	All animals given gross exam; histopathology for: 1) gross lesions/suspect tumors, 2) list of organs for all treated & control animals	All animals given gross exam; full histopathology for all treated & control animals	All animals given thorough examination; consider alternatives to reduce burden of histopathology, such as inverse pyramid & selected inverse pyramid	Gross exam of all; detailed histopathology on high dose & control groups; if no difference is found, then histology can be restricted to examining gross lesions & sites where significant lesions are observed in high dose group; suggests distinguishing between fatal & incidental tumors, especially for statistical analysis	Gross exam for all animals, microscopic exam for: 1) all visible tumors, 2) all animals that died during study, 3) high dose group & controls; if questionable, then examine other exposure groups	Full histopathology on: 1) all animals in control & high dose groups & all that died during study, 2) all gross lesions, 3) target organs in all animals; if there were excessive early deaths or problems with high dose group, use next lower dose group for full histopathology	Full histopathology on: 1) all animals in control & high dose groups & all that died during study, 2) all gross lesions, 3) target organs, 4) lungs, liver & kidneys in all animals; if there were problems with high dose group, use next lower group for full histopathology
Study director	—	Doctorate in toxicology, pathology, veterinary medicine, biochemistry, or chemistry	—	—	—	—	Appropriately educated, trained, & experienced toxicologist
Pathologist	Board-certified w/experience in laboratory animal pathology	Formal training & experience required; board certification desirable	—	—	—	"Individual possessing expertise in laboratory animal pathology"	Board-certified or board-eligible or person w/equivalent training w/expertise
Histology technicians	Supervised by HT/ASCP technician	ASCP-registered technicians	—	—	—	—	Certified by HT/ASCP or having equivalent training & capability

SOURCE: Office of Technology Assessment, based on cited documents.

use information on metabolism and pharmacokinetic studies, if available, to help set dose levels.

The control group should be completely untreated or sham treated, and should otherwise be handled by the lab workers in the same way as the treated animals. Sometimes control animals are treated with the "vehicle" used to administer the test compound, such as the corn oil used in gavage studies (vehicle controls). Sometimes, researchers will also include a group of animals to be exposed to a known animal carcinogen (positive control), to be sure that the animals being used are in fact sensitive to a known carcinogen. Except for the NCI guidelines, the guidelines that mention this possibility generally include it for routine studies. IARC scientists state information should be collected on control animals to evaluate any changes over time.

The basic laboratory alternatives for dosing the animals include adding the substance to the animals' food or water, exposing the animals by contaminating their air in special inhalation chambers, painting the substance onto the animals' skin, or delivering the substance, usually dissolved in corn oil, directly into the animals' stomachs using a special tube (gavage studies). With regard to the dosing regimen, the guidelines provide that the route of administration be as close as possible to the human exposure route, recognizing that sometimes this is not possible, for example, when the suspect compound is so unpalatable that animals will not eat the treated feed. For dosing in food or water, exposure is generally 7 days per week. For inhalation or gavage studies, laboratories generally expose the animals five times per week to match the schedules of laboratory personnel.

The guidelines also specify the age of the animals at the start of the study. The NCI, NTP Statement of Work, FDA "Red Book," TSCA, and FIFRA guidelines all require dosing to begin shortly after the animals have been weaned and before the animals reach 6 to 8 weeks. For food and color additives, FDA often requires the manufacturer to conduct the carcinogenicity study in at least one rodent species with in utero exposures. Parents are exposed to the test compound prior to mating, and exposure continues through pregnancy and throughout the lives of the animals. It is argued that this design is particularly sensitive in detecting carcinogenic effects and is especially appropriate for substances in the food supply because exposures may be continuous for parents and children. PMA guidelines also mention the in utero design as a possibility, although data on this design are lacking. The guidelines suggest use of the design to develop data, especially for drugs that may be used in childbearing women. The NTP Ad Hoc Panel also suggests that the in utero design be considered under certain circumstances.

The guidelines differ concerning when a study should end. The basic principle is that carcinogenicity studies should expose the animals for the "greater part of the animals' lifespans" (251). For rodents, this is generally considered to be 2 years. Thus, the NCI guidelines provide for exposures of 24 months, the NTP Statement of Work provides for 103 weeks (plus up to 2 more weeks to schedule autopsies), and the FDA "Red Book" requires at least 104 weeks (24 months), though such a study may last up to 130 weeks (30 months). PMA, TSCA, and FIFRA guidelines provide for a shorter exposure for mice: 18 months for PMA, at least 18 months for FIFRA, and 18 to 24 months for TSCA. For rats, these guidelines provide for studies of at least 24 months.

NCI guidelines suggest it "may be desirable to hold animals for an additional period of 3-6 months" after exposure has stopped. This time was termed an "observation period." More recently, however, the NTP Ad Hoc Panel concluded that, except when there is concern about exposure to lab personnel, "it does not seem wise to terminate exposure prior to sacrifice." The NTP Statement of Work sets a 1-week period for observation. FIFRA guidelines say that rat studies may be 24 to 30 months long and mouse studies 18 to 24 months long.

IARC scientists define the length of a study in terms of the animals' survival. According to them, a study should be terminated when 75 percent of either the control or low-dose group have died. But in no case should the study extend beyond 130 weeks (30 months) for rats or 120 weeks (28 months) for mice. IARC prefers no observation period, but suggests that if one is desired, treatment can continue 104 weeks for rats and 96 weeks for mice.

NCI guidelines provide for the termination of a study when survival drops to 10 percent in any group. The NTP Statement of Work allows for termination when "cumulative mortality jeopardizes ability to draw appropriate conclusions on carcinogenicity," but requires consultation with NTP before a contractor can sacrifice the animals. The FDA "Red Book" provides that a study may be terminated "under special circumstances" if there are only 10 surviving animals in any group (because these groups contain 50 animals, this is a survival rate of 20 percent) after 24 months in rats and 18 months in mice.

The FDA "Red Book," TSCA guidelines, FIFRA guidelines, and IARC scientists specify the minimum survival necessary for a valid negative study. The FDA "Red Book" requires at least 25 animals per sex at 24 months for rats and at 18 months for mice. In addition, no more than 10 percent of the animals should have been lost due to autolysis (tissue destruction before necropsy), cannibalism, or management problems. TSCA guidelines require at least 50-percent survival in all groups, while FIFRA guidelines specify that survival must not be less than 50 percent at 18 months for rats and 15 months for mice and not less than 25 percent at 24 months for rats and 18 months for mice. IARC scientists suggest that a study is not satisfactory if mortality is greater than 50 percent before week 104 (24 months) for rats and 96 weeks (22 months) for mice.

All the guidelines describe the nature of the necropsies that should be conducted after animals die during the experiment or are sacrificed at the end. In general, all the animals should be examined carefully, including gross visual examination of a number of specified tissues. Instructions on preserving tissues are given, and tissue portions are prepared for microscopic examination to discover tumors and their types. The guidelines differ with regard to the extent of this microscopic examination. In general, it is required for all observed gross lesions and for sections of major body tissues. NCI guidelines require microscopic examination of these tissues for all exposed and control animals. The other guidelines allow for

less comprehensive microscopic examinations, specifically, of all animals that died during the study, all animals in the control group, and all animals in the highest exposure group. Microscopic examination should also be conducted for animals in lower exposure groups on the specific organs (target organs) in which tumors were discovered in the highest exposure group, and in some guidelines, on the lungs, livers, and kidneys of all animals. Microscopic examination of animals in the lower exposure groups may also be necessary if there are excessive early deaths or other problems in the highest dose group.

A large part of the costs of long-term carcinogenicity bioassays owes to examining or reading the large number of microscopic slides. For example, EPA has estimated that a bioassay conducted to meet the requirements of TSCA regulations will generate about 40,000 slides, requiring about three-quarters of a year of a pathologist's time in addition to the costs of technicians and materials. Because of these costs, the NTP Ad Hoc Panel suggested that alternatives be considered to reduce the burden of conducting microscopic pathology. NTP has tried to implement such an approach to reduce the pathology requirements. But when using a reduced pathology system, NTP often found it necessary to go back to the original tissues to obtain additional slides for diagnosis. Consequently, NTP has now returned to examining all tissues at all dose levels. While the reduced pathology system decreased the costs of pathology, it increased the time necessary to complete the study (95,121).

Several of the guidelines also detail the qualifications necessary for principal study personnel. NCI guidelines required that the study pathologist be board-certified and that histology technicians be supervised by a registered technician. The NTP Statement of Work requires that the study director have a doctorate in a relevant discipline, that the pathologist have formal training and experience in animal pathology, and that histologists be registered with the relevant accrediting organization. PMA guidelines specify only that the pathologist should have expertise in laboratory animal pathology.

ANALYSIS OF AGENCY CARCINOGEN ASSESSMENT POLICIES

The following comparison will examine policies* that have been issued by the major regulatory agencies:

- EPA Interim Guidelines—EPA (1976) (ref. 293),
- EPA Water Quality Criteria—EPA (1980) (ref. 323),
- EPA Standard Evaluation Procedure for Pesticides—FIFRA (1985) (ref. 328),
- EPA Carcinogen Risk Assessment Guidelines—EPA (1986) (ref. 284),
- CPSC Interim Carcinogen Policy—CPSC (1978) (ref. 228),
- OSHA Carcinogen Policy—OSHA (1980) (ref. 276), and
- FDA Sensitivity of Method Policy (proposed)—FDA SOM (1985) (ref. 246).

There have also been several interagency collaborative efforts, and efforts by nonregulatory bodies:

- National Cancer Advisory Board, Subcommittee on Environmental Carcinogenesis—NCAB (1977) (ref. 348),
- Interagency Regulatory Liaison Group—IRLG (1979) (ref. 347),
- Office on Science and Technology Policy—OSTP (1979) (ref. 23), and
- Office on Science and Technology Policy—OSTP (1985) (ref. 351).[7]

These policies were issued under a variety of circumstances and are organized in several different ways. In some cases, they appear to have been adopted as relatively informal statements of scientific understanding on how carcinogens might be identified. In other cases, they are formally adopted agency regulations, specifying how the agency will identify carcinogens and attempting to limit the kinds of arguments and evidence to be considered in any specific regulatory proceeding. In between these two extremes, some documents outline an agency's standard procedures and discuss problematic areas of interpretation, often including the inference options the agency will generally use.[8]

This chapter focuses on formal written policies, with only limited attention to actual agency practices on carcinogen risk assessment. The policies themselves will be referred to using agency acronyms and the year of the policy, for example, OSHA (1980), OSTP (1985).

Definitions

Not all of these policies propose a formal definition of "carcinogen," although in most cases the text of the policy outlines the various criteria and considerations that will be used to identify and classify carcinogens. In its simplest form, a carcinogen may be defined as a substance that causes cancer (217). Two more complete definitions of a carcinogen are those of OSHA (1980) and OSTP (1985). OSHA gave the following definition of a potential occupational carcinogen:

> . . . any substance, or combination or mixture of substances, which causes an increased incidence of benign and/or malignant neoplasms, or a substantial decrease in the latency period between exposure and onset of neoplasms in humans or in one or more experimental mammalian species as the result of any oral, respiratory or dermal exposure, or any other exposure which results in the induction of tumors at a site other than the site of administration. This definition also includes any substance which is metabolized into one or more potential occupational carcinogens by mammals.

A more recent definition is offered by OSTP:

> . . . a substance which is capable under appropriate test conditions . . . of increasing the incidence of neoplasms (combining benign and malignant when scientifically defensible) or decreasing the time it takes for them to develop.

OSTP (1985) added the qualification that, before concluding that a chemical is carcinogenic, gen-

---

*See reference numbers in the following two bulleted lists for complete policy citations. Policies are only cited by year elsewhere in this section.

[7]This discussion does not cover one other policy that was prepared by the Committee to Coordinate Environmental and Related Programs (CCERP) of the Department of Health and Human Services (232).

[8]For another comparison of agency policies, see Rushefsky (179,180).

eral principles in evaluating animal test results should be followed (e.g., eliminating experimental artifacts).

Qualitative Risk Assessment: Hazard Identification

Use of Human Epidemiologic Data

Most policies declare that well-conducted positive epidemiologic studies provide conclusive evidence for carcinogenicity. The FIFRA (1984) evaluation procedure did not discuss epidemiology at all, and FDA SOM (1985) treated it only cursorily. In evaluating new pesticides and residues of new animal drugs, it is not likely that there will be relevant epidemiologic studies.

EPA (1986), IRLG (1979), and OSTP (1985) all discuss in some detail several of the important factors to consider in evaluating epidemiologic studies, including the strength of association, level of statistical significance, information on dose-response relationships, biological plausibility, temporal relationships, confounding factors and bias, accuracy of exposure and cause-of-death classifications, adequacy of followup, and whether sufficient time has elapsed to allow for latent effects.

One epidemiologic issue provoking special attention in many of these policies is the role to be played by negative human studies in evaluating chemical hazards. All Federal policies addressing this issue state that negative human studies can only set an upper bound on risk estimates. A negative study cannot prove the absence of a carcinogenic hazard. A negative study indicates, at most, that the true risk is unlikely to exceed the specified upper bound. The magnitude of this upper bound depends on the size of the study and the background incidence of the cancer in question.

OSHA (1980) went even further than this, generally referring to negative studies as "nonpositive" studies. In characteristic fashion, OSHA (1980) also set down explicit and stringent criteria for when a "nonpositive" study would be acceptable evidence for an OSHA rulemaking. Such a study will be considered only if:

1. the study involved at least 20 years of exposure and at least 30 years of observation after initial exposure,
2. documented reasons are provided for predicting human cancer site(s) at which the substance would induce cancer if it were carcinogenic in humans, and
3. the exposed group was large enough to detect a 50 percent excess risk at the predicted sites.

To use a "nonpositive" study to set an upper limit on risk, both of the first two criteria must be met and, in addition, there must be reliable human exposure data.

OSHA (1980) pointed out that there have been negative studies for arsenic, benzene, coke oven emissions, petroleum refinery emissions, and vinyl chloride, substances and mixtures now generally believed to be carcinogenic on the basis of other epidemiologic studies. Even the epidemiologic evidence of the association between asbestos exposure and lung cancer among nonsmokers was "nonpositive" for a long time. Selikoff found no excess of lung cancer among nonsmoking asbestos workers for the first 30 years after exposure, though 5 more years of followup demonstrated a positive effect.

Use of Long-Term Animal Bioassay Data

All policies accept the use of animal data as predictive for human beings. Explicitly or implicitly, all the policies acknowledge that substances shown to be carcinogenic in animals should be presumed to present a carcinogenic hazard to humans.

An often-quoted statement on the value of animal data in assessing human risk is that of IARC. Their principle is based on two points: that a number of chemicals were first identified as animal carcinogens, and then evidence confirmed carcinogenicity in humans. Second, all chemicals accepted as human carcinogens that have been adequately studied in animals are positive in at least one species. (See the discussion in ch. 4.) IARC concluded:

> Although this association cannot establish that all animal carcinogens also cause cancer in humans, nevertheless, in the absence of adequate data on humans, it is biologically plausible and prudent to regard agents for which there is *suffi-*

cient evidence of carcinogenicity in experimental animals as if they presented a carcinogenic risk to humans (99).

However, determining exactly what evidence will be considered sufficient to demonstrate a substance to be an animal carcinogen is a little more complex.

OTA identified the following major issues on use of long-term animal bioassay data for hazard determination in agency policies:

- use of the maximum tolerated dose,
- route of administration,
- criteria for a valid negative study,
- classification of tumors as benign or malignant and deciding which should count as evidence for carcinogenicity,
- evaluation of certain problem tumor types and commonly spontaneous tumors,
- use of historical control data,
- statistical evaluation, and
- performance of overall qualitative evaluation.

Some of the agency policies also gave guidance for the design of bioassays. These points have generally been covered in the earlier section on testing.

Use of Maximum Tolerated Dose.—For reasons of economics and practicality, long-term animal bioassays are much too small to provide experimental data on the hazards of low exposures. Therefore, to maximize the sensitivity of animal bioassays for detecting carcinogenic effects, agency guidelines for designing tests specify use of the MTD. This position was also affirmed by the Ad Hoc Panel convened by NTP to consider issues in carcinogenicity testing (258).

In bioassays, the power of a study to detect a tumor increase reliably depends on the number of animals in each exposure group, spontaneous incidence of the particular tumor increased, and the magnitude of the increase. The probability of missing an increase even though the substance is truly carcinogenic (a false negative) is fairly high. For example, with a standard bioassay design using 50 animals per exposure level, the probability of not detecting an increased tumor incidence from 1 percent in controls to 10 percent in the exposed group is 73 percent or nearly three-quarters of the time (258).[9]

With one exception, all agency policies on interpreting test results accept positive test results using the MTD. One policy (FIFRA 1984) raised the concern that exposures at the MTD may represent a toxic insult qualitatively different than those at much lower exposures. FDA (1985) indicated when test levels turn out to have exceeded the MTD (after conducting the 2-year study), negative results do not remove suspicion about possible carcinogenicity. Several other policies address how to interpret positive results at levels that exceed the MTD or show noncarcinogenic toxic effects.

EPA (1986) suggested that studies be carefully reviewed to determine whether the high exposure levels induce effects that would not be seen at lower levels. OSHA (1980) generally accepts positive results from high-dose testing. OSHA will entertain arguments that high doses are not relevant to human exposures only if documentation shows that:

1. at high doses the test animals produce metabolites that are produced only at high doses,
2. these high-dose metabolites are the ultimate carcinogens and the ones produced at low doses are not, and
3. the carcinogenic metabolites are not produced by humans exposed to low doses.

FDA SOM (1985) required "convincing evidence" to rule out carcinogenic effects seen at exposures above the MTD. OSTP (1979) suggested accepting these results only if the noncarcinogenic toxic effects have not altered metabolism or immune system responses in a way that could have caused the carcinogenic effects, while OSTP (1985) declares it is appropriate to consider animal test results that use exposures exceeding human exposures although, as mentioned above, OSTP (1985) also requires that possible organ damage

---

[9]Using Fisher exact test with a one-sided significance level < 0.025.

and metabolic saturation (experimental artifacts that may occur at high doses) be considered before concluding that a chemical is carcinogenic.

Thus, agency policies accept the use of high-dose testing, but many policies raise concerns about how to interpret test results at levels that exceed the MTD. One difficulty in conducting these tests is making a guess concerning the MTD for a 2-year study based on the results from a 13-week study. Sometimes the researchers estimate poorly what the MTD will be. The result is that the study is conducted with doses that are substantially above or below the MTD.

The decisions about how to use results from these studies are a problem. The policies generally appear to reject use of negative results from such studies, but differ in how to handle positive results. FDA SOM (1985) accepted positive studies, even if the MTD was exceeded, "unless there is convincing evidence to the contrary." OSHA (1980) set a policy of entertaining arguments that high-dose results are not relevant to humans only if documented evidence is presented that shows that the ultimate carcinogenic metabolites are produced only at high doses and not in humans exposed at low doses.

Other policies are more restrictive in interpreting such studies. EPA (1986) asked for careful review of studies at levels above the MTD to determine if there was a response that does not occur at lower exposures. FIFRA (1984) went further in arguing that use of the MTD was "interjecting biases of considerable importance" in evaluating animal studies.

As will be discussed later in this chapter, there are other difficulties in applying test results from high doses in animals to predict human risk, even using high doses that do not exceed the MTD.

Route of Administration. —In animal bioassays, the substance under test may be administered in any of several ways: it may be incorporated into the animals' diet, or into their drinking water; the animals may inhale the substance as they breathe; the substance may be dissolved in corn oil (or similar vehicle) and then administered through a feed-ing tube directly into their stomachs (in gavage); or it may be injected or implanted in the animals or painted onto their skin. Although it is desirable that the exposure route used in the animal study be similar to human exposures, this is not always possible. For example, some substances when mixed with feed or water will alter the taste, leading the animals to refuse to eat or drink enough to receive the desired dose.

The two major issues of interpretation are these: If an exposure route that differs from the human route is used, are the results applicable to humans? If tumors are found only in tests that use "unusual" administration routes (such as injection or implantation) or the tumors are found only at the site of administration, are these results applicable to humans?

One view the policies express is that if the resultant tumors are found in organs or tissues distant from the site of application, then the substance should be considered a carcinogen, irrespective of the route of exposure. EPA (1980), OSHA (1980), and IRLG (1979) clearly express this view. On the other hand, tumors found only at the site of administration or by unusual methods may "raise the possibility" of carcinogenicity (NCAB 1977) or might be used as "concordant evidence" (OSHA 1980). Other policies caution that these results merit additional evaluation in assessing their human relevance (EPA 1984, CPSC 1978) or suggest that more testing is needed to resolve safety concerns (FDA SOM 1985). OSHA (1980) treated as indicative of carcinogenicity any contact tumors (those occurring at the site of exposure), from oral, respiratory, and dermal exposure routes and noncontact site tumors (those found at sites away from the exposure site), regardless of the route of exposure. OSHA (1980) will consider arguments that a tumor response only at the site of administration is not predictive of human hazard if: 1) the exposure route is not oral, respiratory, or dermal (i.e., is through injection or implantation); and 2) the tumor induction is related to physical configuration or formulation of material and not to its chemical properties.

Criteria for Valid Negative Study.—Two of the policies provide criteria that must be satisfied for an animal study to be considered a negative study. (As discussed above, some of the test protocols also provide minimum standards for negative studies.) For the IRLG, negative test results can only be considered evidence of no effect when "minimum requirements have been met." "Accepted procedures" include: a) the observation of all animals in the study . . . until their spontaneous death, b) the sacrifice of animals that show clinical signs of severe illness or impending death . . ., and c) terminal sacrifice at a scheduled date near the end of the lifespan (e.g., after 24 months on test). None of the other policies provide criteria for a valid negative animal study, although, as discussed below, many of these policies do specify the weight to be given to "negative" studies.

EPA's CAG guidelines (1986) do not specify minimum test design requirements, but provide that a substance will be classified in the "no evidence" category if there is no tumor increase in "at least two well-designed and well-conducted animal studies of adequate power and dose in different species."

Use of Data on Benign and Malignant Tumors. —When diagnosing cancer in a human patient, microscopic examination of tumor tissue yields a classification of the tumor as benign or malignant. Cancer is a disease of malignant tumors, that is, ones that have the potential to invade other tissues and spread throughout the body. The cells of benign tumors, on the other hand, remain together and do not invade other parts of the body. The clinician can then formulate a prognosis and develop a therapy based on whether the tumor is benign; malignant, having the potential to spread, but localized; or already widespread. Benign tumors may still be of concern, however, because if they develop in vital organs (e.g., the brain), they can cause serious disability and death.

This classification is also used when examining animals exposed during a bioassay. While there are difficulties in the precise classification of some tumors,[10] the more general controversy

concerns how to count the tumors diagnosed as "benign." If there is an increase in benign tumors in the exposed groups compared to the control group, with no increase in the number of malignant tumors, should that increase in benign tumors be sufficient to classify a substance as a carcinogen? If there is no statistically significant increase in the frequency of malignant tumors, but benign and malignant tumors together increase, should that serve to classify a substance as a carcinogen?

Most of the policies have taken the position that it is appropriate to count both benign and malignant tumors when evaluating the carcinogenicity of a substance, although EPA and NTP policies now provide that benign tumors will not be grouped with malignant tumors when they are of a type that is not known to progress to a malignant stage. This principle of grouping tumors based on their potential for progression appears to be gaining general acceptance. The burden, however, is to demonstrate that progression is not likely. In the absence of such evidence, the benign tumors will be grouped with malignant ones.

As an example of an early policy, NCAB (1977) argued that because compounds that induce benign tumors frequently also induce malignant ones, that because benign tumors may represent a stage in transformation to malignancy, and that because benign tumors may themselves endanger health, "if a substance is found to induce benign neoplasms in experimental animals it should be considered a potential human health hazard which requires further evaluation." If the increase in malignant tumors is of questionable significance, the NCAB (1977) policy provided that a parallel increase in benign tumors in the same tissue adds weight to evidence for carcinogenicity. CPSC (1978), IRLG (1979), OSHA (1980), and FDA SOM (1985) all provide for grouping benign and malignant tumors together. EPA (1976) and EPA (1986) provided for combining benign and malignant tumors, unless the benign tumors are not considered to have the potential to progress to malignancy. FIFRA (1984) cited the 1984 proposal of the EPA (1986) guidelines and also includes a list, prepared for the NTP Ad Hoc Panel, of specific tumor types that should and should not be combined. OSTP (1979) apparently dropped a

---

[10]NCAB (1977), FIFRA (1984), and OSTP (1985) pointed to the lack of standard nomenclature for classifying tumor types and the need for professional judgment in examining tissue slides.

discussion of benign and malignant tumors in their final document because a "wide range of opinion" had been expressed, making it "clear that no consensus exists on this issue."

For cases in which the animal response consists of only "benign" tumors, EPA (1976) and EPA (1986) would classify an increased incidence of benign tumors alone as only "limited" evidence of carcinogenicity. OSHA (1980) did allow for the possibility of a benign-tumor-only response, although it required a substantial amount of proof that this response is truly limited to benign tumors and that the tumors will not progress to malignancy.

For evaluating its bioassays, NTP considers chemically induced benign tumors to be an "important toxicological indicator of a chemical's carcinogenic potential in rodents," and includes these findings in its evaluations. A substantially increased incidence of benign tumors alone may serve to place an experimental result in the category of clear evidence for carcinogenicity category, as discussed below.

Although the weight to be placed on benign tumors remains controversial, the fact remains that very few chemicals testing positive in NCI/NTP studies induced only benign neoplasms. Of 113 chemicals studied by the NTP and reviewed by the NTP Peer Review Panel at the time of the analysis, 56 were found to have evidence of carcinogenicity in rodents. Of these, only four (7 percent) were based entirely on the finding of benign neoplasia. Moreover, none of these four were placed in the category of "clear evidence." Of the 56 chemicals, 20 were carcinogenic in all experiments (17 in all 4 experiments, 1 in 3 of 3 experiments, and 2 in 2 of 2 experiments). Of these 20, all caused malignant neoplasms (92).

Evaluation of Certain Problem Tumor Types, Common Spontaneous Tumors, and Historical Control Data.—Another difficulty in evaluating bioassay results arises when an increase in the frequency of relatively common, "spontaneous" tumors is found. "Spontaneous" tumors are ones that arise, generally for unknown reasons, in animals that are not being deliberately exposed to a carcinogenic agent. The difficulty in interpretation occurs because the incidence of some spontaneous tumors is relatively high and variable.

This variability can create an especially problematic situation for evaluation when the control group in a particular bioassay happens to have had a spontaneous tumor incidence that is on the low side of the range for the particular species or strain of laboratory animal. (This may occur simply as a result of random variations among groups of animals: some will be above average, some below average in the incidence of tumors.) For a particular bioassay, the exposed groups may then have a tumor incidence that is similar to what could be expected for purely spontaneous tumors, but there would appear to be a significant excess just because the control group had an abnormally low incidence. In addition, background frequency may in some cases be affected by the animals' diet and metabolic state. On the other hand, if the study has been conducted properly, the animals will have been assigned randomly to treatment and control groups. A control group with an abnormally low incidence would be accompanied by exposed groups that would be expected to have had a similar incidence if they had not been exposed.

Historical control data consist of information on the incidence of tumors in groups of control animals of particular species and strains within a given laboratory or from a particular source of animals. Historical control data can be useful in two circumstances:

1. judging the likelihood that the difference in tumor incidence between the exposed and concurrent control group (the animals used during the actual study) can be explained as random variations within an unexposed population of animals; and
2. judging whether a rare tumor may be of concern, even though the comparison between exposed and control groups has low statistical power because there are very few cases.

Care must be taken when using historical control data because spontaneous tumor incidence can change over time as a result of genetic changes in the animal population over generations or changes in pathology and tumor diagnosis. In addition, there may be differences of opinion among the pathologists who examined the different sets of animals. Because concurrent controls have been treated to the same conditions, except for chemical exposure, as the exposed group, and have been examined by the same pathologists, they are the best source of information on the spontaneous tumor incidence in the group of animals being studied.

In their policies, OSHA (1980) and CPSC (1978) simply declared that a significant increase in spontaneous tumors would serve to identify carcinogens. NCAB (1977) concluded that an increase in spontaneous tumors would "raise the possibility" that a substance was carcinogenic, and IRLG (1979) urged caution in interpreting bioassay results when the observed increase in a spontaneous tumor type is within the range observed in historical controls from the same colony of animals.

OSTP (1985) acknowledged the problems of evaluating increases in spontaneous tumors, emphasized the need to consider other biological evidence, and suggested that historical controls can aid in evaluation, although "care should be exercised when combining different control groups."

FIFRA (1984) suggested that while the occurrence of spontaneous tumors "complicates" evaluation, "judicious use" of historical control data can be of assistance, although it should not substitute for concurrent control data. However, in an example given in the text, the authors of FIFRA (1984) discounted an apparent tumor excess because the incidence in the exposed group was within the range found in historical controls. FIFRA (1984) also described how historical control data could be used for interpreting the observation of rare tumors, while it cautioned that underlying spontaneous frequency can change, depending, for example, on changes in pathology technique.

According to IRLG (1979), the occurrence of rare tumors raises suspicion and is worthy of careful review, but this occurrence by itself is "not necessarily evidence" of carcinogenicity without additional supporting evidence.

Because an increase in the occurrence of mouse liver tumors is a subject of continuing controversy, some policies have special provisions for dealing with this result. According to EPA (1986), a bioassay that shows an excess only in mouse liver tumors should be considered "sufficient" evidence of carcinogenicity when other conditions for classification of "sufficient" evidence occur. Classification of evidence could be downgraded to "limited," however, if a number of factors among the following are observed:

- the liver tumors occur only in the highest dose group or only at the end of the study,
- there is no substantial dose-related increase in proportion of tumors that are malignant,
- the tumors that occur are predominantly benign,
- there is no dose-related shortening of the time to tumor appearance,
- short-term tests are negative, or
- excess tumors occur only in one sex.

Statistical Evaluation.—Only a few of the policies actually devote any substantial discussion to the topic of statistical analysis of bioassay results. OSHA (1980), EPA (1986), and OSTP (1985) tated that such analyses shall be performed. OSHA (1980) did provide that these analyses would not be used exclusively to evaluate evidence for carcinogenicity. EPA (1986) provided, on the other hand, that evidence for carcinogenicity should be based on statistically significant response in specific organs or tissues, although the weight given to level of significance and other information is "a matter of overall scientific judgment." The policies have not specified the level of statistical significance to be used in evaluation.

Overall Evaluation of Bioassay Results.—Some of the policies provide a list of some general principles for overall evaluation of bioassay results. EPA (1986) stated that the strength of positive evidence increases with:

- an increase in number of tissue sites affected;
- increase in number of species, strains, and sexes showing response;
- "occurrence of clear-cut dose-response relationships" and high level of statistical significance of tumors in treated compared to control animals;
- dose-related shortening of time to tumor occurrence or time to death with tumor; and
- dose-related increase in proportion of malignant tumors.

OSTP (1985) listed several factors that increase confidence in the conclusion that a substance is a carcinogen. These include:

- an observed dose-response relationship,
- a marked increase in tumor incidence in treatment groups,
- tumors being found at multiple sites,
- significant reduction in latent period, and
- information comparing the neoplastic stages of tumors in treatment and control groups.

In addition, information on preneoplastic lesions, target organ effects in prechronic studies, and chemical activity at physiological, cellular, and molecular levels may help.

But OSTP (1985) added a caution:

> . . . the carcinogenic effects of agents may be influenced by non-physiological responses (such as extensive organ damage, radical disruption of hormonal function, saturation of metabolic pathways, formation of stones in the urinary tract, saturation of DNA repair with a functional loss of the system) induced in the model systems.

Tests that produce these responses need to be evaluated for human relevance, according to OSTP (1985). While there has always been concern that high-dose testing might not be relevant to human exposures, the OSTP (1985) caution represents a more explicit discussion of these potential problems than had been seen in most earlier policies.

Use of Short-term Tests and Structure-Activity Relationships

While none of the agency policies provided for using positive short-term test results as the sole basis for identifying or regulating carcinogens, they all indicated that such test results may be used either as supporting information (EPA 1980, EPA 1986, OSHA 1980, IRLG 1979, OSTP 1985) or as an indication that further testing may be warranted (CPSC 1978, NCAB 1977, OSTP 1979).

Few of the policies directly discussed use of structure-activity relationships (SARs) to identify carcinogens. Of those that do, none will use SARs as the sole basis for such identification, although IRLG (1979) states they might be used as corroborative evidence along with other data, or in the absence of other data, as limited, suggestive evidence. EPA (1980) states that SARs will not be used as the sole basis for quantitative risk assessments, while CPSC (1978) suggests that if related chemicals are carcinogenic, the chemical in question should be tested prior to being used in consumer products.

Evaluating Conflicting Data

There are major issues in evaluating substances for carcinogenicity when the evidence is mixed, such as both positive and negative results in animal bioassays, or negative data in human studies and positive animal data. Many of the policies have come to conclusions about what to do in these situations, and provide guidance for the overall qualitative evaluation.

Conflicting Animal Data.—Animal evidence may differ in two ways: it may be positive and negative studies in different species, or positive and negative studies in the same species. Agency policies generally hold that positive results in one species outweigh negative results in another. Thus, in theory, the policies imply that agencies can regulate chemicals based on positive results in a single species. This principle, that positive results supersede negative results, may be extended to cover not just conflicts between results in different species, but also conflicting results in different strains of the same species or between sexes. Of course, arguments will still occur over what constitutes convincing positive evidence. Conflicting positive and negative results in the same species will, in general, provoke a case-by-case evaluation. For example, EPA (1986) provided:

> Positive responses in one species/strain/sex are not generally negated by negative results in other species/strains/sexes. Replicate negative studies that are essentially identical in all other respects to a positive study may indicate that the positive results are spurious.

Conflicting Animal and Human Data.—Similarly, the policies generally provide that positive animal data will outweigh negative human data. OSTP (1979) points out that limitations in the power of human epidemiologic investigations to detect an effect can often explain the apparent discrepancy between positive animal data and negative human data. As noted elsewhere, EPA (1980) and IRLG (1979) provide that in these situations the negative human studies can be used to set an upper bound on the estimate of human risk.

Quantitative Risk Assessment: Dose-Response Determination

The NAS committee divided the more quantitative aspects of risk assessment into dose-response determination, exposure estimation, and risk characterization. Quantitative estimation has been a particularly vexing area because of the need to make a series of often untestable assumptions to perform quantitative extrapolation from animal to human cases, the inadequacy of historical information from epidemiologic studies, and a frequent lack of data on current human exposures. During the 1970s, many argued that quantitative risk assessment was far too imprecise for use by regulatory agencies except for the relative ranking of different hazards for setting priorities. OSHA (1980) in fact adopted this position in its cancer policy, although because of court interpretations of its regulatory authority, OSHA now conducts quantitative risk assessments.

Long-term animal bioassays provide information relating the dose administered to the animals (the exposures) to the proportion of animals that are diagnosed with tumors. In these bioassays, exposure levels are deliberately set at high levels to maximize the probability of detecting a carcinogenic effect.

Information from epidemiologic studies may be used to relate some measure of exposure with the proportion of people incurring cancer, although there are often significant inaccuracies in exposure estimates made many years ago. For epidemiologic studies, the population examined is often a group of workers who were exposed to relatively high levels, often much higher than the levels workers are currently exposed to or that the public may be exposed to.

In many ways, the problems of extrapolating from effects of high doses to those of low doses differ based on whether one is using epidemiologic data or animal data. For example, an epidemiologic study may have examined the incidence of lung cancer among workers exposed to relatively high levels of arsenic inside a plant. Extrapolating from those exposures to the lower exposures found among community residents outside the plant poses difficulties, but the magnitude of the range from high worker exposure to lower ambient environmental exposure is less than that encountered when extrapolating from animal study results. As mentioned earlier, the highest exposed animals are to be exposed at the MTD. The exposures in human study populations are high compared to general environmental exposures, but are not at levels approaching the MTD. Of course, for regulating worker exposures, frequently the exposure levels for the study population are close to those found in workplaces that are to be regulated.

When using either animal or human data, the first issue is to extrapolate from the estimated probability of harm at these higher exposure levels to estimate what the probability of harm is at the lower exposure levels of interest. This high- to low-dose extrapolation is often very uncertain and controversial because there are usually few observed data on health outcomes at the lower levels.

This problem is most severe in the case of animal data because of the use of the MTD in those studies. The dose levels that the animals are exposed to are often several orders of magnitude greater than exposures that most people experience in the general environment.[11]

Risk assessors using epidemiologic data of exposed worker groups must also extrapolate from relatively high exposures to low exposures. Past exposures, before the agent's toxic effects were recognized or before the beginning of concerted public and private efforts to reduce workplace hazards, were often much higher than current exposures, and nearly always much higher than proposed new exposure limits. For example, workers exposed to benzene during the 1950s and 1960s at levels substantially above the current OSHA

---

[11]Orders of magnitude refer to differences that can be expressed in powers of 10. Thus the difference between 10 and 100 is one order of magnitude, while the difference between 10 and 1,000 is two orders of magnitude.

standard of 10 ppm incurred a significantly increased risk of leukemia. But how large is the risk of leukemia among workers currently exposed below 10 ppm?[12]

Unfortunately, there are very often few data on historical exposure levels because quantitative industrial hygiene measurements may never have been taken. In these cases, the risk assessors must make guesses about what the exposure levels were.

The relationship between dose (or exposure) and biological response (in this case the induction of cancer) is one of the most fundamental in the fields of toxicology and epidemiology. If data on exposures and responses are available, they may be plotted on a graph. The line joining these plotted points is called a dose-response curve. Generally, the dose-response curve has a positive slope, that is, the greater the dose, the larger the response.

Even so, the dose-response curve may have several different shapes, ranging from a straight line to differently shaped curves. Figure 2-1 shows several possible dose-response curves. In an ideal world, dose-response assessment would ascertain the shape of the curve and thus make estimates of what human risk is likely to be. In this ideal world, there would be data on the response in the range of human exposures, and there would be enough data points at different levels to distinguish between different possible dose-response curves. Alas, in the real world, the data from animals usually represent dose levels substantially higher than the range of human exposure, often several orders of magnitude higher.

A number of methods have been proposed for extrapolating from high to low doses. They range from the simple technique of drawing a straight line on a graph to sophisticated computer programs that fit the available data to develop a mathematical equation relating exposure to response. As discussed below, the Federal agencies have adopted linear no-threshold models, which, while allowing for nonlinear dose-response curves at higher exposures, extrapolate to low doses using an assumption of low-dose linearity.

[12]See ref. 172 for a recent study of the dose-response for benzene.

Figure 2-1.—A Stylized Dose-Response Curve and Some Extrapolated Curves

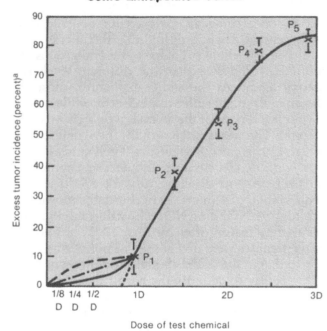

Dose of test chemical

[a]Excess tumor incidence (percent) is defined as:

$$\left(\frac{\text{tumors in exposed population}}{\text{number of exposed population}} - \frac{\text{tumors in control population}}{\text{number of exposed population}} \right) \times 100$$

——— a sigmoid dose-response curve; infralinear between O and P₁
—·—·— linear extrapolation
——— supralinear extrapolation
------ line projected to a threshold

SOURCE: U.S. Congress, Office of Technology Assessment, *Assessment of Technologies for Determining Cancer Risks From the Environment* (Springfield, VA: National Technical Information Service, June 1981).

Agency Policies on Dose-Response Assessment

The different policies of the Federal agencies contain various degrees of discussion concerning some of the problematic areas of quantitative risk estimation. These policies include a series of assumptions about how to estimate human risk based on animal data. Some of the assumptions discussed in this section apply only to risk assessments based on animal data (choice of animal species and species conversion factors), while the others apply to use of both human and animal data.

Do Thresholds Exist?

The first issue in dose-response determination concerns whether there might be a "no effects" or "safe" level of exposure to carcinogens. This particular issue has generated intense regulatory debates.

For noncarcinogenic toxic agents, toxicologists have generally believed that no-effect thresholds could be determined. To do so, several groups of animals would be exposed at different levels to a toxic agent. At the higher exposures, most of the animals might suffer toxic effects, while at the lowest levels none of the animals would show such effects. The researcher would then determine which of the various exposure levels was associated with no toxic effects and declare that to be a NOEL or a "no observed adverse effects level" (NOAEL). To estimate "safe" human exposures, the highest NOEL or NOAEL would be divided by a safety factor, often by 10 or 100. The underlying premises were that human response was similar to that in the tested animals, that humans had some ability to detoxify the harmful agent or recover from its effects, and that the safety factor would provide sufficient protection from incorrect guesses about the degree of toxicity.

However, for carcinogenic effects the general belief in the scientific community is that it is not possible to determine a no-effect threshold for carcinogens. This belief is based on observations, experimental limitations, and theoretical considerations.

Dose-response data from many epidemiologic and toxicologic studies of carcinogens fit mathematical models that are linear without an apparent threshold. Also, because of experimental limitations inherent in the size of bioassays typically used, it is not possible to demonstrate conclusively the existence of a no-effect threshold. This is especially true for risk levels of interest to regulators, which are much smaller than those detectable in these studies.

Certain theoretical considerations about cancer causation also imply the absence of no-effect thresholds. Cancer is a disease of self-replicating cells. Tumors can begin from a single cell, the DNA of which has been damaged by a small amount of a carcinogenic chemical. Unless that damage is repaired, the genetic material of the cell's "daughters" will have been altered. These daughter cells are then irreversibly "initiated" and may eventually develop into a tumor. Various cellular repair processes do exist, but these repair processes will lead to a no-effect threshold only if their efficiency is 100 percent.

Moreover, even if thresholds for individuals could be determined, genetic differences among people would make it difficult to demonstrate a no-effect threshold for a population. Finally, if exposure to a carcinogenic agent is contributing to the background incidence of cancer, the additional effect of the new carcinogen is approximately linear at low doses, without a no-effect threshold.

On the other hand, there are data from epidemiologic and toxicologic studies that are curvilinear. Some scientists interpret these data as revealing possible thresholds. Some scientists also believe that a threshold may exist if the carcinogen acts through an indirect mechanism, although it is currently difficult to distinguish among carcinogens based on mechanism. In addition, if the carcinogen contributes to the background of cancer, the additional effects will be linear and would not show a threshold. Finally, even if carcinogens lack a theoretical no-effects threshold, many believe that at some finite exposure level the additional risk is so low that it may be regarded, for practical purposes, as safe. In this case, the carcinogen might be considered to have a practical threshold.

For policy reasons, in addition to these scientific considerations, the agencies have generally assumed that carcinogenic chemicals lack no-effect thresholds. This reflects the agency's desire to be conservative in assessing risk, that is, to err on the side of safety in protecting public health. Most of the agency policies endorse the view that no-effects threshold levels do not exist for carcinogens. EPA (1980) presents this view:

> Because methods do not now exist to establish the presence of a threshold for carcinogenic effects, EPA's policy is that there is no scientific basis for estimating "safe" levels for carcinogens. The criteria for carcinogens, therefore, state that the recommended concentration for maximum protection of human health is zero.

EPA (1986), CPSC (1978), OSHA (1980), FDA (1985), IRLG (1979), and OSTP (1985) also set forth the position that there is no safe exposure level for carcinogens. EPA (1976) stated that the linear model derived from study of radiation effects is also applicable to chemical carcinogens, but added that the costs of prohibiting exposures to some chemicals might be socially unacceptable.

Two of the policies, however, did not strongly endorse the no safe threshold principle: OSTP (1979) and FIFRA (1984). OSTP took no stand on the issue, citing disagreement within the scientific community. FIFRA argued that the no-effect threshold concept is contrary to the toxicologic principles for other kinds of toxicity, suggests that this concept has inhibited scientific discussion in this field, and provides that if an EPA evaluator believes a threshold exists, this fact should be stated along with its rationale.

Mathematical Models for Fitting Data and Extrapolating From High to Low Doses

Beyond the issue of whether safe thresholds might be discoverable, the agencies have issued some guidance concerning the techniques that will be used to describe, in mathematical terms, the dose-response relationship. This process is important because carcinogens vary a great deal—by a factor of more than 10 million—in their potency (70). Quantitative dose-response estimation is designed to produce estimates of carcinogenic potency, which can then be used to estimate the degree of human risk associated with a given exposure level or the exposure level that corresponds to a preselected level of human risk, for example, what exposure would lead to an increased risk of 1 in 1 million? Information on the degree of risk for given exposure levels can be further combined with information on exposure levels to estimate the number of cases occurring or expected to occur in a population. These estimates could then be used to decide on a particular regulatory action.

The basic issue at this step in risk assessment is to develop estimates of response rates at low exposure levels using dose-response information from the generally much higher doses of animal bioassays or epidemiologic studies. Put another way, the problem involves extrapolating from the high-dose region of the dose-response curve where animal tumor rates are in the range of 5 to 50 per-

cent to the low-dose region corresponding to an estimated human incidence range of between 1 case for every 100,000 people (10^{-5}) to 1 case for every 100 million people (10^{-8}).[13]

A variety of mathematical models have been proposed for this analytic step. These models fall into three basic types: mechanistic models, tolerance distribution models, and time-to-tumor models (351). The major ones used by the agencies include the multistage, one-hit, multihit, logit, probit, and Weibull models.

Animal bioassays produce only a few points on the dose-response curve: the tumor incidence for two or three exposure levels and the control group. A major difficulty in risk assessment is that many of the mathematical models can fit these two or three data points equally well, yet differ by orders of magnitude in the corresponding estimates of human risk at low doses. Figure 2-2 illustrates several possible dose-response curves in the low dose region. All of these curves fit the actual experimental data reasonably well, but the models underlying them utilize different assumptions about the nature of the biological processes that may be involved.

Choosing among these competing models involves questions of risk assessment policy. The major issues include crucial assumptions concerning whether there might be a no-effect threshold, whether the risk of exposure adds to the background cancer rate, whether the dose-response curve at low doses is linear, and the precise mathematical techniques for the calculation algorithm.

As a rule, the agency policies endorse models that assume the absence of no-effect thresholds and that the low-dose portion of the dose-response curve is linear. The assumption of linearity is based on several considerations and is generally thought to be conservative in the sense that it is unlikely to understate the true risk. Recent research, however, indicates that this assumption is not always true. In these cases, the dose-

---

[13]For environmental exposures, there is no consensus on what level of risk might clearly be considered "desirable," "acceptable," or "unacceptable," but agencies often act to regulate exposures that pose a cancer risk greater than 1 death for every million people exposed (or a risk of 10^{-6}).

Figure 2-2.—Possible Types of Dose-Response Curves in the Low-Dose Region

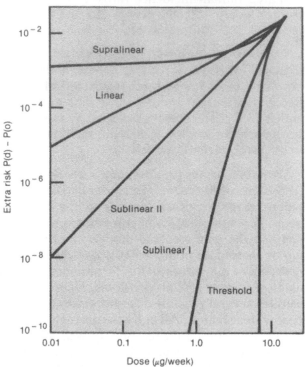

SOURCE: National Academy of Sciences/National Research Council, *Risk Assessment in the Federal Government: Managing the Process* (Washington, DC: National Academy Press, 1983).

response curve may be supralinear and the use of linear models may actually underestimate the true risk (13). On the use of models, OSTP (1985) states:

> No single mathematical procedure is recognized as the most appropriate for low-dose extrapolation in carcinogenesis. When relevant biological evidence on mechanism of action (e.g., pharmacokinetics, target organ dose) exists, the models or procedures employed should be consistent with the evidence. However, when data and information are limited, and when much uncertainty exists regarding the mechanisms of carcinogenic action, models or procedures which incorporate low-dose linearity are preferred when compatible with the limited information.

Over time the agencies have become more convinced that quantitative extrapolation is possible and useful. NCAB (1977) emphasized the uncertainties of extrapolating based on animal data. OSHA (1980) rejected an approach of setting regulatory standards based on quantitative risk

assessment. On the other hand, EPA, with the formation of CAG in the mid-1970s, began applying quantitative extrapolation and the agency has now built up an extensive background in risk assessment.

One important change in the late 1970s was EPA's shift away from the "one-hit" model to the linearized multistage model developed by Kenneth Crump (discussed in EPA 1980). While this model is able to fit dose-response curves in the high-dose region that are very curvilinear, the upper confidence limit of this model is effectively linear in the low-dose region. This model was chosen because it was compatible with the multistage theory of carcinogenesis and because it uses all the data from most animal experiments (123).

Some of the policies also indicated that a variety of models should be used in any particular case (EPA 1976, OSTP 1985). The use of several different plausible models allows the risk assessor to characterize the potential uncertainty that is related to the choice of model. Some policies, in particular OSTP (1985) and EPA (1986), emphasized that selection of the extrapolation model must be chosen case by case. The chosen model should be the one that has the most correspondence with other evidence which relates to the expected mechanism of action and to the biological activity of the chemical in question. The selection should also be based on statistical considerations. In both of these policies, however, there was a preference for models which incorporate an assumption of low-dose linearity. FDA (1985), in contrast, stated that using a variety of models is not likely to provide useful information.

Should Dose-Response Estimates Be Upper Confidence Limits or the Maximum Likelihood Estimate?

There has been some debate on which of two different risk estimates—maximum likelihood estimates or upper confidence limits—should be emphasized in risk assessments. For a given model, a maximum likelihood estimate is the estimated risk at low doses that corresponds to the maximum likelihood curve, which is defined as the mathematical curve that best fits the given high-dose data. The upper confidence limit (often designated as the 95-percent confidence limit), for a given model, is calculated under certain assumptions about the dose-response curve and is linear in the low-dose region.

For example, the results of a bioassay for exposures at 100 and 200 ppm can be fed into the computer. The program generates a risk estimate for exposures at 0.1 ppm. The maximum likelihood estimate might be 1 chance in 10,000 or 1×10^{-4}, while the upper confidence limit might be 1 chance in 100 or 1×10^{-2}.

The estimated risk at the upper confidence limit will be higher than that for the maximum likelihood estimate. For certain dose-response data, the difference will not be important. For other data, the differences may be large. Of course, if the underlying model is wrong, then both the maximum likelihood estimate and upper confidence limit will also be wrong.

Because the upper confidence limit is forced to be linear in the low-dose region, it allows for the dominant view that carcinogens lack no-effect thresholds and have dose-response curves that are linear at low doses. It is usually stated that the true risk is unlikely to exceed the upper confidence limit, and it is possible that the true risk is actually less. Although this is the case for most bioassay data, for some data sets this is not always true.

Some argue that the maximum likelihood estimate is the "best estimate" for a given model and ought to be used in preference to the upper confidence limit, which is possibly too high and thought to represent unnecessary conservatism.[14] In fact, the maximum likelihood estimate may be misnamed. It is the extrapolated risk estimate for a particular dose on the maximum likelihood curve, which, under certain assumptions about the mathematical form of the dose-response curve, is the curve that most closely fits the actual experimental data.

But it is possible, with certain high-dose animal data, to develop a maximum likelihood estimate for the low doses of regulatory interest that does not have a linear term and would thus not

exhibit low-dose linearity. Because, for reasons discussed above, it is generally presumed that the dose-response curve for carcinogens is linear at low doses, the maximum likelihood estimate in these cases would systematically understate the "most likely" risk based on our understanding of cancer causation. Moreover, small fluctuations in the underlying data at high doses in the bioassay can dramatically change the maximum likelihood estimate at the low doses of interest. The upper confidence limit is a more stable number. Finally, because in most cases the true risk is not likely to exceed the upper confidence limit, a regulation based on this estimate will be sufficiently protective. Thus, the estimates based on upper confidence limits are used not only to be conservative in assessing risk, but because a dose-response curve that is linear in the low-dose region is plausible on biological grounds.

Only two of the policies explicitly chose between the maximum likelihood estimate and the upper confidence limit. Both of these are in discussions prepared by CAG. For EPA's water quality criteria documents (323), the discussion supports the adoption of the linearized multistage procedure developed by Kenneth Crump. Although EPA's CAG still uses the upper confidence limit, EPA (1986) seems to respond to critics of the upper confidence limit with a different view:

> Such an estimate [based on the upper confidence limit] . . . does not necessarily give a realistic prediction of the risk. The true value of the risk is unknown, and may be as low as zero. The range of risks, defined by the upper limit given by the chosen model and the lower limit which may be as low as zero, should be explicitly stated.

EPA (1986) argued that current data and procedures do not allow calculations of "best estimates," but promises to use them if they become available, most likely "when human data are available and when exposures are in the dose range of the data."

Choice of Data: Using the Most Sensitive Species

If data on carcinogenic response are available from more than one animal study or from both animal and human studies, which of these data

---

[14]Because different models give different estimates, the best estimate depends on the model selected.

sets should be used? The general principle in the agency policies is to use the data from the most sensitive species, that is, the study showing the highest response for a given exposure level, provided the study is of acceptable quality (EPA 1980, OSHA 1980, IRLG 1979).[15] The general rationale is that little is known about the relative sensitivities of different species and in the absence of evidence on the effects of a chemical in human beings, it is not possible to know human sensitivity. Because it is possible that humans are in fact the most sensitive species, the approach taken is to use data for the species, strain, and sex that is most sensitive, and not to reduce the risk estimates by combining data from a less sensitive species, strain, and sex.

EPA (1986) suggests use of animal data from a species that responds most like humans, if information on this correlation exists. In the more likely event that it does not exist, then the policy suggests use of all biologically and statistically acceptable data from all animal studies to identify a range of risk estimates. However, emphasis is to be placed on results from the "animal studies showing the greatest sensitivity . . . with due regard to biological and statistical considerations." When human and animal exposure routes differ, the policy states that the risk assessment should consider uncertainties about doses delivered to target organs and outline the assumptions used.

Species Conversion Factors

The second important mathematical step in dose-response assessment is estimating human doses or exposures equivalent to those used in the animal studies (converting "from mouse to man"). There are several different ways this can be done: using the ratio of body weights, the ratio of body surface areas, daily or lifetime doses; or by assuming equivalence in terms of exposure concentrations in food, air, or water. Depending on the method and whether rat or mouse data are being used, the resulting risk estimates can vary by up to a factor of 40 (217).

The two most debated methods for cross-species scaling use the ratio of body weights and the ratio of body surface areas.[16] The assumption that equivalent doses may be calculated using body surface areas was based on studies of the effects of certain drugs in different species. Studies of some other drugs and chemicals show effects that are proportional to the ratio of body weights. The choice of the body weight conversion (using mg/kg/day), will lead to risk estimates that are one-fifth (using rat data) to one-twelfth (using mouse data) of those developed using the surface area conversion (mg/m³). Thus the risk estimates scaled using body surface area will indicate higher estimated risks for a given exposure level than those based on body weights. Both methods have their proponents.

EPA's policies (1980 and 1986) assumed that response is proportional to daily dose per unit of body surface area, although the latter policy allows for the use of other methods if information is available.[17] IRLG (1979) argued that there is no one single factor to capture the differences in animal and human susceptibility and suggests that "several species-conversion factors should be considered in estimating risk levels for humans . . . ." The other policies did not specify what species scaling factor to use.

Agency Practice of Quantitative Risk Assessment

In practice, many of the choices in performing a risk assessment for a particular chemical are specific to that chemical. These choices focus on which particular study to use; for instance, if there are several animal bioassays, which are of acceptable scientific quality, and of those that are acceptable, which particular study should be used? Also at issue is which data set to use; for example, if several tumor sites are affected, which one should be used? While several policies suggest using the most sensitive animal and tumor sites, often the

---

[15]If, however, human data are also available, those data may be used to set an upper limit on the risk estimates, as discussed above.

[16]In practice, surface area is approximated by raising the body weight to the 2/3rd power.

[17]Specifically, comparative toxicologic, physiological, metabolic and pharmacokinetic information might be useful for directly developing a cross-species extrapolation (58). However, most of the time, these data are limited or not available.

policies also suggest incorporating information on the human relevance of these tumor sites, if that information is available.

Choices like these are inescapable in risk assessment. General guidelines will not obviate these choices. It may not be desirable to do so either. If the process is made routine and a "cookbook" approach has been formulated, talented individuals with the most to contribute toward developing new approaches may be dissuaded from entering the process (86).

In discussions with OTA, agency staff often indicated that they use a flexible approach to risk assessment, incorporating new knowledge when it becomes available, and allowing choices to be made case by case to develop a risk assessment that is appropriate for a given chemical. Still, the agencies also tend to use certain "default" assumptions in the absence of other data or considerations.[18] Some of these default assumptions are discussed in the written policies, although not all of them are. To develop an understanding of the agencies' approaches to these issues, OTA asked agency staff about the default assumptions used for extrapolating from high to low doses and for converting animal data to estimate human risk.

To extrapolate from high to low doses, EPA, OSHA, and CPSC all use the multistage model, and more specifically, the same computer program for the actual mathematical manipulations. This program uses a specific mathematical algorithm to develop an equation for the dose-response curve and then uses that equation to estimate the risk at low doses. In addition, these three agencies often run other models (e.g., one-hit, multihit, probit, logit, or Weibull models) to obtain a range of possible estimates (33,58,118).

Again, these three agencies differ in whether they use the upper confidence limit or maximum likelihood estimate. As discussed above, EPA uses the upper confidence limit of the multistage model, also known as the "linearized multistage model," in the absence of information that would

indicate the use of another model. In published risk assessments using the multistage model, OSHA has used the maximum likelihood estimate (118). CPSC also uses the maximum likelihood estimate, although only if the data appear to be linear at low doses (33).

FDA uses a different procedure, which usually gives results similar to those of the linearized multistage model. This method represents a modification of the Gaylor-Kodell linear interpolation method. In the FDA modification, the estimated response rate at the lowest dose, where curvilinearity is no longer discernible from the data, is extrapolated linearly to the background response rate (the tumor incidence when the exposure level is zero) to estimate risk at low doses. With data that are considered insufficient or of questionable quality, the upper confidence limit on the response rate is used as the starting point for this linear extrapolation. FDA has used upper limits on other models (e.g. multihit, logit, and Weibull models) and found the results comparable to the Gaylor-Kodell procedure (188).

In setting a species conversion factor for converting animal data to human risk estimates, the four agencies split evenly on the default factor to be used—EPA and CPSC convert on the basis of body surface area, while FDA and OSHA convert on the basis of body weights (33,58,118,182).

For this step in risk assessment, agency staffs are actively exploring the use of pharmacokinetic models; with appropriate data for both humans and animals, it is theoretically possible to perform the cross-species conversion directly rather than by relying on an assumption. CAG has developed a draft risk assessment for tetrachloroethylene that uses pharmacokinetic information (58). The other agencies are exploring ways to use this kind of information, both qualitatively and quantitatively, but have not yet published risk assessments that used pharmacokinetic modeling.

Thus, the four main regulatory agencies have chosen four different approaches to these three issues: the method of extrapolating from high to low doses, use of the Upper Confidence Limit or the Maximum Likelihood Estimate, and converting data from one species to predict another specie's response. How much these choices affect the

---

[18]Computer users will recognize this use of the term "default." In computer terminology, "default" usually refers to the variables, values, or parameters that will be used unless the user specifies otherwise.

resulting risk assessments depends on the precise nature of the data used, such as the shape of the animal dose-response curve. In some cases, these different approaches could lead to important differences in estimated risk; in other cases, there would be little difference. However, compared to the use of models that assume a no-effects threshold, the models chosen by the four agencies will tend to be relatively close to each other.

Agency Policies on Human Exposure Estimation

After the dose-response characterization, the next step is to combine the resulting mathematical representation of hazard with data on actual human exposures. According to OSTP (1985), a risk assessment is only as good as the human exposure estimates, although far less attention has been paid to this aspect of the risk assessment process. In fact, exposure is often the crucial step in determining whether human risk is substantial or trivial. But this is often the area where the data are weakest. The agency policy documents generally devote only limited attention to the issues of exposure assessment, often merely presenting questions that should be answered.

OSTP (1979) recommended that research be done on exposures and CPSC (1978) stated that its staff analyses consider the nature and extent of human exposure to products containing the regulated substance and their potential for human uptake.

OSTP (1985) gave some attention to exposure assessment. Exposure assessments rely largely on monitoring data (e.g., actual measurements of chemical concentrations in water) and modeling (e.g., computer programs designed to predict exposure levels under a variety of assumptions). In the appendixes to a chapter on exposure (OSTP 1985), participating agencies presented descriptions of a number of the data bases, such as on food consumption, food additive use, modeling techniques, and other pertinent topics.

In its summary principles, OSTP (1985) argued that "a single generally applicable procedure for a complete exposure assessment does not exist." Exposure assessments should be tailored to pro-

vide information relevant for the risk assessment and should describe the "strengths, limitations, and uncertainties of the available data and models and should indicate the assumptions made to derive the exposure estimates." A range or array of exposure values is generally preferred to a single numerical estimate.

EPA (1976) presented a list of exposure variables to identify and factors to consider in risk assessment: known and possible exposures, data on factors relevant to effective dose, physical and chemical parameters, possible interaction of agents, likely exposure levels, both time pattern and weighted averages for total population and subgroups with different exposures, size of groups (and whether exposures involve children and pregnant women), adequacy of exposure estimation methods, and uncertainty.

EPA's Carcinogen Risk Assessment Guidelines (EPA 1986) called for a case-by-case selection of methods to match data and level of required sophistication and, unless there is evidence to the contrary, for basing risk estimates on cumulative doses received over lifetimes, expressed as average daily exposure prorated over a lifetime. Furthermore, analysts should assess the level of uncertainty in exposure assessment.

EPA has also issued separate guidelines on exposure assessment (285). The intent was that these guidelines, "by laying out a set of questions to be considered in carrying out an exposure assessment, should help avoid inadvertent mistakes of omission" (285). Thus, consistency among exposure assessments would be promoted and the information developed would be in a form compatible with dose-response assessments. The text of the guidelines, only 11 pages in the *Federal Register*, is largely an outline of points that analysts should cover in exposure assessment. These include information on the properties of the chemical in question, the sources of production and distribution, exposure pathways and environmental fate, information on measured or estimated concentrations, a description of the exposed population, and an "integrated exposure analysis." The last item consists of the actual calculation of exposures, information on human dosimetry, development of exposure scenarios (occupational, consumer, transportation, disposal, food, drinking water, ambient), and discussion of uncertainty. An issue of emerging importance for exposure estimation involves potential human exposures to toxic chemicals from different routes. For example, a carcinogenic chemical in drinking water might also present a dermal and inhalation hazard when people are taking showers.

The EPA guidelines are general and do not specify particular methods, procedures, or assumptions to use in the absence of data. The guidelines express a preference for measured data, but recognize the need to use mathematical modeling in many cases. In actual practice, exposure estimation has involved extensive use of computer models in the absence of exposure measurements.[19] The EPA guidelines encourage "the development of realistic assessments based on the best data available," rather than the use of "worst-case assessments." But EPA "will err on the side of public health when evaluating uncertainties, when data are limited or nonexistent" (285). EPA's Office of Toxic Substances has developed nine volumes presenting methods for assessing exposure in the ambient environment, chemical disposal, drinking water, occupational exposure, consumer exposure, food contamination, transportation-related spills, and on methods for enumerating and characterizing exposed populations (342).

Two of the guidelines present specific assumptions used in risk assessment calculations. EPA (1980), for the water quality criteria documents, provided an assumed average drinking water consumption of 2 liters of water per day and average fish consumption of 6.5 grams of fish per day. FDA (1985), for the SOM guidelines, states that allowable drug residue level will be set after correcting for food intake in total human diet. For these calculations, FDA specifies various "food factors" which imply that up to one-third of diet might be cattle, pig, sheep, or poultry muscle or poultry eggs, and 100 percent of diet might be cow's milk.

---

[19]EPA, for example, has been criticized by its Science Advisory Board and others for an overreliance on computer modeling for exposure assessment (156).

Risk Characterization

The first issue in risk characterization is whether to make the characterization quantitative. The Federal agencies are all using quantitative risk assessment today, although some observers urge caution in the use of these quantitative approaches (8,156).

Different approaches to whether quantitative estimates are possible or desirable are found in the policies. CPSC (1978) and OSHA (1980) argued that quantitative risk assessment would be used at those agencies only for setting priorities. Court decisions and the general regulatory environment have superseded that stand, and both agencies today prepare quantitative risk assessments.

OSTP (1979) admitted that "extrapolation from the animal model to humans represents something of a leap of faith," but nevertheless recommends that quantitative potency estimates should be used in determining the human risk posed by a carcinogen and that assessment of relative potencies "will aid agencies in the establishment of regulatory priorities and in the selection of appropriate regulatory action." EPA (1976, as well as subsequent EPA policies) accept quantitative estimation as important for setting regulatory standards, but indicate it "should be regarded only as rough indications of effect."

EPA (1986) presents several options for quantitative characterization of risk:

- unit risk—"excess lifetime risk due to continuous constant lifetime exposure of one unit of carcinogen concentration,"
- dose corresponding to a given level of risk,
- individual risks—excess individual lifetime risk, and
- population risk—excess number of cancers in an exposed population.

Individual and population risks are those most used in policy debates. Both of these estimates incorporate information on potency from the dose-response characterization with estimates of exposures. Individual risk is the increase in the probability of disease or death for an individual in a lifetime. It is often expressed as the number of deaths per thousand or million similarly exposed persons. Thus, exposure to a particular carcinogen might present a 1 in 1,000 lifetime risk. Population risk involves the number of excess cases of disease found in the exposed population. Thus among 70,000 people exposed to a 1 in 1,000 lifetime risk, there would be 70 excess cancer deaths associated with this exposure. Some exposures might present high individual risks among a relatively small subgroup in the population, yet also present a low population risk (because of the small number of people exposed). What to do in these situations is an important regulatory issue. Most of the policies are silent on this issue, although OSTP (1985) suggests that agencies consider identifying high-risk populations.

EPA (1986) also cautioned against using more than one significant figure in the quantitative estimates, although their published potency estimates in the past frequently had three significant figures. (See table 3-24 in ch. 3.)

EPA (1980) used a boilerplate for its risk summary of the water quality criteria documents: [name of chemical] is a carcinogen, exposures to carcinogens should be zero, but may not be attainable. "Therefore, the levels which may result in incremental increase of cancer risk over the lifetime are estimated at 10^{-5}, 10^{-6}, 10^{-7}." The corresponding estimates for the particular chemical were then presented.

Treatment of Uncertainty

It is also important to describe uncertainties in the characterization of the risk. There is little opposition to discussing uncertainties and assumptions in risk assessments. The agency policies differ, however, on the utility of developing a range of estimates using different extrapolation models.[20] For example, EPA (1976) urges that "where appropriate, a range of estimates should be given on the basis of several modes of extrapolation." OSTP (1985) states that it is important to discuss the various sources of uncertainty, including statistical uncertainty, variability introduced by the chosen extrapolation model, and variability asso-

ciated with interspecies scaling. The uncertainty in the choice of model can be characterized by indicating the range of estimated risks that can be developed using different plausible models.

EPA (1980) and FDA (1985) on the other hand argued that little is gained by adding estimates from several different models. EPA (1980) went so far as to state that this would "add no additional scientific information while at the same time would create confusion and thereby undermine the utility of risk estimates." OSTP (1985) advises agencies to distinguish clearly among facts, consensus, assumptions, and science policy decisions. Although it is not clear that this effort will actually reduce regulatory controversies, it may clarify the issues in dispute and outline the areas of greatest uncertainty.

New Areas for Risk Assessment Policy

Two topics have been given increased attention in recent years: possible distinctions among carcinogens based on their mechanisms of action and consideration of the pharmacokinetics of toxic chemicals within the body. The former received considerable attention and argument early in the Reagan Administration when suggestions were made that regulatory distinctions could be made based on carcinogenic mechanisms.

Distinctions Based on Mechanism

The development of cancer consists of stages. These are typically called initiation, promotion, and progression. Initiation involves an alteration in a cell's genetic material, an alteration that can remain latent (without apparent disease) for years. Promotion involves the expression of genetic information and the transformation of latent initiated cells into tumors. Progression consists of the growth of tumors and the development of metastases in distant tissues.

Some chemicals are primarily initiators; others act only as promoters. Many chemicals are both initiators and promoters and are termed complete carcinogens.

Because they can directly damage genetic material, leading to creation of initiated cells, and

---

[20]Although there are other sources of uncertainty in risk assessments, much of the discussion in policies on this point concerns the choice of extrapolation model.

because such damage might be from an interaction with a very small amount of the chemical and may not be reversible, it is generally felt that initiators would not exhibit a no-effects threshold.

It is possible that the mechanism of promotion involves alteration in body chemistry, cellular growth and repair, and other processes. Because these alterations may be reversible and may not be harmful at relatively low doses, it has been suggested that there may be safe thresholds for these agents (358,359).

Weisburger and Williams have suggested the terms genotoxic and epigenetic to distinguish carcinogens based on their mechanisms (358,359). Other distinctions made, though with different meanings, are genotoxic and nongenotoxic, and direct and indirect carcinogens. But while such distinctions have been hypothesized, most scientists do not believe that chemicals in these groups can be reliably distinguished (98,157,258,351,356). This is particularly the case because many carcinogenic chemicals affect both initiation and promotion. While research on mechanisms is moving rapidly, there is currently no accepted group of tests for determining the mechanism of action for carcinogenic chemicals.

Moreover, promoters may themselves be very potent. "Dioxin" (2,3,7,8-TCDD), the most potent animal carcinogen known, may be acting as a promoter. Finally, some argue that the dose-response curve for indirect carcinogens will be linear at low doses, just as it is for direct carcinogens (86). An important argument on this point is that if it is assumed that exposure to a carcinogen adds to the background risk of cancer, the dose-response curve will be linear in the low-dose region (37).

While the issue has received enhanced attention in recent years, it is one that has been addressed in some of the earlier policies as well. In general, agency policies have refused to give a blanket endorsement of regulatory distinctions based on mechanism, although several appear willing to entertain an argument, with supporting documentation, that a substance acts only through an indirect mechanism.

EPA (1980) saw no currently satisfactory way of estimating risk for "epigenetic" agents and until mechanisms are better understood, will continue to use the linear no-threshold model. EPA (1986) generally considers substances positive in bioassays to be complete carcinogens, "unless there is evidence to the contrary." Individual consideration will be given to cases where the substance is positive in special tests for initiation, promotion, or cocarcinogenicity, but negative in long-term bioassays. NCAB (1977) wanted additional tests before extrapolating to humans when a substance tests positive in a bioassay in which the animals were also treated with a known carcinogen or cocarcinogen. OSHA (1980) allowed consideration of the argument that an indirect mechanism is involved and that this would not occur under conditions of human exposure. IRLG (1979) required "rigorous documentation" that a positive animal bioassay does not represent a complete carcinogenic process, but is due solely to an enhancing factor. IRLG (1979) also noted that in considering indirect mechanisms, promoters, and metabolic pathways, it would consider evidence, but expressed concern that false-negative judgments be avoided.

Incorporation of Pharmacokinetics

Exposure to a drug or chemical can lead to a variety of chemical and biological reactions in the body. The substance can be absorbed into the body, metabolized into other substances, distributed to other organs and tissues, or removed from the body. The general term for all of these processes is pharmacokinetics (171).

The various biochemical pathways in the body that activate, metabolize, detoxify, transport, and excrete chemicals will determine the relationship between the administered doses and the effective dose. As shown in figure 2-3, this relationship might be linear, but may not be. Either activation or detoxification mechanisms could become saturated or overwhelmed. In these cases, the administered dose (the dose externally administered to the animal) will not be proportionate to the effective dose (the amount of the chemical or its metabolites) that actually reach the target tissue, and qualitatively different effects may occur. The shape of the curve relating administered dose

Figure 2-3.—Possible Types of Relationships Between Administered and Effective Dose

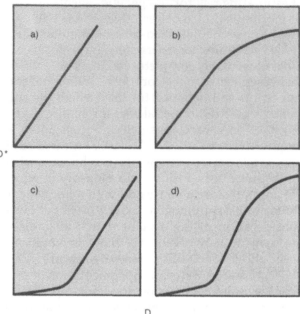

Possible relations between administered dose, D, and effective dose at the target (e.g. DNA), D*, for several kinetic models: a) simple first-order kinetics; b) saturation of the activation system; c) saturation of detoxification or repair systems; and d) combination of b) and c).

SOURCE: D.B. Hoel, N.L. Kaplan, and M.W. Anderson, "Implication of Nonlinear Kinetics on Risk Assessment in Carcinogenesis," *Science* 219:1032-37, 1983.

with effective dose depends on the particular pathways used and the levels at which each becomes saturated.[21]

While some observers hope that analysis of pharmacokinetics will refine risk assessments and improve regulatory decisions, there is much that is not understood, and there are practical difficulties in undertaking the experiments and analyzing the data. Until recently, the agencies have not attempted to include quantitative modeling of a chemical's pharmacokinetics in their quantitative risk assessments. However, even some of the earlier policies place a value on information concerning the metabolism of carcinogenic compounds.

EPA (1976) stated that a risk analysis should describe a substance's metabolic characteristics

and similarities to other known classes of carcinogens at high and low doses and in different species. EPA (1980) stated that pharmacokinetic information can be useful for interspecies comparisons and estimating human risk. Further, EPA stated that relevant metabolic differences among species should be considered. EPA (1986) mandated a summary of relevant metabolic and pharmacokinetic data and suggested that this information might affect choice of a high to low dose extrapolation model. With this information or other evidence on the cancer mechanism, an extrapolation using a model other than the linearized multistage model "might be considered more appropriate on biological grounds."

FIFRA (1984) also desired a summary of available pharmacokinetic data. In an example, the policy suggested that an evaluator should then consider the available metabolic and pharmacodynamic data for an explanation of the shape of the dose-response curve. Although not stated in the text, the data in this example appear to be from a study of rats exposed to formaldehyde. FIFRA (1984) suggested that a "threshold dose has been exceeded" in describing a very curvilinear dose-response curve.

OSHA (1980) set forth detailed requirements for arguments that metabolic differences between animals and humans justify the conclusion that a positive animal carcinogen does not pose a human health risk. This policy stated that OSHA would also consider a substance to be a carcinogen if it is metabolized into one or more potential occupational carcinogens. IRLG (1979) stated that while knowing the dose at the target organ is the ideal and that study of a substance's pharmacokinetics can in theory provide that information, there are still many uncertainties about metabolic pathways in humans and considerable variation within the human population. OSTP (1985) urged that the extrapolation model chosen needs to be consistent with available information on pharmacokinetics and target tissue dose.

Use of pharmacokinetic modeling is frequently hampered by lack of data and by our incomplete knowledge about which metabolites are the ultimate carcinogens. Obtaining these data in humans

---

[21]Note that each of the four curves in figure 2-3 assumes the absence of a no-effects threshold.

for known animal carcinogens would require deliberately exposing people to suspect carcinogens —an enterprise that is ethically objectionable.

Classification of Carcinogens

Some policies provide for classifying substances by the nature or strength of the evidence for carcinogenicity. Regulatory agencies that provide such classifications on major statements include OSHA (1980), CPSC (1978), and EPA (1986, and the proposed EPA airborne carcinogen policy in 1979). NTP also has a classification of "levels of evidence" for the results of the long-term carcinogenicity studies. Finally, there is the carcinogen classification scheme of IARC.

Boxes 2-B through 2-G summarize the various classification systems. Except for the NTP "levels of evidence," which is designed solely for evaluating animal test results, all of the systems accept human evidence and accord the highest overall classification to substances shown by human epidemiologic studies to be carcinogenic. For example, for IARC this is Group 1, for EPA (1986), Group A.

All of the policies accept the use of animal data alone for suggesting a carcinogenic hazard to humans, although there are some differences in the nature of the required evidence. A significant increase in malignant tumors in two or more species or two or more independently conducted studies in the same species is considered "sufficient" evidence by IARC and EPA (1986), "strong" evidence by CPSC (1978), and enough to bring the substance into "Category I" of OSHA (1980). Positive results in only one animal species are considered to be "limited" evidence by IARC and EPA (1986), but would be adequate to place the substance in the "high probability" category of the EPA airborne carcinogen policy (1979). Both CPSC (1978) and OSHA (1980) indicated that positive results in one species with supporting or concordant short-term test results would also lead to classification as a carcinogen (CPSC: strong evidence; OSHA: Category I). In addition, both

agencies indicate that in certain circumstances they could decide to classify a substance as a carcinogen even without such supporting data. Thus, for both CPSC and OSHA, as well as EPA (1979), a substance could be classed as a carcinogen of regulatory interest based on a positive bioassay in single species.

All the schemes will classify substances as carcinogens based on both benign and malignant tumors. Both IARC and EPA (1986) include the possibility of downgrading a substance to the "limited evidence" category if the response is an increase in tumors that have a high spontaneous background rate. The classic example of this is an increased incidence of mouse liver tumors.

The IARC classification scheme presented in box 2-B represents the results of a recent modification. The major changes involved creation of a new Group 4 for agents probably not carcinogenic to humans and the criteria for Group 2. Group 2A (probably carcinogenic) will generally be used for agents with limited evidence in humans and sufficient evidence in animals. Group 2B (possibly carcinogenic) will be used for agents that have only limited evidence in humans without sufficient evidence in animals, inadequate or no data in humans but sufficient evidence in animals, and inadequate or no data in humans and limited evidence in animals when there is other supporting data. Generally agents will be classified in Group 4 based on combined evidence from animals and humans which indicate a lack of carcinogenicity (99).

The EPA weight-of-the-evidence classification system was developed as an adaptation of an earlier version of the IARC classification system. EPA (1986) quotes extensively the IARC definitions of "sufficient" and "limited" evidence. The EPA classification has regulatory meaning as well. The regulation of chemicals in drinking water depends on the weight-of-the-evidence classification. In addition, the weight-of-the-evidence classifications were used in developing adjustments of the reportable quantities of chemicals covered under CERCLA. (See ch. 3.)

Box 2-B.—1987 Classification of Carcinogens by the International Agency for Research on Cancer (99)

Degree of Evidence for Carcinogenicity to Humans and to Experimental Animals and Supporting Evidence

It should be noted that these categories refer only to the strength of the evidence that these agents are carcinogenic and not to the extent of their carcinogenic activity (potency) nor to the mechanism involved. The classification of some agents may change as new information becomes available.

Human Carcinogenicity Data.—The evidence relevant to carcinogenicity from studies in humans is classified into one of the following categories:

Sufficient Evidence of Carcinogenicity.—The Working Group considers that a causal relationship has been established between exposure to the agent and human cancer. That is, a positive relationship has been observed between exposure to the agent and cancer in studies in which chance, bias, and confounding could be ruled out with reasonable confidence.

Limited Evidence of Carcinogenicity.—A positive association has been observed between exposure to the agent and cancer for which a causal interpretation is considered by the Working Group to be credible, but chance, bias, or confounding could not be ruled out with reasonable confidence.

Inadequate Evidence of Carcinogenicity.—The available studies are of insufficient quality, consistency, or statistical power to permit a conclusion regarding the presence or absence of a causal association.

Evidence Suggesting Lack of Carcinogenicity.—There are several adequate studies covering the full range of doses to which human beings are known to be exposed, which are mutually consistent in not showing a positive association between exposure to the agent and any studied cancer at any observed level of exposure. A conclusion of evidence suggesting lack of carcinogenicity is inevitably limited to the cancer sites, circumstances and doses of exposure, and length of observation covered by the available studies. In addition, the possibility of a very small risk at the levels of exposure studies can never be excluded.

In some instances, the above categories may be used to classify the degree of evidence for the carcinogenicity of the agent for specific organs or tissues.

Experimental Carcinogenicity Data.—The evidence relevant to carcinogenicity in experimental animals is classified into one of the following categories:

Sufficient Evidence of Carcinogenicity.—The Working Group considers that a causal relationship has been established between the agent and an increased incidence of malignant neoplasms or of an appropriate combination of benign and malignant neoplasms . . . in (a) two or more species of animals or (b) in two or more independent studies in one species carried out at different times or in different laboratories or under different protocols.

Exceptionally, a single study in one species might be considered to provide sufficient evidence of carcinogenicity when malignant neoplasms occur to an unusual degree with regard to incidence, site, type of tumor, or age at onset.

In the absence of adequate data on humans, it is biologically plausible and prudent to regard agents for which there is *sufficient evidence* of carcinogenicity in experimental animals as if they presented a carcinogenic risk to humans.

Limited Evidence of Carcinogenicity.—The data suggest a carcinogenic effect but are limited for making a definitive evaluation because, e.g., (a) the evidence of carcinogenicity is restricted to a single experiment; or (b) there are unresolved questions regarding the adequacy of the design, conduct, or interpretation of the study; or (c) the agent increases the incidence only of benign neoplasms or lesions of uncertain neoplastic potential, or of certain neoplasms which may occur spontaneously in high incidences in certain strains.

Inadequate Evidence of Carcinogenicity.—The studies cannot be interpreted as showing either the presence or absence of a carcinogenic effect because of major qualitative or quantitative limitations.

Evidence Suggesting Lack of Carcinogenicity.— Adequate studies involving at least two species are available which show that, within the limits of the tests used, the agent is not carcinogenic. A conclusion of evidence suggesting lack of carcinogenicity is inevitably limited to the species, tumor sites, and doses of exposure studied.

Supporting Evidence of Carcinogenicity.—The other relevant data judged to be of sufficient importance as to affect the making of the overall evaluation are indicated.

Overall Evaluation

Finally, the total body of evidence is taken into account; the agent is described according to the wording of one of the following categories, and the designated group is given. The categorization of an agent is a matter of scientific judgment, reflecting the strength of the evidence derived from studies in humans and in experimental animals and from other relevant data.

Group 1: The Agent Is Carcinogenic to Humans.—This category is used only when there is *sufficient evidence* of carcinogenicity in humans.

Group 2.—This category includes agents for which, at one extreme, the degree of evidence of carcinogenicity in humans is almost sufficient, as well as agents for which, at the other extreme, there are no human data but for which there is experimental evidence of carcinogenicity. Agents are assigned to either 2A (probably carcinogenic) or 2B (possibly carcinogenic) on the basis of epidemiological, experimental, and other relevant data.

Group 2A: The Agent Is Probably Carcinogenic to Humans.—This category is used when there is *limited evidence* of carcinogenicity in humans and *sufficient evidence* of carcinogenicity in experimental animals. Exceptionally, an agent may be classified into this category solely on the basis of *limited evidence* of carcinogenicity in humans or of *sufficient evidence* of carcinogenicity in experimental animals strengthened by supporting evidence from other relevant data.

Group 2B: The Agent Is Possibly Carcinogenic to Humans.—This category is generally used for agents for which there is *limited evidence* in humans in the absence of *sufficient evidence* in experimental animals. It may also be used when there is *inadequate evidence* of carcinogenicity in humans or when human data are nonexistent but there is *sufficient evidence* of carcinogenicity in experimental animals. In some instances, an agent for which there is inadequate evidence or no data in humans but *limited evidence* of carcinogenicity in experimental animals together with supporting evidence from other relevant data may be placed in this group.

Group 3: The Agent Is Not Classifiable as to Its Carcinogenicity to Humans.—Agents are placed in this category when they do not fall into any other group.

Group 4: The Agent Is Probably Not Carcinogenic to Humans.—This category is used for agents for which there is *evidence suggesting lack of carcinogenicity* in humans together with *evidence suggesting lack of carcinogenicity* in experimental animals. In some circumstances, agents for which there is *inadequate evidence* of or no data on carcinogenicity in humans but *evidence suggesting lack of carcinogenicity* in experimental animals, consistently and strongly supported by a broad range of other relevant data, may be classified in this group.

Box 2-C.—1986 Classification of Carcinogens by National Toxicology Program (255)

- **Clear Evidence** of Carcinogenic Activity is demonstrated by studies that are interpreted as showing a dose-related: (i) increase of malignant neoplasms, (ii) increase of a combination of malignant and benign neoplasms, or (iii) marked increase of benign neoplasms if there is an indication from this or other studies of the ability of such tumors to progress to malignancy.

- **Some Evidence** of Carcinogenic Activity is demonstrated by studies that are interpreted as showing a chemically related increased incidence of neoplasms (malignant, benign, or combined) in which the strength of the response is less than that required for clear evidence.

- **Equivocal Evidence** of Carcinogenic Activity is demonstrated by studies that are interpreted as showing a marginal increase of neoplasms that may be chemically related.

- **No Evidence** of Carcinogenic Activity is demonstrated by studies that are interpreted as showing no chemically related increases in malignant or benign neoplasms.

- **Inadequate Study** of Carcinogenic Activity is demonstrated by studies that because of major qualitative or quantitative limitations cannot be interpreted as valid for showing either the presence or absence of carcinogenic activity.

When a conclusion statement for a particular experiment is selected, consideration must be given to key factors that would extend the actual boundary of an individual category of evidence. This should allow for incorporation of scientific experience and current understanding of long-term carcinogenesis studies in laboratory animals, especially for those evaluations that may be on the borderline between two adjacent levels. These considerations should include:

- the adequacy of the experimental design and conduct;
- occurrence of common versus uncommon neoplasia;
- progression (or lack thereof) from benign to malignant neoplasia as well as from preneoplastic lesions;
- some benign neoplasms have the capacity to regress but others (of the same morphologic type) progress. At present, it is impossible to identify the difference. Therefore, where progression is known to be a possibility, the most prudent course is to assume that benign neoplasms of those types have the potential to become malignant;
- combining benign and malignant tumor incidences known or thought to represent stages of progression in the same organ or tissue;
- latency in tumor induction;
- multiplicity in site-specific neoplasia;
- metastases;
- supporting information from proliferative lesions (hyperplasia) in the same site of neoplasia or in other experiments (same lesion in another sex or species);
- the presence or absence of dose relationships;
- the statistical significance of the observed tumor increase;
- the concurrent control tumor incidence as well as the historical control rate and variability for a specific neoplasm;
- survival-adjusted analyses and false positive or false negative concerns;
- structure-activity correlations; and
- in some cases, genetic toxicology.

These considerations together with the definitions as written should be used as composite guidelines for selecting one of the five categories. Additionally, the following concepts (as patterned from the International Agency for Research on Cancer Monographs) have been adopted by the NTP to give further clarification of these issues:

The term *chemical carcinogenesis* generally means the induction by chemicals of neoplasms not usually observed, the induction by chemicals of more neoplasms than are generally found, or the earlier induction by chemicals of neoplasms that are commonly observed. Different mechanisms may be involved in these situations. Etymologically, the term *carcinogenesis* means induction of cancer, that is, of malignant neoplasms; however, the commonly accepted meaning is the induction of various types of neoplasms or of a combination of malignant and benign neoplasms. In the Technical Reports, the words *tumor* and *neoplasm* are used interchangeably.

Box 2-D.—1986 Classification of Carcinogens by the Environmental Protection Agency (284)

Group A—Human Carcinogen:

This group is used only when there is sufficient evidence from epidemiologic studies to support a causal association between exposure to the agents and cancer.

Group B—Probable Human Carcinogen:

This group includes agents for which the weight of evidence of human carcinogenicity based on epidemiologic studies is "limited" and also includes agents for which the weight of evidence of carcinogenicity based on animal studies is "sufficient." The group is divided into two subgroups. Usually, Group B1 is reserved for agents for which there is limited evidence of carcinogenicity from epidemiologic studies. It is reasonable, for practical purposes, to regard an agent for which there is "sufficient" evidence of carcinogenicity in animals as if it presented a carcinogenic risk to humans. Therefore, agents for which there is "sufficient" evidence from animal studies and for which there is "inadequate evidence" or "no data" from epidemiologic studies would usually be categorized under Group B2.

Group C—Possible Human Carcinogen:

This group is used for agents with limited evidence of carcinogenicity in animals in the absence of human data. It includes a wide variety of evidence, e.g., (a) a malignant tumor response in a single well-conducted experiment that does not meet conditions for sufficient evidence, (b) tumor responses of marginal statistical significance in studies having inadequate design or reporting, (c) benign but not malignant tumors with an agent showing no response in a variety of short-term tests for mutagenicity, and (d) responses of marginal statistical significance in a tissue known to have a high or variable background rate.

Group D—Not Classifiable as to Human Carcinogenicity:

This group is generally used for agents with inadequate human and animal evidence of carcinogenicity or for which no data are available.

Group E—Evidence of Non-Carcinogenicity for Humans:

This group is used for agents that show no evidence for carcinogenicity in at least two adequate animal tests in different species or in both adequate epidemiologic and animal studies.

The designation of an agent as being in Group E is based on the available evidence and should not be interpreted as a definitive conclusion that the agent will not be a carcinogen under any circumstances.

Box 2-E.— 1979 Classification of Carcinogens in the EPA Airborne Carcinogen Policy (324)

Identify carcinogens based on EPA guidelines, supplemented by IRLG guidelines, judgments based on quality and weight of evidence, classify into high, moderate, or low, based on probability of human carcinogenicity

- high probability—"best" or "substantial" evidence exists from epidemiologic and/or at least one mammalian study;
- moderate probability—"suggestive" evidence exists from epidemiologic, animal, or "short-term" studies; and
- low probability—only "ancillary" evidence exists, such as from structural correlations, or for which epidemiologic or animal results are judged to indicate low probability.

Box 2-F.—1982 Classification of Carcinogens by the Occupational Safety and Health Administration (276)

Category I Potential Carcinogens.—If substance meets definition of potential occupational carcinogen in: 1) humans, 2) a single mammalian species in a long-term bioassay where the results are in concordance with some other scientifically evaluated evidence of a potential carcinogenic hazard, 3) in a single mammalian species in an adequately conducted long-term bioassay, in appropriate circumstances where [OSHA] determines the requirement for concordance is not necessary. Evidence of concordance is any of the following: positive results from independent testing in the same or other species, positive results in short-term tests, or induction of tumors at injection or implantation sites.

Category II Potential Carcinogens.—1) Meets criteria for category I, but the evidence is only "suggestive" or 2) meets criteria for category I in a single mammalian species without evidence of concordance.

The requirement for concordance may be waived in "cases where the evidence has been carefully scrutinized and found to be unusually compelling," including the induction of many unusual tumors or early deaths for most of the exposed animals.

Box 2-G.—1978 Classification of Carcinogens by the CPSC (228)

Category A—Strong Evidence:
1. NCI has issued a finding that the substance is an animal or human carcinogen,
2. substance significantly increases incidence or reduces time to onset of benign or malignant neoplasms in humans in exposed compared to nonexposed, or
3. substance significantly increases incidence or reduces time to onset of one or more types of benign or malignant neoplasms in treated compared to control groups [of experimental animals].

Ordinarily, positive animal results must derive from systemic distribution of substance and must be obtained in:
- two animal species; or
- one species when replicated in a second experiment using independent control groups; or
- one species of test animal when supported by a battery of well designed and soundly conducted relevant short-term tests; or
- CPSC finds that there is other evidence sufficiently compelling to classify substance in Category A. Thus classification may be based on single, unreplicated long-term animal study.

Category B—Evidence is Suggestive:
1. human or animal data are suggestive but not conclusive because they are statistically inconclusive or methodologically deficient but nonetheless tend to support carcinogenicity;
2. positive results in one or more short-term tests, but not confirmed in human or animal studies;
3. positive results in only one unreplicated long-term animal study which is not compelling enough to classify in Category A; absence of positive short-term results.

Category C:
Substances which are members of chemical classes which include known carcinogens and other substances about which questions have been raised, but with very limited evidence.

Category D:
Reclassified substances.

AGENCY POLICIES ON CARCINOGEN RISK MANAGEMENT

Regulatory Procedures

Compared to the problems of identifying and assessing carcinogenic risk, relatively little attention is given in the various agency policies to the topic of how to reduce, eliminate, or control the risks posed by carcinogens. Only CPSC (1978), EPA (1979), OSHA (1980), and FDA (1985) give any details beyond a summary of the agency's statutory mandate on the kind of regulatory action that can be anticipated after identifying a carcinogen. In addition, OSTP (1979) suggests focusing regulatory action on particular exposures to improve the ratio of benefits to costs.

Most of these statements are in documents that have the official status of proposals (EPA 1979 and FDA 1985), or that have been suspended (OSHA) or withdrawn (CPSC 1978). However, in communication with OTA, EPA staff suggest that the airborne carcinogen policy (1979) reflects the broad outlines of their approach, FDA staff stated in their proposed SOM (1985) that they would follow the procedures outlined in it until it was published in final form. OSHA has not taken action to revoke their cancer policy.

The basic policy outlined by three of the documents is a very protective approach for eliminating or substantially reducing carcinogen exposures. CPSC (1978) states a general policy of not permitting "known carcinogens to be intentionally added to consumer products if they can be absorbed, inhaled or ingested . . . ." CPSC (1978) required the use of substitutes for identified carcinogens or reduction to the lowest attainable level" until substitutes can be found. However, actual practice at CPSC since 1978 demonstrates that CPSC has followed a less conservative course in addressing chemical hazards. As discussed in chapter 3, CPSC has often deferred to voluntary industry action and CPSC-mandated labeling of hazardous consumer products to reduce exposures (81).

For Category I carcinogens (for which the evidence is clear), OSHA (1980) originally required that exposures be reduced to the lowest feasible level, using engineering controls and work practices (and not through use of respirators). If suit-able, safer substitutes are found, OSHA will set permissible exposures at zero to encourage substitution. The EPA airborne carcinogen policy (288) mandates that for identified carcinogens, the standards issued under section 112 of the Clean Air Act will, at a minimum, require use of best available technology. If the risk remaining after application of the best available technology is still unreasonable, EPA will consider mandating further control.

In all three cases, the agencies state that they do not believe that safe thresholds exist for carcinogens. This view, combined with their interpretations of statutory mandates, leads them to require that exposures be reduced as much as possible through the use of technology and substitution.

In its SOM paper, FDA argues that it follows a protective approach in setting a maximum lifetime risk cutoff of 1 in 1 million and in using upper bound risk estimates for determining the added risk of one permitted animal drug residue. In the U.S. population of 240 million, assuming that everyone is exposed, this would imply a maximum of 240 deaths. FDA suggests that because of the assumptions behind the development of their risk estimates, the actual risks are lower than that. In fact, they argue that it is likely that no one will actually die as a result of these exposures. The FDA policy is in fact the only one to adopt a risk level that it deems to be "safe." Other agencies may do this informally and implicitly; FDA alone has done this explicitly.

OSHA's 1980 policy is the most detailed in describing regulatory procedures that will be followed if OSHA identifies a substance as an occupational carcinogen. In addition to regulating Category I carcinogens to the lowest feasible level (see above), OSHA will regulate what it calls Category II carcinogens (for which the evidence is only suggestive) in a manner consistent with statutory requirements. OSHA does not seem willing to force substitution, set a permissible exposure limit, or require compliance plans, hygiene facilities, and regulated areas in the case of substances with only suggestive evidence of carcinogenicity. For Category II substances, OSHA is also to ask for additional research from the appropriate agency.

The OSHA policy further provided for the publication of lists—annual candidate lists of substances under scientific review and semiannual priority lists of the substances with the agency's highest priority ranking. Only one of these lists was ever published (277). In 1981, OSHA suspended the publication of these lists.

Finally, in an attempt to increase the timeliness of agency action, the OSHA policy set a number of specific deadlines for agency action. For example, a final standard was to be published within 120 days from the end of a hearing or 90 days from the end of a posthearing comment period, whichever is earlier. This time may be extended for one more 120-day period, unless important new evidence is found.

Regulatory and Research Priorities

Some of the policies also give guidance on setting regulatory and research priorities. In particular, the EPA airborne carcinogen policy and the OSHA cancer policy provide some general statements about setting priorities. These are very general and probably provide only limited insight into decisionmaking at either agency. The magnitude of the exposed population and the availability of controls or the low cost of applicable controls appears in both policies. The EPA airborne carcinogen policy explicitly refers to estimated carcinogenic potency and the upper bound incidence associated with exposures, both presumably derived from EPA's quantitative risk assessments. The policy refers to risk estimates for the most highly exposed individuals and the population risk.

OSHA (1980) rejected quantitative risk assessment for setting the exposure level in an occupational health standard; instead the policy was to regulate down to the lowest level feasible. However, it did suggest that quantitative risk assessment could be used in setting priorities. As discussed in other sections of this background paper, this policy was changed in response to court decisions. OSHA (1980) also mentioned that they would consider a substance's molecular structure, the potential for controls to prevent other adverse occupational and environmental effects, pending actions by other agencies, and OSHA's other responsibilities before taking action on a chemical.

OFFICE OF MANAGEMENT AND BUDGET POLICY ON CARCINOGEN RISK ASSESSMENTS

Since 1981, the Office of Management and Budget (OMB) has had an important, if not central, role in many decisions on regulatory policy through its review of agency proposals and final rules. Each of those actions takes place case by case as OMB interprets the requirement for cost-benefit analysis contained in Executive order 12291 and judges the desirability of particular regulations. Summarizing such case-by-case interpretation is difficult, in part because much of it takes place in private meetings between OMB officials, agency officials, and others. In this section, OTA will not attempt such a summary.

OMB has publicly indicated its concern over several areas of carcinogen risk assessments, and its expressed opinions on these matters are contrary to the general consensus that has evolved in the agency policies. OMB's general position is that the use of many "conservative" assumptions (in these cases, assumptions designed to err on the side of caution and minimize the chances of understating the true risk), will compound each other, guaranteeing that the estimated risk is overstated. If the estimated risks are overstated, regulatory decisions will not be as efficient or as cost-effective as they could be.[22]

For hazard identification and dose-response assessment, OMB concern involves assumptions about the treatment of benign tumors, selection of the most sensitive species and sex for risk assess-

___
[22]For an academic discussion of this issue, see ref. 149.

ments, and the use of conservative high- to low-dose extrapolation techniques. On benign and malignant tumors, OMB suggests that because not all benign tumors become malignant, use of the benign tumor data in the risk assessment "can overstate the real risk present." Regarding choice of data for the risk assessment, OMB argues that use of the most sensitive species and sex will bias the risk assessment and that a "more accurate" estimate would be derived from a "weighted average of all the scientifically valid, available information." On the choice of extrapolation technique, OMB is worried about use of the upper confidence limit, suggesting that "such an extrapolation has a 95 percent chance of overstating the true risk." In fact, OMB is misinterpreting the meaning of the upper bound estimates prepared by the regulatory agencies.

In regard to exposure assessment, OMB expresses opinions on the use of worst-case environmental scenarios, the assumption of lifetime exposure, and the focus of regulation being placed on the most highly exposed individuals. The worst-case scenarios are used to simplify the tasks of estimating risks and setting standards. OMB is worried that these worst-case conditions are not representative of actual conditions throughout the Nation. It is usually assumed that people might be exposed to a lifetime of drinking contaminated water or might spend their entire lives working with the same hazards. OMB thinks that this assumption can bias upward the estimates of risk because people move and change jobs. Finally, OMB expresses concern about basing regulations on the maximally exposed individual who has the worst combination of exposures (350).

One example of the implementation of OMB's approach may be found in OMB's comments to OSHA about OSHA's proposed regulation of occupational formaldehyde exposures. In those comments, OMB surveyed the epidemiologic literature on formaldehyde exposures and cancer and concluded that formaldehyde is not a human carcinogen because there is little consistency in the tumor sites among 19 different studies and little evidence that the observed excesses are actually related to the level of exposures. They cited, in addition, a recent epidemiologic study conducted by NCI, which was also negative. The approach

embodied in most agency policies is to use negative epidemiologic studies only to estimate an upper bound on estimated risk and not to use these studies to dispute positive animal evidence. In fact, the position of the agency policies is that positive animal evidence outweighs negative human evidence. In contrast to this, OMB used the human evidence to cast doubt on the validity of the animal evidence. OMB also argued that the pharmacokinetics of formaldehyde exposure predict that carcinogenic effects will occur only at high doses that overwhelm the body's protective mechanisms at the exposure sites.

For developing its own quantitative dose-response assessment, OMB follows an approach to selecting animal data and extrapolating from high to low doses that is different from that used by the regulatory agencies. Instead of selecting the most sensitive sex, species, and strain or choosing one with the biologically most plausible response, OMB combined the animal responses from six different studies in three different species. In the case of formaldehyde this has a large effect on estimated risk because the studies are clearly positive only in rats; for mice and hamsters, the studies are largely negative. Thus, the OMB approach uses both positive and negative data and calls it a weight-of-evidence approach.

OMB then recalculated the extrapolation from high to low doses using several mathematical models—the multistage model and several models that do not incorporate low-dose linearity assumptions—and also incorporating different exposure scenarios and estimates of effective or delivered dose. On the basis of these calculations OMB concluded that the "carcinogenic risk of formaldehyde to workers is likely to be de minimis," that is, less than 1 in a billion (349).

OSHA has not issued its final formaldehyde rule. In this particular instance, it is not clear how persuasive OMB's arguments will be to OSHA. Nor is it clear how strongly OMB will push its position about risk assessment assumptions and de minimis risks on the agencies in particular cases or if it will expect the agencies to rewrite their general guidelines.

CONCLUSIONS

Federal agency guidelines are generally consistent on major features of animal bioassay design, specifying testing in two animal species and generally requiring use of three dose groups and a control group. The guidelines agree that a study must set the highest dose as high as possible without shortening the animals' lives because of noncarcinogenic toxic effects.

Agency policies value epidemiologic studies as the most conclusive evidence for human carcinogenicity, generally presume that substances carcinogenic to animals in long-term bioassays should be treated as presenting a hazard to humans, and treat short-term test results as supportive information. Analyses of structure-activity relationships are used mostly when no other data are available.

The policies state that the agencies will use animal data derived from use of the maximum tolerated dose and will treat the appearance of malignant or benign tumors as evidence for carcinogenicity, except when the benign tumors are of a type that does not progress to malignancy. Policies usually state that positive results in animals outweigh negative epidemiologic results, and that positive results in one species outweigh negative results in another.

During the 1970s and 1980s, the agencies began using quantitative risk assessments for carcinogens. Today, while there are still considerable uncertainties in quantitative risk assessment, all the agencies use it. They assume that human risk estimates can be derived from animal data, that carcinogenic chemicals lack no-effect thresholds, and that risk estimates should be based on the most sensitive animal species. All the agencies use extrapolation models that assume low-dose linearity, although they differ on the mathematical technique to use, whether the focus should be on the "upper confidence limit" or "maximum likelihood estimate," and the method to convert animal doses into human doses. The agency policies do not distinguish chemicals based on their mechanisms of action and are only beginning to explore the use of pharmacokinetic modeling techniques.

Agency policies give much less detailed guidance on estimating human exposures to specific chemicals. Instead, they rely on case-by-case evaluations. Nevertheless, the lack of detailed guidelines does not diminish the great importance of this factor in estimating human risk.

Despite some differences, the approaches of the regulatory agencies to carcinogen risk assessment have many similarities. OMB, on the other hand, has indicated that it does not agree with parts of the common approach.

PUBLISHER'S NOTE:

The recommended carcinogenicity testing required under NSF Standards 60 and 61 (direct and indirect additives, respectively) will be triggered by contaminant concentration in the finished drinking water. Barring evidence of mutagenicity or preneoplastic changes in sub-chronic toxicity studies, chronic toxicity/carcinogenicity testing will not be required unless the contaminant is present at the tap in concentrations exceeding 1000 ppb. Recommended testing protocols follow those recommended by FDA, EPA, or comparable international agency. Risk assessment models are comparable to those employed by EPA. The maximum level allowed, under the standards, to be contributed to the drinking water by any single contaminant (Maximum Allowable Level or MAL) is based on EPA MCLs for regulated contaminants, calculated as described by EPA procedures.

For unregulated contaminants, the MAL is calculated using EPA procedures, safety factors, and daily water consumption figures.

Chapter 3

Federal Agency Assessment and Regulation of Carcinogens

CONTENTS

Chapter 3
Federal Agency Assessment and Regulation of Carcinogens

This chapter describes the major statutes and agency actions regulating human exposure to carcinogens. Most of these statutes do not single out carcinogens for specific consideration, but merely regulate them as a species of toxic substance. The exception to this is the Food, Drug, and Cosmetic Act (FDCA), which, in its "Delaney clause," prohibits the intentional use of carcinogens as food or color additives. In the Clean Water Act (CWA), the Toxic Substances Control Act (TSCA), and the Resource Conservation and Recovery Act (RCRA), Congress did not specify any particular type of regulatory action for carcinogens beyond what is required for other toxic effects. But carcinogens were mentioned as agents of particular congressional concern.

These laws established different regulatory mechanisms and specified the considerations on which the agencies are to make regulatory decisions and the range of allowable discretion. Under some provisions of FDCA, TSCA, and the Federal Insecticide, Fungicide, and Rodenticide Act (FIFRA), premarket review of a substance is required before it can enter into commerce. Most of the statutes, including other provisions of FDCA, TSCA, and FIFRA, however, provide for postmarket regulation of substances after they have been in commerce and people have been exposed to them. Once a health problem has been identified an agency would be required to either propose a regulation based on that finding, such as, under the Clean Air Act (CAA), CWA, the Safe Drinking Water Act (SDWA), or the Occupational Safety and Health Act, or establish this fact in court, and seek a judicial remedy on that basis. Some sections of FDCA and the statutes administered by the Consumer Product Safety Commission (CPSC) follow this approach.

To regulate carcinogenic substances, Federal agencies follow rulemaking procedures that are set by their statutes or those established by the Administrative Procedure Act (5 U.S.C. 553), which may be either informal or formal. Under the latter procedures the agency must issue a "Notice of Proposed Rulemaking" (NPRM), which describes the proposed regulation, explains the basis for the proposal, and announces an opportunity for comment by interested parties. After written comments are received, a hearing may be held to obtain public comments. After considering the comments, a final rule is published in the *Federal Register*. The proposal may also be altered or withdrawn. Formal rulemaking procedures differ from informal ones by the nature of the evidence presented during the comment period, the opportunity for cross-examination of witnesses in hearings, which may resemble the proceedings in a court of law, and, some argue, closer scrutiny by the courts.

The nature of the evidence an agency may consider as the basis for carcinogen regulation reveals various attitudes toward the acceptability of risk. In the most general terms, regulatory approaches are of several types:

- risk-based, such as the Delaney clause of the FDCA which requires the ban of a food additive shown to cause cancer in humans or in animal tests;
- *technology-based*, which might require the use of "best available technology" (BAT) or "best practical technology" (BPT) to control emissions from a particular source; or
- *risk-benefit or cost-benefit balancing*, which could permit the consideration of competing health risks and benefits—such as in the case of cancer-causing drugs used to treat fatal illnesses—or the costs of control and other economic impacts.

There are also different types of agency actions. Some statutes set exposure standards for air (e.g., the Occupational Safety and Health Administration's permissible exposure limits) or water (e.g., exposure limits in water under SDWA). Others require emission standards for air (under CAA) or water (under CWA). Under other statutes, Federal agencies issue rules concerning the "safe use" of a product (under the Consumer Product Safety Act (CPSA), FDCA, or FIFRA), and some permit or require outright banning of substances or products containing them (under FDCA, FIFRA, and CPSA).[1]

Agency actions in trying to control carcinogens have been as varied as the statutes under which

---

[1]This background paper only describes the nature of the standards or regulations issued, and does not discuss implementation and enforcement of these regulations.

they work. Some agencies have regulated as many as 191 potential carcinogens (under the Comprehensive Environmental Response, Compensation and Liability Act (CERCLA, or Superfund) and 29 carcinogenic chemical classes (of the 65 classes required to be regulated in a judicial order under CWA), while others have regulated only a few (e.g., under the authority of CAA, the Environmental Protection Agency (EPA) has regulated 4 carcinogens in 16 years). For the most part, it appears that the agencies do not act quickly when they learn that a substance is carcinogenic, since there is usually a considerable delay between the time when the outcome of human epidemiologic studies or animal bioassays becomes known and final regulations are published, as well as between the issuing of proposed and final rules.

OSHA REGULATORY ACTIONS

The Occupational Safety and Health Act of 1970 established the Occupational Safety and Health Administration (OSHA) and the National Institute for Occupational Safety and Health (NIOSH). OSHA is a regulatory agency which, among its other duties, issues and enforces regulations that limit exposure to carcinogens in the workplace. NIOSH is a research agency that has supported epidemiologic and toxicologic research and makes recommendations to OSHA concerning changes in occupational health standards.

The act provides three statutory mechanisms for establishing standards for protection from hazardous substances:

1. Section 6(a), which authorized OSHA to adopt the standards already established by Federal agencies or adopted as national consensus standards, as "startup standards" during the first 2 years after the act went into effect. The original source for most of these standards was the list of threshold limit values (TLVs) published by a professional society, the American Conference of Government Industrial Hygienists (ACGIH).
2. Section 6(b), which authorizes OSHA to issue new permanent exposure standards in rulemaking proceedings.

3. Section 6(c), which authorizes OSHA to issue emergency temporary standards that require immediate action to reduce a workplace hazard when employees are exposed to a "grave danger." A section 6(c) action also initiates the process of establishing a permanent standard under 6(b).

The standards issued may require monitoring and medical surveillance, modification of workplace procedures and practices, requirements for recordkeeping, and new or modified Permissible Exposure Limits (PELs), which are the maximum concentrations of toxic substances permitted in the workplace air.

Standards adopted under section 6(b) must "adequately assure" that "to the extent feasible . . . no employee will suffer material impairment of health or functional capacity even if such employee has regular exposure to the hazard dealt with by such standard for the period of his working life." Interpreting this mandate has been at the center of legal disputes around OSHA's regulation of toxic workplace exposures. In a major decision, the Supreme Court invalidated OSHA's 1978 benzene standard, ruling that OSHA had to demonstrate that exposure at the current permissible levels presents a "significant risk" to workers before lower exposure standards can be issued.

Prior to this decision, OSHA had prepared quantitative risk assessments concerning substances it regulated. The first one of these was prepared in 1976 for the proposed standard regulating coke oven emissions (118). In the case of benzene, however, OSHA concluded that because of the uncertainties, it would not conduct quantitative risk assessment. The subsequent legal battle involved whether OSHA would be required to conduct such assessments. Following the benzene decision, OSHA has conducted quantitative risk assessments to demonstrate that current permissible exposure levels present a "significant risk" to workers. The first quantitative risk assessment conducted after the Supreme Court decision involved worker exposure to arsenic.

In 1971, OSHA adopted exposure limits on approximately 400 specific chemical substances as startup standards as required by section 6(a). Although these exposure limits had been developed primarily to protect against noncarcinogenic toxicities, some of the substances are also carcinogens.

From 1972 to 1986 OSHA issued more stringent health standards covering 22 carcinogens (see table 3-1). Nine of OSHA's final actions on health standards established new PELs and other requirements on individual carcinogens (asbestos (1972), vinyl chloride (1974), coke oven emissions (1976), benzene (1978), 1,2-dibromo-3-chloro-propane (DBCP) (1978), arsenic (1978), acrylonitrile (1978), ethylene oxide (1984), and a second regulation on asbestos (1986)). One OSHA standard regulating a group of "14 carcinogens" did not institute or change a PEL, but created new requirements for work practices and medical surveillance for this group of carcinogens and mandated use of "closed system operations."[2] One final action in 1983 clarified that asphalt fumes, which contain known carcinogens, are not regulated under

___

[2]According to OSHA's preamble to the final rule, 13 of these 14 substances were derived from ACGIH Appendix A. Alpha-naphthylamine was added because it is often found together with beta-naphthylamine. Dimethyl sulfate appears in the ACGIH appendix but was excluded for inadequate documentation of carcinogenicity (272).

the OSHA startup standard for coal tar pitch volatiles (219). The reason for this is that asphalt fumes contain a significantly lower percentage of the polyaromatic hydrocarbons listed in the coal tar pitch volatile standard (278).

Thus, OSHA has used two different approaches for limiting exposures: setting permissible exposure limits and requiring specific process technology and procedures. The latter was used for the group of 14 carcinogens and, while a permissible exposure limit for coke oven emissions was set, the standard also included relatively specific requirements concerning the types of engineering controls that were to be used. For either approach, OSHA has mandated the use of engineering controls as the primary method of compliance, in contrast to use of gas masks and respirators (219). The focus on "closed systems" and stringent work procedures instead of setting PELs for the 14 carcinogens reflected the facts that closed system operations were possible and that many of these chemicals had readily available substitutes.

OSHA has also issued "generic" standards that apply to large groups of chemical exposures. Two of these affect workers exposed to carcinogens—the Hazard Communication Standard and the Access to Medical and Exposure Records Standard. The former requires that hazardous chemicals in the workplace be labeled, that employers set up training programs for workers on chemical hazards in the workplace, and that chemical manufacturers prepare and that employers keep copies of material safety data sheets for hazardous chemicals. The standard defines hazardous chemicals to include carcinogens (29 CFR 1910.1200).

The Access to Medical and Exposure Records Standard (29 CFR 1910.20) requires that employers allow employees and their representatives access to medical and exposure records and requires that these records be preserved for specified time periods. However, the standard does not require that employers conduct exposure monitoring or medical surveillance. The standard does require that if the employer conducts such monitoring or surveillance that records be kept and made available to workers. Again, exposures to carcinogenic chemicals are covered by this standard.

Table 3-1.—OSHA Regulation of Carcinogens

Substance	Type of evidence	Petition	NPRM	Final	Court action challenge
2-Acetylaminofluorene	Animal, 6 species—mice, rats, rabbits, dogs, fowl, hamsters	1-4-73 (OCAW/HRG)	9-7-73[b]	1-29-74	"Fourteen Carcinogens"[a]
Acrylonitrile	Human; animal, 1 species—rats	3-78 (Manufacturing Chemists Assoc.)	1-17-78[b]	10-3-78	—
4-Aminobiphenyl	Human; animal, 3 species—mice, dogs rabbits	1-4-73 (OCAW/HRG)	9-7-73[b]	1-29-74	"Fourteen Carcinogens"[a]
Asbestos (I)	Human	—	1-12-72; 10-9-75	6-7-72	1972 standard affirmed—*Industrial Union Dept., AFL-CIO v. Hodgson*, 499 F.2d 467 (D.C. Circuit, Apr. 15, 1974)
Asbestos (II)[b]	Human	—	4-10-84	6-20-86	Standard lowered from 2 to 0.2 fibers/cm^3.
Benzene (I)	Human-suggestive evidence	—	5-27-77	2-10-78	Vacated, *American Petroleum Institute v. OSHA*, 581 F.2d 493 (5th Cir., Oct. 5, 1978) & *Industrial Union Department, AFL-CIO v. American Petroleum Institute*, 448 U.S. 607 (Supreme Court, July 2, 1980)
Benzene (II)	Animal, 2 species—mice, rats	4-83	12-10-85	—	—
Benzidine	Human; animal, 4 species—mice, rats, dogs, hamsters	1-4-73 OCAW/HRG	9-7-73[b]	1-29-74	"Fourteen Carcinogens"[a]
Beryllium	Human-suggestive evidence; animal, 3 species—rats, rabbits, monkeys	—	10-17-75	—	—
bis-Chloromethyl ether	Animal, 2 species—mice, rats	1-4-73 OCAW/HRG	9-7-73[b]	1-29-74	"Fourteen Carcinogens"[a]
Coal tar pitch volatiles	—	1980, 4-81 (Asphalt Institute)		1-21-83	Petroleum derived asphalt fumes removed from definition
Coke oven emissions	Human	—	7-31-75	10-22-76	Affirmed, *American Iron & Steel Inst. v. OSHA*, 577 F.2d 825 (3d Cir., 3-28-78)
1,2-Dibromo-3-chloro-propane	Animal, 2 species—rats, mice	8-23-77 (OCAW request)	11-1-77	3-17-78	—
1,2 Dibromoethane (EDB)	Animal, 2 species—mice, rats	9-81, 2-3-84 (Teamsters)	10-7-83	—	—
3,3'-Dichlorobenzidine	Animal, 3 species—rats, mice, hamsters	1-4-73 (OCAW/HRG)	9-7-73[b]	1-29-74	"Fourteen Carcinogens"[a]
4-Dimethylaminoazobenzene	Animal, 4 species—rats, mice, dogs, trout	1-4-73 (OCAW/HRG)	9-7-73[b]	1-29-74	"Fourteen Carcinogens"[a]

continued on next page

Table 3-1.—OSHA Regulation of Carcinogens—Continued

Substance	Type of evidence	Petition	NPRM	Final	Court action challenge
Ethyleneimine	Animal, 2 species—rats, mice	—	9-7-73[b]	1-29-74	"Fourteen Carcinogens"[a]
Ethylene oxide	Human, strongly suggestive evidence; animal, 1 species—rats	8-31-81 (HRG/AFSCME)	4-21-83	6-22-84	HRG et al. v. Auchter, 554 F. Supp. 242 (D.C. Dist. Ct. 1-5-83) & 702 F.2d 1150 (D.C. Cir. 3-15-83)
Formaldehyde	Animal, 2 species—mice, rats	10-26-81 (UAW)	12-10-85	—	—
Inorganic arsenic	Human		1-21-75	5-5-78	Affirmed, ASARCO Inc. et al. v. OSHA, 746 F.2d 483 (9th Cir. 9-13-84)
Methyl chloromethyl ether	Inconclusive—contains bis-chloromethyl ether	—	9-7-73[b]	1-29-74	"Fourteen Carcinogens"[a]
4,4-Methylenebis (2-Chloroaniline) (MBOCA)	Animal, 2 species—rats, mice	—	9-7-73[b] 2-3-75	1-29-74	"Fourteen Carcinogens"[a] —vacated 12-17-74
alpha-Naphthylamine	Found in association with beta-Naphthylamine & has carcinogenic derivatives	—	9-7-73[b]	1-29-74	"Fourteen Carcinogens"[a]
beta-Naphthylamine	Human; animal, 5 species—rats, mice, hamsters, dogs, monkeys	1-4-73 (OCAW/HRG)	9-7-73[b]	1-29-74	"Fourteen Carcinogens"[a]
4-Nitrobiphenyl	Animal, 1 species—dog; forms 4-Amino-diphenyl which is carcinogenic in humans	1-4-73 (OCAW/HRG)	9-7-73[b]	1-29-74	"Fourteen Carcinogens"[a]
beta-Propiolactone	Animal, 3 species—mice, rats, hamsters	1-4-73 (OCAW/HRG)	9-7-73[b]	1-29-74	"Fourteen Carcinogens"[a]
N-Nitrosodimethylamine	Animal, 6 species—rats, mice, rabbits, hamsters, guinea pigs, fish	1-4-73 (OCAW/HRG)	9-7-73[b]	1-29-74	"Fourteen Carcinogens"[a]
Trichloroethylene	Animal, 1 species—mice: data inconclusive	—	10-20-75	—	—
Vinyl chloride	Human; animal, 3 species—rats, mice, hamsters	3-14-74 (United Rubber Workers, IUD, OCAW)	4-5-74(1)	10-4-74	Affirmed, Society of the Plastics Industry Inc. v. OSHA, 509 F.2d 1301 (2d Cir. 1-31-75)

[a]Standards were issued simultaneously for 14 carcinogens. Initially issued as emergency temporary standards, 2 of the 14 were struck down in Dry Color Manufacturing v. Department of Labor, 486 F.2d 98 (3rd Circuit, Oct. 4, 1973). Final standards were then issued. The Court upheld all final standards except 4,4-Methylenebis (2-chloroaniline) in Synthetic Organic Chemical Manufacturers Association v. Brennan, 506 F.2d 385 (3rd Circuit, Dec. 17, 1974). The ethyleneimine standard was affirmed in Synthetic Organic Chemical Manufacturers Association v. Brennan, 503 F.2d 1155 (3d Cir., 8-26-74).

[b]Also an emergency temporary standard.

SOURCES: OSHA response to OTA request; Federal Register notices cited in the agency response; Preventing Illness and Injury in the Workplace, OTA, 1985; "Summary of OSHA Regulations & NIOSH Recommendations for Occupational Safety & Health Standards, 1986" published in MMWR vol. 35, No. 1S.

Both standards are important. The Hazard Communication Standard, in particular, has led to a major effort devoted toward updating material safety information and communicating this information to workers. In addition, OSHA is now in the process of expanding the scope of the Hazard Communication Standard to all industries. The standard currently applies to the manufacturing sector, including chemical companies that supply the relevant hazard information, but nonmanufacturing industries are not required to implement the program. The effects of these standards on actual exposures in the workplace, however, are indirect. OSHA is still working on standards setting specific exposure levels and mandating other types of health and safety activities, even for chemicals covered by the Hazard Communication and Access to Records standards.

OSHA's regulation of carcinogens has been controversial. Of nine final actions on individual carcinogens, seven have resulted in court challenges: on asbestos (twice), vinyl chloride, coke oven emissions, benzene, arsenic, and ethylene oxide. The rules on DBCP and acrylonitrile were not challenged as final standards. For the standard regulating 14 carcinogens as a group, the ethyleneimine and 4,4-methylenebis(2-chloroaniline) (MBOCA) standards were challenged. In all, two permanent standards were struck down as a result of those challenges: MBOCA and benzene. Thus, standards are in effect for 20 of the 22 chemicals regulated as carcinogens.[3]

In the 16 years since the OSH Act passed, NIOSH, part of the Centers for Disease Control, issued 93 recommendations concerning carcinogens. As summarized in table 3-2, these consisted of 32 criteria documents, 6 revised criteria documents, 19 current intelligence bulletins, 10 special hazard reviews, 3 health hazard alerts, 1 occupational hazard assessment, and a list of 22 substances (including the "14 carcinogens") for which recommendations were made in testimony presented in 9 OSHA regulatory proceedings. Criteria documents and regulatory testimony identify substances that pose potential health problems and recommend exposure levels to OSHA.

The other documents represent NIOSH efforts to communicate research findings and warnings to workers and employers.

NIOSH has changed its policy on criteria documents. While a large number were produced in the early to mid-1970s, OSHA criticized the quality of the documents and rarely responded (see table 3-2). Starting under NIOSH Director Anthony Robbins, the agency placed greater emphasis on epidemiologic studies and health hazard evaluations (evaluations of reported worker health problems in particular workplaces), and has deemphasized production of criteria documents. From 1971 to 1980, NIOSH issued 77 recommendations, about 7.7 per year, while from 1981 to 1986, it has issued 16 criteria documents, current intelligence bulletins, or special hazard alerts, about 2.6 per year.

In all, the NIOSH recommendations cover 71 different chemicals or processes that were determined by NIOSH to be carcinogenic. Of these 71, OSHA has responded by issuing health standards for 21 chemicals or processes: asbestos (twice), arsenic, coke oven emissions, vinyl chloride, benzene, acrylonitrile, ethylene oxide, and the 14 carcinogens regulated together. However, for two chemicals (benzene and MBOCA), the OSHA standard was vacated by the courts. Thus 19 chemicals and processes have actually been regulated for carcinogenic effects based on NIOSH recommendations. OSHA regulated one chemical, DBCP, for carcinogenic effects although the NIOSH recommendation was only based on adverse toxic effects. As of early 1987, OSHA has an active proposal pending on benzene exposures. However, no activity is currently being considered for MBOCA exposures, although this substance is still being imported and used in the United States.

The remaining 50 chemicals or processes have not been the subjects of final OSHA 6(b) standards, although many are being regulated under 6(a) standards that were adopted in 1971 based on recommendations concerning noncarcinogenic effects. Of these 50 chemicals or processes, OSHA has proposed regulations for 4: formaldehyde, ethylene dibromide, trichloroethylene, and beryllium, although the latter two were proposed in

---

[3]A recent report of the Administrative Conference of the United States discusses OSHA rulemaking. See (1).

Table 3-2.—NIOSH Identification of Carcinogens

Substance	Type of recommendation	Date	OSHA proposed or final action	Type of evidence
2-Acetylaminofluorene	T	9-73	1-29-74	Animal
Acrylonitrile	CD	9-77	10-3-78	Animal, human
Aldrin/Dieldrin	SHR	9-78	—	Animal
4-Aminobiphenyl	T	9-73	1-29-74	Human
Arsenic	CD	9-74	—	Human (not conclusive)
	CD (rev.)	6-75	5-5-78	Human
Arsine	CIB	8-79	—	Human[a]
Asbestos	CD	1-72	1-21-75*	Human
	CD (rev.)	12-76	—	Human
	T	6-84	6-20-86	Human
Benzene...............................	CD	7-74	—	Human (limited evidence)
	CD (rev.)	8-76	—	Human
	T	7-77	2-10-78**	Human
	T	3-86	12-10-85*	Human
Benzidine	T	9-73	1-29-74	Human
Benzidine-based dyes	SHR	11-79	—	Human
Beryllium..............................	CD	6-72	10-17-75*	Animal
	T	8-77	—	Animal, human
bis-Chloromethyl ether	T	9-73	1-29-74	Human
1,3-Butadiene	CIB	2-84	—	Animal, human
Cadmium..............................	CD	8-76	—	Human (toxic effects)
	CIB	9-84	—	Animal, human
Carbon black	CD	9-78	—	Animal
Carbon tetrachloride	CD	12-75	—	Animal
	CD (rev.)	6-76	—	Animal
Chloroform	CD	9-74	—	Animal
	CD (rev.)	6-76	—	Animal
Chloroprene	CD	8-77	—	Human
Chromium	CD	12-75	—	Animal, human
Chrysene..............................	SHR	6-78	—	Animal
Coal gasification plants	CD	9-78	—	Animal, human
Coal liquefaction	OHA	3-81	—	Animal, human
Coal tar products.......................	CD	9-77	—	Animal, human
Coke oven emissions	CD	2-73	—	Human
	T	11-75	10-22-76	Animal
DDT	SHR	9-78	—	Animal
2,4-Diaminoanisole	CIB	1-78	—	Animal, human (suggestive evidence)
3,3'- Dichloro-benzidine..................	T	9-73	1-29-74	Animal
Di-2-ethylhexyl Phthalate (DEPH)	SHR	3-83	—	Animal
4-Dimethylaminoazobenzene	T	9-73	1-29-74	Animal
Dinitrotoluenes	CIB	7-85	—	Animal
Dioxane	CD	9-77	—	Animal
Epichlorohydrin	CD	9-76	—	Animal
	CIB	10-78	—	Animal, human
Ethylene dibromide	CD	8-77	—	Animal
	T	11-83	—	Animal
Ethylene dichloride	CD	3-76	—	Human (toxic effects)
	CD (rev.)	9-78	—	Animal
Ethyleneimine..........................	T	9-73	1-29-74	Animal
Ethylene oxide	SHR	9-77	—	Human (potential)
	T	7-83	6-22-84	Animal, human
Ethylene thiourea.......................	SHR	10-78	—	Animal
Formaldehyde..........................	CD	12-76	12-10-85*	Human (toxic effects)
	CIB	4-81	—	Animal
	T	5-86	—	Animal, human (potential)
Foundries	CD	9-85	—	Human
Glycidyl ethers.........................	CD	6-78	—	Animal
Hexachloroethane	CIB	8-78	—	Animal
Hydrazines	CD	6-78	—	Animal

continued·on next page

Table 3-2.—NIOSH Identification of Carcinogens—Continued

Substance	Type of recommendation	Date	OSHA proposed or final action	Type of evidence
Isopropyl alcohol	CD	3-76	—	Human[b]
Kepone	CD	1-76	—	Animal
Methyl chloromethyl ether	T	9-73	1-29-74	Animal
4,4'-Methylenebis-(2-chloroaniline) (MOCA)	T	9-73	1-29-74**	Animal
	SHR	1978	2-3-75*	Animal
Methylene chloride	CD	3-76	—	Human (toxic effects)
	CIB	4-86	—	Animal
4,4-Methylenedianiline	CIB	7-86	—	Animal, human
Monohalomethanes	CIB	9-84	—	Animal
alpha-Naphthylamine	T	9-73	1-29-74	Human
beta-Naphthylamine	T	9-73	1-29-74	Human
Nickel carbonyl	SHR	5-77	—	Animal
Nickel, inorganic compounds	CD	5-77	—	Animal, human
4-Nitrobiphenyl	T	9-73	1-29-74	Animal
2-Nitronapthalene	CIB	12-76	—	Animal (caused by metabolite)
2-Nitropropane	CIB	4-77	—	Animal
	HHA	10-80	—	Animal
N-Nitrosodimethylamine	T	9-73	1-29-74	Animal
O-Dianisidine-based dyes	HHA	12-80	—	Animal, human[c]
O-Tolidine-based dyes	HHA	12-80	—	Animal, human[c]
O-Tolidine	CD	1-74	—	Animal
Phenyl-beta-naphthylamine	CIB	10-76	—	Animal[d]
Polychlorinated biphenyls	CD	9-77	—	Animal
beta-Propiolactone	T	9-73	1-29-74	Animal
2,3,7,8-Tetrachloro-dibenzo-p-dioxin (TCDD)	CIB	1-84	—	Animal
1,1,2,2-Tetrachloro-ethane	CD	12-76	—	Animal, human (toxic effects)
1,1,2,2-Tetrachloroethane	CIB	8-78	—	Animal
Tetrachloromethylene	CD	7-76	—	Human (toxic effects)
	CIB	1-78	—	Animal
Trichlorethylene	CD	6-73	10-20-75* no final action	Human (toxicity data)
	SHR	1-78	10-20-75*	Animal
1,1,2-Trichloroethane	CIB	8-78	—	Animal
Vinyl chloride	CD	3-74	10-4-76	Animal
Vinyl halides	CIB	9-78	—	Animal, human

*Date of notice of proposed rulemaking (NPRM).
**Vacated.
[a]Based on evidence for arsenic because Arsine metabolizes to inorganic arsenic in the human body.
[b]Carcinogenic effect considered derived from strong-acid in production process, no evidence for carcinogenicity of isopropyl alcohol itself.
[c]Based on evidence for benzidine because it converts to benzidine in humans.
[d]Based on evidence for beta-naphthylamine because it metabolizes to beta-naphthylamine in humans.
ABBREVIATIONS: CD = Criteria Document; CIB = Current Intelligence Bulletin; HHA = Health Hazard Alert; OHA = Occupational Hazard Assessment; (rev.) = Revised; SHR = Special Hazard Review; T = Testimony to Dept. of Labor.

SOURCES: NIOSH/OSHA responses to OTA request including summary of NIOSH Documents for Carcinogenic Agents; "Summary of OSHA Regulations and NIOSH Recommendations for Occupational Safety and Health Standards," 1986, published in MMWR vol. 35, No.1S; Preventing Illness and Injury in the Workplace, OTA, 1985.

1975 and can now be considered dormant. For the remaining chemicals or processes, no OSHA proposals have been issued.

OSHA's regulatory agenda has increasingly been set by outsiders, through petitions, court orders, referrals from EPA and congressional directives. The formaldehyde standard involved a petition, a referral from EPA and a court order after OSHA delays. OSHA has also received formal or informal referrals from EPA under section 9 of TSCA, for 4,4'-methylene dianiline (MDA),

1,3-butadiene, chloromethane, MBOCA, toluenediamine, glycol ethers, acetaldehyde, and acrylamide. In the Superfund amendments of 1986, Congress directed OSHA to issue standards on health and safety protection for workers at hazardous waste sites.

During the first 10 years of its existence, OSHA issued 8 final rules, while from 1981 to 1986, OSHA has issued 2 final rules to reduce exposures to specific carcinogens, a rate of 3 years per regulation. On average, 15 months have elapsed between OSHA's issuance of an NPRM and publication of a final rule on a particular substance. Further, during the first 10 years of OSHA's existence, the time from criteria document to final rule was about 32 months, while during the last 6 years, the average has been about 60 months. There are still a number of regulations yet to be acted on that could modify this last average, which is based on just two final rules. One final rule issued since 1981 was a more stringent regulation of asbestos, previously regulated by OSHA, while the final rule for ethylene oxide was issued under court order (see the discussion of judicial action in app. A). Benzene, first identified as a health hazard by NIOSH in a criteria document in 1974, was initially regulated in 1978. That final rule was overturned by the Supreme Court in 1981. Subsequently, NIOSH has issued several revised recommendations (in 1976, 1977, and 1986), but OSHA has yet to issue a final rule on benzene.

Although OSHA's cancer policy permits the regulation of carcinogens on the basis of animal evidence alone, for the most part the agency has regulated on the basis of human evidence, evidence which in most cases was confirmed by animal evidence. Not considering the 14 carcinogens regulated as a group, OSHA has issued final rules for 8 individual substances based on at least some evidence of human carcinogenicity. DBCP, however, was regulated primarily because it caused infertility in men, although animal carcinogenicity data was available.

Of the "14 carcinogens," 3 are human carcinogens, 8 are positive in 2 or more species (a number of these 11 carcinogens are positive in 3 to 5 mammalian species), and 1 was positive in a single species. The remaining 2 chemicals (alpha-napththylamine and methyl chloromethyl ether) are found in association with other carcinogens on the list of 14 and were themselves thought to pose potential risks. Thus 11 of the 22 chemicals were regulated based partly on human evidence, and most of the animal carcinogens were regulated early in OSHA's history—in the 1974 standard on 14 carcinogens.

MSHA REGULATORY ACTIONS

The Mine Safety and Health Administration (MSHA) regulates the exposure of miners to carcinogens. The Federal Mine Safety and Health Amendments Act (1977) consolidated the regulation of mine health and safety under one statute, and transferred responsibility from the Department of the Interior to the Labor Department. Safety and health in coal mines had previously been regulated under the Federal Coal Mine Health and Safety Act of 1969 by the Department of the Interior's Mining Enforcement and Safety Administration (MESA), while safety and health in metal and nonmetal mines had been regulated by MESA under the Federal Metal and Nonmetallic Mine Safety Act of 1966 (150).

Separate standards are issued for surface and underground coal mines and for surface and underground metal and nonmetal mines. For the most part, standards adopt the 1972 and 1973 recommendations of the ACGIH. Other than in radiation and asbestos, MSHA has found relatively few exposures to carcinogens in mines.

When MSHA's predecessor agency issued the surface coal mining regulations in 1972 it specified that exposures to toxic substances should not exceed those recommended by ACGIH in 1972. At that time only some potential carcinogens in diesel fumes and polychlorinated biphenyls (PCBs) were thought to present possible problems in the surface mining of coal.

For underground coal mining, MSHA's regulations require mines to keep exposures at least as low as the "current" ACGIH recommended exposures, with the idea that the exposure levels would be updated each time ACGIH changed its list of toxic substances. Although MSHA legal staff interpret "current" to be the 1972 list, which was current when the regulations were issued, MSHA staff stated to OTA that inspectors enforce those that are actually current. The issue has not been resolved in part because there have been very few citations. There is debate within the agency over whether it would be in violation of the Administrative Procedure Act for the regulations to be automatically updated by ACGIH because it would not provide opportunity for the public to comment on the updated exposure levels (190).

Regulations that govern exposure to carcinogens in metal and non-metal mines incorporate by reference the 1973 ACGIH recommendations. In 1978, for metal and non-metal mines, MSHA regulated the 14 carcinogens that had been regulated by OSHA in 1974, and 2 other substances, by restricting their use to approved laboratory conditions and competent personnel (see table 3-3).

Apart from incorporating by reference the ACGIH list of carcinogens, MSHA issued regulations in the late 1970s on asbestos, lowering the permissible exposure level as specified in the 1973 ACGIH list from 5 fibers to 2 fibers greater than 5 microns in length per cubic centimeter of air (161). The regulations applied to metal and non-metal mines and surface coal mines. There are no regulations however governing exposure to asbestos in underground coal mines. Finally, MSHA has regulated exposure to radon daughters and ionizing radiation (1969) and updated these standards once (1976) (see table 3-3).

Following a petition and a lawsuit filed by the Oil Chemical and Atomic Workers Union and the Health Research Group for a more stringent emergency temporary standard, and in response to a court order, MSHA has also proposed to revise the standards for radiation exposure to uranium miners (269). The proposed new standard is 4 working level months (WLM) which results in a cumulative lifetime exposure of 120 WLM over a 30-year period.[4] This standard is similar to the

existing standard except that it also establishes a combined exposure limit for radon daughters, gamma radiation, and thoron daughters. Previously, thoron daughters were excluded from coverage altogether.

These standards merit special attention because uranium miners receive relatively high levels of radiation exposure compared to other nuclear workers and have high rates of occupational mortality (119). Radiation protection standards in general are based on the principle that exposures should be kept "As Low As Reasonably Achievable" (ALARA) (269). The proposed standard is consistent with the Federal Radiation Guidance, which is based on the recommendations of the International Commission of Radiological Protection (ICRP). A recent NIOSH evaluation however found a significant excess of lung cancer deaths from cumulative radon daughter exposures well below 100 WLM and concluded that a standard of 1 WLM a year is achievable (252). NIOSH is in the final stages of preparing a criteria document recommending a permissible exposure level.

Increased use of diesel engines in mines has raised concern about mine worker exposures to those fumes. Although MSHA routinely samples and regulates gaseous diesel emissions in accordance with the 1972 and 1973 TLVs, these are not based on carcinogenic effects. There is no standard regulating particulate diesel emissions except the limit on respirable dust, which also does not take into account carcinogenic effects. Finally, in 1986 OSHA revised its standard for asbestos. It is not clear whether MSHA will take additional action on regulating exposures to radiation, diesel fumes, or asbestos.

---

[4]A working level refers to a specified concentration of radon daughters in the air. A working level month is defined as the exposure to one working level for 170 hours. The commonly used dose-equivalent of 1 WLM is approximately 1 rem or 4.8 WLM = 5 rem (a rem is a measure of absorbed dose).

FDA REGULATORY ACTIONS

Carcinogens in foods and cosmetics are regulated by the Food and Drug Administration (FDA) under the FDCA. Of the regulatory statutes considered in this report, this act has by far the longest history. Its requirements have evolved over time. Prior to the 1950s, the statutory approach was to prohibit the sale of adulterated food. Food was considered adulterated if FDA could show

Table 3-3.—MESA/MSHA Regulation of Carcinogens: Metal & Non-Metal Mines, Underground & Open Pit (other than coal), Sand, Gravel, and Crushed Stone Operations, Including Uranium Mine Radiation Standards

Substance	Adoption of 1973 ACGIH recommendations			Revisions[a]		
	NPRM	Final	Evidence	NPRM	Final	Basis
Asbestos	8-29-73	7-1-74[b]	Human	7-7-77	11-17-78	OSHA Standard
bis-(Chloromethyl) ether	8-29-73	7-1-74[b]	Human	7-7-77	11-17-78	OSHA Standard
Chromates	8-29-73	7-1-74[b]	Human	—	—	—
Coal tar pitch volatiles	8-29-73	7-1-74[b]	Human	—	—	—
Nickel carbonyl	8-29-73	7-1-74[b]	Human	—	—	—
4-Aminobiphenyl (P-xenylamine)	8-29-73	7-1-74[c]	Human	7-7-77	11-17-78	OSHA Standard
Benzidine & its salts	8-29-73	7-1-74[c]	Human	7-7-77	11-17-78	OSHA Standard
Beta-naphthylamine	8-29-73	7-1-74[c]	Human	7-7-77	11-17-78	OSHA Standard
4-Nitrodiphenyl	8-29-73	7-1-74[c]	Human	7-7-77	11-17-78	OSHA Standard
Beryllium	8-29-73	7-1-74[d]	Animal	—	—	—
Chloromethyl methyl ether	8-29-73	7-1-74[d]	Animal	7-7-77	11-17-78	OSHA Standard
3,3'-Dichlorobenzidine	8-29-73	7-1-74[d]	Animal	7-7-77	11-17-78	OSHA Standard
Dimethyl sulfate	8-29-73	7-1-74[d]	Animal	—	—	—
Ethylenimine	8-29-73	7-1-74[d]	Animal	—	—	—
4,4'-Methylene bis (2-chloroaniline)	8-29-73	7-1-74[d]	Animal	7-7-77	11-17-78	OSHA Standard
N-Nitrosodimethylamine	8-29-73	7-1-74[d]	Animal	7-7-77	11-17-78	OSHA Standard
beta-Propiolactone	8-29-73	7-1-74[d]	Animal	7-7-77	11-17-78	OSHA Standard
2-Acetylaminofluorene	—	—	—	7-7-77	11-17-78	OSHA Standard
Carbon tetrachloride	—	—	—	7-7-77	11-17-78	[e]
4-Dimethylaminoazobenzene	—	—	—	7-7-77	11-17-78	OSHA Standard
Alpha-naphthylamine	—	—	—	7-7-77	11-17-78	OSHA Standard
Phenol	—	—	—	7-7-77	11-17-78	[f]

Uranium mining radiation exposure standards:

Type of radiation	NPRM	Final	Basis	Court challenge/petitions
Radon daughters	1-16-69[g]	2-25-70	FRC Guidance, 1967	
	6-24-70	12-8-70	—	
	9-25-75	6-10-76	EPA Guidance 5-25-71 (derived from FRC recommendation 1-15-69)[h]	
Gamma radiation	1-28-77	6-8-77	NCRP/ICRP recommendations	
Radon & thoron daughters and gamma radiation[i]	12-19-86	—	ICRP 26	4-21-80, Petition (OCAW /HRG) for 1.7 WLM/yr ETS 3-13-84; *OCAW & Public Citizen HRG* v. *David Zeeger*, 768 F.2d 1480, (D.C. Cir. 8-2-85); 1-29-85 denied ETS, issued ANPRM

[a]In the revised regulations (except for asbestos for which the exposure limit was reduced), MSHA restricted the use of the listed substances to approved laboratory conditions by competent personnel. The list was based on technical and medical evidence compiled by OSHA for the regulation of certain carcinogens. According to MSHA, however, "most if not all, of the listed chemicals are seldom if ever used in the mineral industries subject to the Metal and Nonmetal Mine Safety Act."
[b]Regulation consists of ACGIH TLV.
[c]ACGIH did not issue a TLV; regulation adopts ACGIH recommendation that no exposure or contact be permitted.
[d]ACGIH did not issue a TLV; regulation adopts ACGIH recommendation that exposure should be reduced to a minimum.
[e]Carbon tetrachloride was included on the list of restricted chemicals even though it was not regulated as a carcinogen by OSHA. It had already been promulgated as a mandatory standard by MSHA according to information provided in response to the OTA request (Agenda to 9th Annual Meeting) but there is no indication when this occurred.
[f]Phenol was listed for acute toxicity and because it is a suspect carcinogen according to NIOSH even though it was not regulated by OSHA.
[g]Prior to the 1970 final standards, health and safety in uranium mines was regulated by the States. Several Federal agencies provided advice and assistance and formed an interagency committee but

held that they lacked the proper statutory authority to regulate radiation exposure in privately owned uranium mines. In 1950 the U.S. Public Health Service began to investigate the levels of radiation exposure and the health effects of uranium mining at the request of the Colorado health department. In 1959, the Atomic Energy Commission did begin to inspect mines and mills on Federal lands and require plans for compliance with the 1957 radiation standards. By doing so, the AEC demonstrated the feasibility of controlling radiation levels in mines which prompted the interagency committee to convene a conference of governors from uranium mining States and representatives of the concerned Federal agencies on Dec. 12, 1960. It was at this meeting that the USPHS reported preliminary data showing that uranium miners had a lung cancer rate 5 times higher than the U.S. white male population.
[h]The 1967 FRC (Federal Radiation Council) recommendations were based on the recommendations of the NCRP (National Commission on Radiological Protection). The revised 1969 FRC recommendations were based on a NAS report prepared for the Interagency Uranium Mining Radiation Review Group (IUMRRG) entitled "Epidemiologic Studies of Uranium Miners" which recognized an increased incidence of lung cancer at 120-359 cumulative WLM.
[i]The existing radiation exposure standards cover only radon daughters and gamma radiation in underground mines. The NPRM for new standards also includes thoron daughters.

SOURCES: MSHA response to OTA request including *Federal Register* notices and agenda for the 19th Meeting of the Federal Metal & Non Metal Mine Safety Advisory Committee, June 29, 30 and July 1, 1976; MSHA, "Ionizing Radiation Standards for Metal & Non-Metal Mines—Preproposal Draft," November 1985; Comments on MSHA Preproposal Draft submitted by Public Citizen Health Research Group (HRG) and the Oil Chemical and Atomic Workers International Union (OCAW), Feb. 18, 1986; Southwest Research & Information Center; NIOSH, "Evaluation of Epidemiologic Studies Examining the Lung Cancer Mortality of Underground Miners," prepared for MSHA, 5-9-86; George T. Mazuzan and J. Samuel Walker, "Control of the Atom" University of California Press, 1984.

that it was "ordinarily injurious to health" or that it contained an added substance that "may render it injurious to health" (section 402(a)(1)).

Starting in 1954, Congress enacted special statutory provisions for particular groups of substances that might be added to foods or cosmetics, first, pesticide residues (in 1954, now regulated by EPA), then food additives (1958), color additives (1960), and animal drug residues (1962).

A "food additive" is any substance, the intended use of which leads it to become a component of food either directly or indirectly. The 1958 Food Additives Amendment established a premarket approval process for food additives, although excluded from this are substances "generally recognized as safe" (GRAS) and substances that had received Federal sanction prior to 1958 ("prior sanction" substances). Food additives include substances used to alter the taste or composition of food, and packaging materials that may migrate into the food.

To approve a food additive, FDA must be convinced of the safety of the additive. On this point, the well-known "Delaney clause" provides "that no additive shall be deemed to be safe if it is found to induce cancer when ingested by man or animal, or if it is found, after tests which are appropriate for the evaluation of the safety of food additives, to induce cancer in man or animal."

For "unavoidable contaminants" found in food (such as aflatoxins in peanuts, mercury in fish, and PCBs in milk and fish), the FDA may set tolerance levels through formal rulemaking procedures. Alternatively, FDA has used more informal "action levels," which are established without going through rulemaking. Because they are not regulations, action levels are easier to change than tolerances. Action levels and tolerances are both levels of a contaminant which, if exceeded, would render the food "adulterated" and lead FDA to bring court action to seize the food.[5]

FDA's regulation of carcinogens was affected by an important case, *Monsanto* v. *Kennedy* (129), which involved a plastic bottle that FDA

---

[5]The authority to establish tolerance levels for pesticide residues on raw agricultural products was transferred to EPA in 1970 and will be discussed under the regulation of pesticides.

believed might release a suspected carcinogen into its contents and thus present a risk to consumers. The court ruled that FDA need not determine that a substance was a food additive based on a theoretical prediction of migration, although this is permissible if there is a safety concern. The court then went on to declare that FDA could exempt small amounts of substances in "de minimis situations" that "clearly present no public health or safety concerns." A 1984 court decision, *Scott* v. *FDA* (184), specifically supported FDA's decision to allow the use of color additives that contain carcinogenic impurities, when FDA believes such use represents an insignificant risk. As discussed below, FDA has started to apply this by approving certain color additives even though they contain known carcinogens, arguing that the risks are insignificant.

As a result, FDA's carcinogen regulation has changed. Prior to 1982, food and color additives that contained carcinogenic impurities (formerly called "constituents") were banned, but since then the policy for such impurities is that, if a food or color additive itself does not cause cancer, but an impurity of the additive is a known carcinogen, the agency will use risk assessment to determine whether "under the general safety clause, there is a reasonable certainty that no harm will result from the proposed use of the additive" (245).

Role of Quantitative Risk Assessment

Over time, FDA has given quantitative risk assessment more importance. The first uses of quantitative risk assessment were for carcinogenic chemicals not subject to the Delaney clause. In 1973, FDA proposed to use risk assessment to specify the sensitivity of the analytic method used to determine the level of potentially carcinogenic animal drug residues (see ch. 2). The next application was for setting tolerances and action levels on food containing unavoidable environmental carcinogenic contaminants. The first uses of risk assessment for such contaminants appear to have been for aflatoxins in 1978 and PCBs in 1979 (table 3-4). In 1982, it was applied to carcinogenic impurities in color additives, which were not regarded as subject to the Delaney clause under FDA's impurities policy. Finally, risk assessment

Table 3-4.—FDA: Carcinogens Associated With Foods and Cosmetics

Substance	When identified	Type of evidence	Petition	NPRM	Final	Type of action/comments
Acrylonitrile—as contaminant of acrylonitrile/styrene copolymer resin in the manufacture of beverage bottles	6-22-80; 11-24-81, risk assessment	Human, respiratory only; animal, 1 species—rats	3-4-83 (Monsanto) requesting rule for safe use	NA	9-19-84	The agency had revoked all beverage container uses of acrylonitrile polymers 9-23-77. Action in 1984 allows its use based on new agency policy and development of a bottle with lower contaminant levels[a]
Aflatoxin	1-19-78, risk assessment	Human, animal	—	12-6-74, proposed tolerance 15 ppb in peanut products; 3-3-78, reopened comment period	—	In 1965 aflatoxin contaminated foods were declared adulterated at the detection limit of 30 ppb. In 1969 the level changed to 20 ppb which remains as the current action level[b]
2-Aminoanthraquinone	2-21-80	—	—	—	—	Hypothesized impurity in several color additives; not found
4-Aminoazobenzene —in FD&C Yellow No. 5	12-20-83, risk assessment	Animal, 1 species—rats	3-27-65	—	9-4-85	FD&C Yellow No. 5 permanently listed for use in cosmetics and externally applied drugs; contains 4-aminoazobenzene impurity[a]
—in FD&C Yellow No. 6	—	—	11-20-68	—	11-19-86	Permanently listed Yellow No. 6; contains 4-aminoazobenzene impurity[a]
4-Aminobiphenyl —in Ext. D&C Yellow No. 1	1954 evidence cited	Human; animal, 1 species—dogs	11-23-76, received comments about substance as contaminant of azo dyes	9-23-76	12-13-77	Terminated prov...nal listing of color additive for use in externally applied drugs and cosmetics because of possible benzidine impurity and 4-aminobiphenyl impurity
—in FD&C Yellow No. 5	12-20-83, risk assessment	—	3-27-65	—	9-4-85	Permanently listed Yellow No. 5 for use in cosmetics and externally applied drugs; contains 4-aminobiphenyl impurity[a]
	12-20-83, risk assessment	—	—	9-4-85	7-7-86	Adopted uniform specifications, including specifications on impurities for all uses of Yellow No. 5
—in FD&C Yellow No. 6	—	—	11-20-68	—	11-19-86	Permanently listed Yellow No. 6; contains 4-aminobiphenyl impurity[a]
4-Amino-2-nitrophenol	4-29-80	—	—	—	—	Coal tar hair dye ingredient[c]
2-Amino-5-nitro-thiazole	1-8-79	Animal, 1 species—rats	—	—	—	Metabolite of animal drug; manufacturer withdrew drug from market after notified about carcinogenic metabolite by FDA

[a]The rules for safe use are based on a 1984 court decision that allowed safe exposure levels to be established for non-carcinogenic additives that contain carcinogenic impurities if the additive itself is not shown to be carcinogenic. (Scott v. FDA, 728 F.2d 322 (6th Cir. 1984).) See discussion in text.
[b]Action levels are not required to be published for comment before they are established.
[c]Coal tar hair dyes are exempt from the adulteration and color additive provisions found in sections 601 & 706 of the FDCA (21 U.S.C. 361.376) provided the label bears a caution statement & patch test instructions to determine if the product causes skin irritation (Fed. Reg. 43:1101. 1978).

continued on next page

Table 3-4.—FDA: Carcinogens Associated With Foods and Cosmetics—Continued

Substance	When identified	Type of evidence	Petition	NPRM	Final	Type of action/comments
Aniline						
—in FD&C Yellow No. 5	4-6-79; 12-20-83, risk assessment	Animal, 1 species—rats	3-7-65	—	9-4-85	Permanently listed Yellow No. 5 for use in cosmetics and externally applied drugs; contains aniline impurity[a]
—in FD&C Yellow No. 6	—	—	11-20-68	—	11-19-86	Permanently listed Yellow No. 6; contains aniline impurity[a]
Asbestos—in talc	6-6-85, risk assessment	—	—	—	—	Cosmetic use of talc; risk assessment indicated lifetime risk is less than 1 in 1 million
Azobenzene						
—in FD&C Yellow No. 5	12-20-83, risk assessment	Animal, 1 species—rats	3-27-65	—	9-4-85	FD&C Yellow No. 5 permanently listed for use in cosmetics and externally applied drugs; contains azobenzene impurity[a]
	—	—	—	9-4-85	7-7-86	Adopted uniform specifications, including specifications on impurities for all uses of Yellow No. 5
—in FD&C Yellow No. 6	—	—	11-20-68	—	11-19-86	Permanently listed Yellow No. 6; contains azobenzene impurity[a]
Benzidine						
—in Ext. D&C Yellow No. 1	Cites 1973 studies	Human; animal, 4 species—dogs, rats, mice, hamsters	11-23-76, received comments about substance as contaminant of azo dyes	9-23-76	12-13-77	Terminated provisional listing of color additive for use in externally applied drugs and cosmetics because of possible benzidine impurity and 4-aminobiphenyl impurity
—in FD&C Yellow No. 5	12-20-83, risk assessment	Human; animal, 4 species—dogs, rats, mice, hamsters	3-27-65	—	9-4-85	Permanently listed Yellow No. 5 for use in cosmetics and externally applied drugs; contains benzidine impurity[a]
—in FD&C Yellow No. 5	12-20-83, risk assessment	Human; animal, 4 species—dogs, rats, mice, hamsters	—	9-4-85	7-7-86	Adopted uniform specifications, including specifications on impurities, for all uses of Yellow No. 5
Chloroform (Trichloromethane)	3-1-76, received NCI report	Animal, 2 species	12-30-75 (HRG)	4-9-76	6-29-76	Prohibited in human drugs & cosmetics; proposal to prohibit use in food contact articles
4-Chloro-m-phenylene-diamine	7-10-79	—	—	—	—	Coal tar hair dye ingredient[c]
4-Chloro-o-phenylene-diamine	6-10-80	—	—	—	—	Coal tar hair dye ingredient[c]
5-Chloro-o-toluidine	5-27-80	—	—	—	—	Hypothesized color additive impurity; not found
Cinnamyl anthranilate	6-19-80; 6-21-81, risk assessment	Animal, 1 species—mice	—	5-25-82	10-23-85	Prohibited for use in food
C.I. Vat Yellow No. 4—in C.I. Vat Orange No. 1	7-20-83, risk assessment	Animal, 1 species—mice	6-17-83 (Custom Tint Laboratories)	—	5-16-85	Listed C.I. Vat Orange No. 1 for coloring contact lenses; contains C.I. Vat Yellow No. 4 impurity[a]

continued on next page

Table 3-4.—FDA: Carcinogens Associated With Foods and Cosmetics—Continued

Substance	When identified	Type of evidence	Petition	NPRM	Final	Type of action/comments
p-Cresidine	4-29-80	—	—	—	—	Hypothesized impurity in FD&C Red No. 40
Dapsone (4,4'-Diaminodiphenyl-sulfone)	3-25-80	—	—	—	—	Hypothesized impurity in FD&C Yellow Nos. 5 & 6; not found
D&C Orange no. 17	1-20-83	Animal, 2 species—mice, rats	4-16-69	—	4-1-83	Provisional listing expired for use in ingested drugs and cosmetics
					8-7-86	Permanently listed for use in externally applied drugs and cosmetics
					10-6-86	Response to objections
					2-19-87	Clarification to preamble
D&C Red Nos. 19 & 37	8-12-82	Animal, 2 species—rats, mice	4-14-69	—	2-4-83	Terminated provisional listing for ingested uses
					8-6-86	Terminated provisional listing for D&C Red No. 37
					8-7-86	Permanently listed D&C Red No. 19 for use in externally applied drugs and cosmetics
					10-6-86	Response to objections
					2-19-87	Clarification to preamble
D&C Red Nos. 8 & 9	8-26-82	Animal, 1 species—rats	5-17-65	—	12-5-86	Permanently listed for use in ingested drug and cosmetic lip products in limited amounts and in externally applied drugs and cosmetics
2,4-Diaminoanisole & 2,4-Diaminoanisole sulfate—as ingredients in coal tar hair dyes	10-18-77, NCI study sent to FDA; 9-7-78, risk assessment	Animal, 2 species—rats, mice	10-19-77 (EDF); 12-14-77 (GAO recommendation)	1-6-78	10-16-79	Product warning statements required in absence of statutory authority to ban(2); 9-18-80, stayed by consent order, U.S. Dist. Ct. for the Southern Dist. of Georgia (Civil Action No. CV 480-71) and remanded to FDA for reconsideration & further rulemaking
2,4-Diaminotoluene	9-20-79; 6-30-83, risk assessment	—	—	—	1-15-80	Denied use of adhesive in which this impurity formed
Dibutyltin diacetate—as contaminant of 2,2-oxamidobis[ethyl 3-(3,5-di-tert-butyl-4-hydroxyphenyl) propionate]	10-18-79; 5-23-83, risk assessment	Animal, 1 species—mice	—	—	8-19-83; 4-2-84, supplement to final rule	Rule for safe use of non-carcinogenic food packaging additive that may contain dibutyltin diacetate.[a] The supplement discloses the identity of the carcinogenic constituent which previously was withheld as a trade secret
Di-(2-ethylhexyl) adipate	11-12-80; 7-21-81, risk assessment	Animal, 2 species—rats, mice	—	—	12-4-85	Request for hearing was denied
					—	Under study—Indirect mechanism of action suspected

[a]The rules for safe use are based on a 1984 court decision that allowed safe exposure levels to be established for non-carcinogenic additives that contain carcinogenic impurities if the additive itself is not shown to be carcinogenic. (Scott v. FDA, 728 F.2d 322 (6th Cir.1984).) See discussion in text.

[c]Coal tar hair dyes are exempt from the adulteration and color additive provisions found in sections 601 & 706 of the FDCA (21 U.S.C. 361,376) provided the label bears a caution statement & patch test instructions to determine if the product causes skin irritation (Fed. Reg. 43:1101, 1978).

continued on next page

Table 3-4.—FDA: Carcinogens Associated With Foods and Cosmetics—Continued

Substance	When identified	Type of evidence	Petition	NPRM	Final	Type of action/comments
Di-(2-ethylhexyl) phthalate	11-12-80; 7-21-81, risk assessment	Animal, 2 species—rats, mice	—	—	—	Under study, indirect mechanism of action suspected
1,4-Dioxane—as impurity in 10 food contact additives & 1 sanitizing solution used on food contact surfaces[d]	8-21-79; 11-19-81, risk assessment	Animal, 1 species—rats	3-19-82, 4-22-83, 6-14-84, 2-3-81, 4-20-84, 2-25-85, 12-28-79, 6-3-83, 9-8-83, 9-3-82, 3-5-86	—	9-28-83, 9-10-85, 8-13-86, 9-5-86, 9-19-86, 9-24-86, 9-24-86, 9-24-86, 12-31-86, 1-6-87, 1-7-87	Rules for safe use of food-contact additives that contain 1,4-Dioxane
1,3-Diphenyltriazene —in FD&C Yellow No. 5	12-20-83, risk assessment	Animal, 1 species—mice	3-27-65	—	9-4-85	FD&C Yellow No.5 permanently listed for use in cosmetics and externally applied drugs; contains 1,3-diphenyltriazene impurity[a]
	—	—	—	—	9-4-85	Adopted uniform specifications including specifications on impurities, for all uses of FD&C Yellow No. 5
—in FD&C Yellow No. 6	—	—	11-20-68	—	11-19-86	Permanently listed Yellow No. 6; contains 1,3-diphenyltriazene impurity[a]
Dulcin (sucrol), (4-ethoxy-phenylurea)	no date	Chronic toxicity data in rats	—	—	1-19-50	Banned as food additive
Epichlorohydrin—contained in polyamide-epichlorohydrin water-soluble thermosetting resins (retention aid in paper coating)	11-29-82, risk assessment		4-9-82 (Sandox Colors & Chemicals)	—	8-19-83	Amended regulations to remove upper viscosity limit of the retention aid—(used in paper coating as indirect food additive); contains epichlorohydrin impurity[a]
	—	—	—	—	4-24-84	Clarification of final rule
	—	—	—	—	12-4-85	Denied request for hearing
17-b-Estradiol	11-14-79; 5-23-83, risk assessment	—	—	—	4-9-84	Animal drug; hormone naturally occurring in animals
Ethylene oxide—as impurity in 9 food contact additives[e]	1-5-82; 4-7-86, risk assessment	—	6-14-84; 2-3-81; 4-20-84; 2-25-85; 12-28-79; 6-3-83; 9-8-83; 9-3-82; 3-5-86	—	8-13-85; 9-5-86; 9-19-86; 9-24-86; 9-24-86; 9-24-86; 12-31-86; 1-6-87; 1-7-87	Rules for safe use of several food contact additives that may contain ethylene oxide impurity[a]
Ethylene thiourea (impurity in mercaptoimidazoline)	No date	"Known carcinogen"	—	4-24-73	11-30-73	Banned mercaptoimidazoline and 2-mercaptoimidazoline in food contact articles due to possible impurity
Flectol H (1,2-dihydro-2,2,4-tri-methylquinoline, polymerized)	No date	Animal, 1 species	—	—	4-7-67	Banned in food contact articles
Hydrazine	6-12-79, (cites IARC); 6-8-82, risk assessment	Animal, 2 species—mice, rats	—	6-12-79	—	To prohibit food additive use; permitted as a boiler water additive with limitation of zero in steam contacting food
Lactitol	5-2-85	—	5-19-83	—	—	Not permitted as a food ingredient; petition requests food additive regulation

continued on next page

Table 3-4.—FDA: Carcinogens Associated With Foods and Cosmetics—Continued

Substance	When identified	Type of evidence	Petition	NPRM	Final	Type of action/comments
Lactose	5-2-85	—	—	—	—	Natural component of food including milk, FASEB report (9-86) states that lactose is not significantly tumorigenic in humans; believes it is necessary to repeat animal study
Lead acetate	3-6-79, found to be absorbed by skin; 5-6-80; 5-23-80, risk assessment	Animal, 2 species—rats, mice (by ingestion)	6-29-73 (Comm. of Progressive Hair Dye Industry) requesting permanent listing[f]	12-10-63	10-31-80; 3-6-81, response to objections & removal of stay	Permanently listed for use as color additive in hair dyes
4,4'-Methylenebis (2-Chloroaniline)	No date	Animal, 1 species	—	—	12-2-69	Banned in food contact articles
Methylene chloride	1-20-83; 4-23-85, risk assessment	Animal, 1 species—mice (suggestive in rats)	—	12-18-85	—	Proposed to ban in cosmetics but not in decaffeinated coffee because risk was determined negligible
4,4'-Methylene dianiline	4-7-83; 5-16-83, risk assessment	—	—	—	—	—
b-Naphthylamine —in FD&C Yellow Nos. 3 & 4	No date	No carcinogenic evidence cited	(Dye Stuffs & Chemicals)	—	10-12-60	Denied petition of Dye Stuffs & Chemicals to restore color additives to provisional list with 25 ppm tolerance in food, possible impurity in these color additives
—in FD&C Yellow Nos. 3 & 4	Cites 1974 IARC report	Human, animal, 4 species—mouse, hamster, dog, monkey	—	—	10-12-60	Provisional listing of these color additives terminated because b-naphthylamine is possible impurity
—in Red Nos. 10, 11, 12, & 13	Cites 1974 IARC report	Human, animal	8-6-73 (for permanent listing), withdrawn 10-21-77	9-23-76	12-13-77	Prohibited these color additives from use in drugs & cosmetics because b-naphthylamine is a possible impurity; terminated provisional listing
—in Orange-B	1-19-78, identified as contaminant of Orange-B	Human, animal	—	10-3-78	—	To revoke listing of Orange-B for use in food; not certified for use in food since 10-78
Nitrilo triacetic acid	10-18-79	—	—	—	—	—
N-Nitrosodimethylamine —as contaminant in malt beverages	1956	Animal, more than 2 species—rats, mice, others	—	—	10-25-79, action level announced in press release[b]; notice of compliance guide published 6-10-80	Action level established at lowest level at which presence can be confirmed in malt beverages

[a]The rules for safe use are based on a 1984 court decision that allowed safe exposure levels to be established for non-carcinogenic additives that contain carcinogenic impurities if the additive itself is not shown to be carcinogenic. (*Scott v. FDA*, 728 F.2d 322 (6th Cir. 1984).) See discussion in text.

[b]Action levels are not required to be published for comment before they are established.

[d]Food contact additives which contain 1,4-Dioxane for which safe use rules were issued: ethylene oxide adduct of 2,4,7,9,-tetramethyl-5-decyn-4,7-diol; polyoxyethylated (5 moles) tallow amine and alpha-alkyl-omega-hydroxypoly (oxyethylene); ethoxylated octadecylamine reacted with octadecanoic acid; *alpha sulfo-omega-*(dodecyloxy)poly(oxyethylene) ammonium salt; impact modified Nylon MXD-6; sulfosuccinic acid 4-ester with polyethylene glycol nonylphenyl ether; disodium salt; *alpha-alkyl(C10-C14)-omega*-hydroxypoly (oxyethylene) poly(oxypropylene) and *alpha-alkyl(C12-C18)-omega-*hydroxy-poly(oxyethylene); polyoxyethylene cetyl alcohols and polyoxyethylene oleyl ether; *alpha-*(p-nonylphenyl)-*omega*-hydroxypoly(oxyethylene); and diethyleneglycol dibenzoate. Sanitizing solution used on food contact surfaces: *alpha-alkyl* (C11-C15)-*omega-*hydroxypoly(oxytheylene) and *alpha-*(p-nonylphenyl)-*omega-*hydroxypoly(oxyethylene).

[e]Food contact additives which may contain ethylene oxide for which safe use rules were issued: ethoxylated octadecylamine reacted with octadecanoic acid; *alpha sulfo-omega-*(dodecyloxy)poly(oxyethylene) ammonium salt; impact modified Nylon MXD-6; sulfosuccinic acid 4-ester with polyethylene glycol nonylphenyl ether; disodium salt; *alpha-alkyl(C10-C14)-omega-*hydroxypoly (oxyethylene) poly(oxypropylene) and *alpha-alkyl(C12-C18)-omega-*hydroxy-poly(oxyethylene) poly(oxypropylene); polyoxyethylene cetyl alcohols and polyoxyethylene oleyl ether; *alpha-*(p-nonylphenyl)-*omega*-hydroxypolyoxyethylene); and diethyleneglycol dibenzoate.

[f]The petition was filed in response to 1-31-73 notice (*Fed. Reg.* 38:2996. 1973)that metallic salts or vegetable colorants could no longer be marketed unless a petition was filed by 7-30-73. In the NPRM, manufacturers were advised that they were not eligible for coal tar hair dye exemptions & data was requested to make final determination. The closing date was postponed several times pending study completions.

continued on next page

Table 3-4.—FDA: Carcinogens Associated With Foods and Cosmetics—Continued

Substance	When identified	Type of evidence	Petition	NPRM	Final	Type of action/comments
—as contaminant in 5-chloro-2-methyl-4-isothiazolin-3-one & 2-methyl-4-isothiazolin-3-one	—	—	1-19-80, 2-22-80 to establish rule for safe use	—	2-1-85	Rule established for safe use of additive containing DMNA impurity in adhesives and paper coating which contact food[a]
5-Nitro-o-toluidine	3-4-80	—	—	—	—	Hypothesized impurity in several color additives; not found
2-Nitro-p-phenylenediamine	11-27-79	—	—	—	—	Coal tar hair dye ingredient[c]
2-Nitropropane	4-25-77 NIOSH Bulletin; 4-7-86, risk assessment	Animal, 2 species—rats, rabbits	—	12-1-78	—	To prohibit in food packaging adhesives
N-Nitrosamines—as contaminants in rubber baby bottle nipples	1981, found as a problem in rubber baby bottle nipples	—	—	—	12-27-83, action level announced; revised 6-26-84	Action level established at lowest avoidable level (10 ppb in any specific N-nitrosamine)
Oil of Calamus	No date	Animal, 1 species	—	—	5-9-68	Banned use of calamus and its derivatives in food
PCB's (Arochlor)	9-23-80	—	—	3-18-72 (& others)	7-6-73 (& others)	Tolerances established to limit exposure
P-4,000 (1-n-propoxy-2-amino-4-nitrobenzene)	No date	Chronic toxicity data in rats	—	—	1-19-50	Banned as food additive
o-Phenylphenol	5-31-84	—	—	—	—	—
Polynuclear Aromatic Hydrocarbons (PAH) —in carbon black	ACS Monograph #173	Certain PAHs are known carcinogens	7-24-73	—	9-23-76	Petition denied and provisional listing terminated for use of carbon black as color additive because PAH possible impurity
—in graphite	—	—	8-6-73	—	11-29-77	Provisional listing terminated for use of graphite in externally applied cosmetics because contains PAH impurities
Radioactive Contamination of Food (accidental)	—	Based on guidance issued by Federal Radiation Council, 7-64, 5-65	—	12-15-78 (proposed "Protective Action Guidelines" for State and local agencies)	10-22-82, prop. recommendations withdrawn, issued notice of recommendations (not codified)	Recommended protective action at 0.5 rem whole body, 1.5 rem thyroid; emergency protective action (when there is high dietary & social cost or impact) at 5 rem whole body, 15 rem thyroid
Saccharin	—	Animal, 1 species—rats	—	4-15-77	Withdrawn	Proposal to revoke as ingredient in food was withdrawn because of congressional action
Safrole & Isosafrole	6-16-78	Long term studies	—	—	12-3-60	Banned addition of safrole, oil of sassafras and related substances to foods; some GRAS substances may contain miniscule amounts of safrole. Cosmetics: low priority for further evaluation because low potency carcinogen with limited exposure

continued on next page

Table 3-4.—FDA: Carcinogens Associated With Foods and Cosmetics—Continued

Substance	When identified	Type of evidence	Petition	NPRM	Final	Type of action/comments
2,3,7,8-Tetrachlorodibenzo-p-dioxin (TCDD)	10-26-78, risk assessment	—	—	—	—	Unavoidable contaminant; advisory issued to Great Lakes States on TCDD in fish in 1981
Tetrachloroethylene	5-13-80	Animal, 1 species—mice	—	—	—	—
Thiourea	6-16-78	—	—	—	—	Possible cosmetic ingredient; no longer used in cosmetics to FDA's knowledge
o-Toluidine	4-1-80	—	—	—	—	Impurity in several color additives found along with p-toluidine, but at lower levels; p-toluidine specification also limits o-toluidine
p-Toluidine —as contaminant of D&C Green No. 6	2-24-81; 2-24-81, risk assessment	Animal, 1 species—mice	—	3-21-86	10-27-86	Adopted uniform specifications, including specification on p-toluidine impurity, for all suture uses of D&C Green No. 6
			11-20-68	—	4-2-82	Permanently listed D&C Green No. 6 for use in externally applied drugs & cosmetics; contains p-toluidine impurity[a]
			1-14-83, 1-28-83 (3 petitions)	—	3-29-83	Listed D&C Green No. 6 for use in contact lenses
			2-14-85, 3-29-83	—	3-21-86	Provided for additional uses of D&C Green No. 6 for coloring absorbable sutures
			—	3-21-86	10-27-86	Adopt uniform specifications, including specification on p-toluidine impurity, for all suture uses of D&C Green No. 6
—as contaminant of D&C Green No. 5			11-20-68	—	6-4-82, 11-2-82, stay removed	Permanently listed D&C Green No. 5 for use in drugs and cosmetics; contains p-toluidine impurity[a]
—as contaminant of D&C Red Nos. 6 & 7			8-6-73	—	12-28-82; 7-29-83, stay removed	Permanently listed D&C Red Nos. 6 & 7 for use in drugs & cosmetics; contains p-Toluidine impurity[a]
1,1,2-Trichloroethane	11-20-79	Animal, 1 species—mice	—	—	—	Permitted for use in food packaging adhesives
1,1,2-Trichloroethylene	3-21-75 NCI report	Animal, 1 species—mice	6-24-75 (HRG)	9-27-77	—	Agency information indicates coffee decaffeination and cosmetic uses discontinued

[a]The rules for safe use are based on a 1984 court decision that allowed safe exposure levels to be established for non-carcinogenic additives that contain carcinogenic impurities if the additive itself is not shown to be carcinogenic. (Scott v. FDA, 728 F.2d 322 (6th Cir. 1984).) See discussion in text.

[c]Coal tar hair dyes are exempt from the adulteration and color additive provisions found in sections 601 & 706 of the FDCA (21 U.S.C. 361.376) provided the label bears a caution statement & patch test instructions to determine if the product causes skin irritation (Fed. Reg. 43:1101. 1978).

continued on next page

Table 3-4.—FDA—Carcinogens Associated with Foods and Cosmetics—Continued

Substance	When identified	Type of evidence	Petition	NPRM	Final	Type of action/comments
Trimethylphosphate	Includes risk assessment	Animal, 2 species—rats, mice	1-28-80 (Sunkyong Fibers)	—	—	Petition for use of trimethylphosphate in manufacture of polyethylene phthalate polymers; withdrawn 4-4-80; impurity in pesticide used in animal feed
Urethane —impurity in diethylpyrocarbonate	No date	Not regarded as safe, no evidence cited[g]	—	2-11-72	8-2-72	Banned diethylpyrocarbonate because urethane is possible impurity
—in alcoholic beverages	—	—	11-24-86	—	—	CSPI is requesting recall of alcoholic beverages adulterated with high levels of urethane
Vinyl chloride (impurity in vinyl chloride polymers)	1-4-73, identified in alcoholic beverages stored in vinyl chloride polymer containers	—	—	5-17-73	Withdrawn 9-3-75	Re: restriction from use of vinyl chloride polymers in alcoholic beverage containers
	1974	Human	—	4-2-74	8-26-74	Prohibited use of vinyl chloride as ingredient of drug & cosmetic aerosol products, NDA required for use in drugs
	—	Animal	—	9-3-75	Withdrawn 2-3-86	Re: restriction on use of vinyl chloride polymers in contact with food
	Cites 1979 IARC Monographs; 5-27-82 (risk assessment)	Animals, 3 species— mice, rats, hamsters	—	2-3-86	Pending	Rule for safe use of vinyl chloride polymers based on new technical capability to reduce levels of vinyl chloride monomer in vinyl chloride polymer resin and on new interpretation of legal requirements[a]

[a]The rules for safe use are based on a 1984 court decision that allowed safe exposure levels to be established for non-carcinogenic additives that contain carcinogenic impurities if the additive itself is not shown to be carcinogenic. (Scott v. FDA. 728 F 2d 322 (6th Cir 1984).) See discussion in text.

[g]No carcinogenic evidence is cited by the FDA for this action. However it is listed as a carcinogenic substance in the NTP Annual Report.

was extended to substances covered by the Delaney clause, under a de minimis interpretation of that clause. In 1985, FDA proposed to allow the continued use of methylene chloride as a direct food additive based on the results of a risk assessment. In 1986, some carcinogenic color additives were permanently listed based on the results of risk assessments (see below).

Regulatory Actions on Carcinogens in Foods and Cosmetics

FDA actions concerning carcinogens in foods and cosmetics can be grouped based on the type of material and the kind of FDA action. The types of materials are direct food additives, indirect food additives (generally from packaging materials and food-processing equipment), color additives (both for ingestion and for external use only), contaminants or potential contaminants of food or color additives, unavoidable environmental contaminants in foods, and cosmetic ingredients. FDA actions may include banning the substance (or terminating a provisional listing), setting a rule for safe use of an additive, requiring warning labels, and setting tolerances and action levels.

Prior to 1958, the FDA had prohibited the food additive use of two carcinogens: Dulcin (4-ethoxyphenylurea) and P-4000 (1-n-propoxy-2-amino-4-nitrobenzene), both in 1950 (see table 3-4). Since then, FDA has banned four direct food additives: safrole (1960); oil of calamus (1968); diethylpyrocarbonate (DEPC) (1972), which forms the carcinogen urethane);[6] and cinnamyl anthranilate (1985). FDA has also proposed to prohibit the use of hydrazine, trichloroethylene, and Saccharin as food additives. The proposal for Saccharin was issued in 1977 but was not made final because of congressional action mandating that Saccharin remain on the market. For hydrazine and trichloroethylene, no final action has been taken.

FDA is also proposing a rule to allow the continued use of methylene chloride to decaffeinate coffee (in contrast to banning its use in cosmetics). This use for decaffeination is being justified on

the grounds that an assessment reveals that the risk from this use is de minimis.[7]

Indirect food additives are generally packaging materials in contact with food and processing equipment. FDA has banned outright two indirect food additives: flectol-H (1967) and MBOCA (1968). In the mid-1970s, FDA had also banned certain uses of bottles made from acrylonitrile copolymers and polyvinyl chloride because of concern that residual acrylonitrile and vinyl chloride might leach into the liquids contained in the bottles. In the 1980s, FDA issued a rule to allow acrylonitrile copolymer bottles and proposed to allow polyvinyl chloride bottles. FDA's argument is that new manufacturing technology can assure that leaching of the residual chemicals from these bottles will be minimal and that FDA has the authority to establish specifications for carcinogenic impurities in regulations for the safe use of additives. FDA has also issued a safe use rule for Epichlorohydrin (1983), dibutyltin diacetate (1983), 1,4-dioxane (1985), dimethylnitrosamine (1985), and ethylene oxide (1986), which are carcinogenic impurities of certain packaging materials. FDA has proposed, but not taken final action on, other indirect food additives in packaging materials: 2-nitropropane and chloroform.

Under the Color Additives Amendment of 1960, FDA established a provisional listing of color additives then in use with the aim of reviewing their suitability for permanent listing for use in foods, drugs, and cosmetics. The original amendments gave industry $2\frac{1}{2}$ years (until the end of 1963) to demonstrate the safety of provisionally listed color additives. However, FDA has extended this deadline a number of times since then. Although the provisional color list was established shortly after enactment of the Color

---

[6]Urethane has been in the news lately because it is found as a byproduct of fermentation in several types of wines and distilled spirits. FDA has been petitioned to take action on urethane levels.

[7]Two other well-publicized actions involved some concern for carcinogenicity, although FDA's final decisions were ultimately based on other grounds. In 1969 FDA removed cyclamates from its lists of GRAS substances. While initially there were concerns about a positive animal study, after review FDA concluded that cyclamates were not carcinogenic. But the listing termination was continued because it was not shown to be safe. In 1976, FDA removed FD&C Red No. 2 from its provisional list of color additives because FDA was not convinced of the safety of this additive. This decision was based partly on an inconclusive study on carcinogenicity. Furthermore, no studies were under way to provide the data necessary to establish safety. Because these chemicals were additives, the manufacturer has the burden of providing this evidence.

Additives Amendment in 1960, the process of obtaining the necessary toxicity data and making regulatory decisions has lasted until now.

Today, 25 years later, there are only a few substances left on this provisional list. For a given color additive, FDA regulatory actions described in table 3-4 include terminating the provisional listing of a color or placing it on the "permanent" list. The provisional list originally included over 200 color additives. As of 1985, from the original list, 126 have been permanently listed, 63 have been removed from the market, and 10 have remained on the "provisional" list (210).

FDA has terminated the provisional listing of several color additives, effectively banning them, because they were carcinogenic or potentially contaminated with a carcinogen. These include Ext. D&C Yellow Nos. 9 and 10; D&C Red Nos. 10, 11, 12, and 13; orange-B (which may contain beta-naphthylamine); Ext. D&C Yellow No. 1 (which may contain benzidine and 4-aminobiphenyl), and carbon black and graphite (containing polynuclear aromatic hydrocarbons). FDA permanently listed lead acetate as a color additive for use in hair dyes in 1980, arguing that the lead exposure was small compared to background exposure and that the estimated risk was insignificant.

Since 1982, following its "impurities policy" (described above), FDA has permanently listed several colors even though they contain known carcinogens. These include D&C Green Nos. 5 and 6, and D&C Red Nos. 6 and 7 (all containing p-toluidine), FD&C Yellow No. 5 (containing 4-aminoazobenzene, 4-aminobiphenyl, aniline, azobenzene, benzidine, and 1,3-diphenyltriazene), FD&C Yellow No. 6 (containing 4-aminoazobenzene, 4-aminobiphenyl, aniline, azobenzene, and 1,3-diphenyltriazene), and CI Vat Orange No. 1 (containing CI Vat Yellow No. 4 as a contaminant).

In 1986, FDA applied a de minimis policy to colors that themselves are carcinogenic in animal studies. D&C Red Nos. 8, 9, and 19, and D&C Orange No. 17 were identified by FDA as carcinogenic in 1982 and 1983. D&C Red No. 9, in fact, tested positive in a National Toxicology Program (NTP) bioassay. In theory, based on the traditional interpretation of FDCA, these carcinogenic color additives should be candidates for banning from drugs and cosmetics. In August and December 1986, FDA permanently listed these colors, arguing that the estimated risk was low. In a February 1987 *Federal Register* notice, FDA went on to argue that because the estimated risk in humans was low, the color additives in question were, for purposes of the Delaney clause, not animal carcinogens either, notwithstanding the bioassay results. In the August notices, FDA had stated that each of these colors (D&C Orange No. 17 and D&C Red No. 19) "induces" cancer in animals. In the new notice it explained:

This statement reflected FDA's policy, as a matter of scientific analysis . . . that any chemical shown to induce cancer even in only one strain, gender, and species, at one dose in one experiment, is an animal carcinogen. This statement did not represent a conclusion that this substance induces cancer in animals within the meaning of [the Delaney clause]. . . .

. . . a conclusion for purposes of the Delaney clause that a substance at a given level poses a de minimis risk to humans implicitly includes the conclusion that a de minimis level of risk at a comparable level of exposure is presented to animals. Accordingly, D&C Orange No. 17 [and D&C Red No. 19] can not be said to induce cancer in animals, as well as in man, within the meaning of the Delaney clause. When a substance causes only a de minimis level of risk in animals, it cannot be said to induce cancer in animals within the meaning of the Delaney clause [239,240].

FDA has also acted on certain other ingredients of cosmetics. It has banned vinyl chloride in aerosol products, prohibited chloroform in cosmetics, and proposed to ban methylene chloride from cosmetics. FDA also attempted in 1979 to require a label on coal tar dyes (which contain 2,4-diaminoanisole and its sulfate) warning of animal evidence for carcinogenicity. In 1980, a Georgia court, in a consent decree, remanded this regulation back to FDA for further consideration, including development of a risk assessment. No further action has occurred.

For potentially carcinogenic unavoidable environmental contaminants, FDA has set regulatory tolerance levels for fish contaminated with PCBs. FDA has also set the more informal action levels for aflatoxins, dimethylnitrosamines (in malt beverages), and N-nitrosamines (in baby bottle nipples).

The direct food additives that FDA has banned are generally carcinogenic in a single species. For the color additives that FDA has banned, the carcinogenic substances at issue were beta-naphthylamine, benzidine, 4-aminobiphenyl, and polyaromatic hydrocarbons—all of which are known human carcinogens and carcinogenic in several animal species. For most other color additives and indirect additives for which safe use rules were issued, the original determination of potential carcinogenicity from impurities or substance migration was usually based on animal data, often in one species.

Table 3-4 also indicates a number of substances that FDA has identified as carcinogenic, but for which no final regulations have been issued. In FDA's view, most of these regulations were not needed because the hypothesized impurity was never actually found or the risk was determined to be insignificant. FDA still might take action on a case-by-case basis if a problem was discovered (182).

Regulation of Animal Drug Residues

Prior to 1962, animal drug residues in food were subject to the Delaney clause, but in 1962 Congress enacted the "DES (Diethylstilbestrol) Proviso," which permits the use of carcinogenic drugs in animals, providing that their residues could not be detected in edible portions of tissue or foods derived from living animals according to methods approved by FDA. (See chapter 2 for a discussion of the sensitivity of method for detecting such substances.)

FDA identified (in its response to OTA) 16 carcinogens that are or were administered as drugs to animals, or are potential drug contaminants, and that might leave residues in animal tissues to which people may be exposed. FDA banned the use of DES in animal drugs, denied approval of one substance (which was however overturned in court) (Gentian violet), and has proposed to withdraw seven other substances. FDA has required residue studies on six substances, including four

that it has proposed to withdraw. Three of those four were instead regulated under FDA's policy for endogenous or naturally occurring hormones under which a certain amount of increase over the naturally occurring levels is permitted.[8] For two substances, residue studies are all that FDA has required. In the case of Reserpine, the sponsor withdrew the application for approval, while for three other animal drugs, no actions are expected (see table 3-5). For one substance, aniline hydrochloride, which is a contaminant of an animal drug, FDA required that it's levels be reduced.

The only animal drug successfully banned is DES, on which there is human evidence from human uses of this drug. The lack of action on potentially carcinogenic animal drugs has also been criticized by the House Committee on Government Operations. In particular, the committee indicated that even though FDA has determined several animal drugs were carcinogenic (dimetridazole, ipronidazole, and carbadox), FDA has not removed these drugs from the market nor has it required that adequate residue monitoring methods be developed. In addition, while FDA had never given premarket approval to Gentian violet and its seizure orders had been supported in a number of courts, FDA has temporarily stopped seizing products containing Gentian violet because one lay jury determined this use to be "generally regarded as safe." Later animal tests revealed Gentian violet to be an animal carcinogen. The Committee criticized FDA for failing to take some action based on this evidence (211).

Regulation of Carcinogenic Human Drugs

Human drugs are subject to premarketing approval based on risk-benefit criteria for the intended condition of use (233). When a drug has a significant effect on an incapacitating or fatal

[8]The hormone policy was announced in guidelines for toxicological testing published in conjunction with FDA's proposed sensitivity of method procedures (mentioned above). Under the hormone policy, FDA distinguishes between endogenous and synthetic hormones. Endogenous hormones are regulated based on the increase over endogenous levels of the hormone found in the residue studies. For synthetic hormones, a study is required to determine the hormone no-effect level. If they cause tumors only in the endocrine target organs, the safety standard is based on the no-effect level. Otherwise the standard is based on the carcinogenicity study (102).

Table 3-5.—Potentially Carcinogenic Animal Drugs Considered for Regulation by FDA[a]

Drug	When identified (type of evidence)	Petition	Proposal to withdraw	Final	Court action/challenge
Aniline hydrochloride	—	—	—	Levels of contaminant reduced	—
Carbadox (& metabolites)	1978 (animal—rats, mice)	5-9-86	—	Required residue & metabolism studies	—
DES[b]	1964 (animal—mice); no date (human)	None	3-11-72 (premixes)	8-4-72, (premixes)	Reversed & reinstated: *Hess & Clark v. FDA,* 161 U.S. App. D.C. 395, 495 F.2d 975 (DC Cir. 1974)
			1-12-76 (hearing granted 11-26-76)	9-21-79	DES is no longer permitted for use in animal drugs, upheld *Rhone Poulenc, Inc., Hess & Clark Division v. FDA,* 636 F.2d 750 (D.C. Cir. 1980)
			6-21-72 (premixes & implants)	4-27-73 (implants)	
			3-27-74 (to revoke method of analysis)	—	*Chemetron Corp. v. U.S. DHEW,* 95 F.2d 995, 997 (D.C. Cir. 1974)
Dimetridazole	1971 (animal—rats)	5-9-86	12-17-86	—	—
3,5-Dinitrobenzamide	1970 (animal—rodent)[c]	—	—	Level of human exposure from use in animals to determine further testing for hazard	—
Estradiol benzoate	1974 (IARC, literature reviews re: sex hormones)	None	1-5-79 (9-22-72, requested more residue data)	Residue studies provided	Regulated under FDA hormone policy
Estradiol monopalmitate	1974 (IARC, literature reviews re: sex hormones)	None	1-5-79 (requested more residue data 9-22-72)	No residue studies provided	—
Furazolidone	1964 (animal—rats, mice: tumorigenic evidence); 1974 (carcinogenic evidence)	None	8-4-71; 5-13-76; 9-4-84	—	Administrative Law Judge initial decision recommends withdrawal of NADAs, 11-12-86

[a]Other potentially carcinogenic animal drugs were mentioned in hearings on the "Regulation of Animal Drugs by the FDA" before the House Committee on Government Operations which were not mentioned in the information provided by the agency in response to the OTA request. These include Albendazole which was found carcinogenic in rats and mice by the CVM Cancer Assessment Committee in a July 1984 meeting. In 1979 it had been regarded as a suspect carcinogen and was not approved except for emergency and investigational use. It was ordered off the market in a letter dated Nov. 8, 1984 (Hearings, p. 424). A proposal to withdraw dibutyltin dilaurate was issued Aug. 29, 1978 but, according to an agency memo, products that contain it are still marketed under other NDAs. Dibutyltin dilaurate is a suspect carcinogen because it is related to dibutyltin diacetate which is carcinogenic according to NTP results. (Hearings, p. 187; FDA memorandum "Re-evaluation of the Status of Certain Marketed Drugs.") The memo also mentions Ronnel which is potentially contaminated with dioxin which is carcinogenic in rats. According to the memo however, the agency did not have enough evidence to take regulatory action and recommended requesting more data on the level of dioxin contamination in Ronnel (Hearings, pp. 188-189).

It was also noticed that iron dextran complex and several estrogens are regulated according to information in the *Annual Report on Carcinogens* but do not appear on the list provided by the FDA. It also appears from the hearing documents that several substances may have been regulated through informal procedures, such as Albendazole which was regulated by correspondence between the agency and the sponsor so the listing may not be complete.

[b]In 1962 Congress enacted the "DES Exception to the Delaney Clause" permitting the use of carcinogenic drugs in animals providing that they could not be detected in edible portions of tissue or foods derived from living animals according to methods determined by the FDA. In 1962, regulations were promulgated permitting the use of DES and establishing detection methods. In the early 1970s however, the USDA found residues at levels below the sensitivity of the prescribed method.

[c]The drug is regulated as a carcinogen under the "DES Exception" (see footnote b). Information about the identification of the drug as a carcinogen and the type of evidence relied on was not provided.

continued on next page

Table 3-5.—Potentially Carcinogenic Animal Drugs Considered for Regulation by FDA[a]—Continued

Drug	When identified (type of evidence)	Petition	Proposal to withdraw	Final	Court action/challenge
Gentian violet	1985 (preliminary animal evidence—mice)[c,d]	—	Denied approval, 3-30-79	—	Overturned, approved for use as mold inhibitor in poultry feed. *Marshall Minerals* v. *FDA* 661 F.2d 409 (5th Cir., 1981)
Ipronidazole[e]	1978 (animal)	5-9-86	—	—	—
Melengestrol acetate	Animal [c]		Proposal to withdraw recommended by CVM in 1984	—	—
Nitrofurazone	1964 (animal—rats, mice: tumorigenic evidence); 1974 (carcinogenic evidence)	None	3-31-71; 8-17-76; 9-4-84	—	Same as furazolidone
Progesterone	c	—	1-5-79	Residue studies provided	Regulated under FDA's hormone policy
Reserpine	—	—	—	5-16-84	—
Testosterone propionate	c	—	1-5-79	Residue studies provided	Regulated under FDA's hormone policy
Zeranol	Animal[f]	—	—	Required residue studies and chronic bioassay	—

[a]Other potentially carcinogenic animal drugs were mentioned in hearings on the ''Regulation of Animal Drugs by the FDA'' before the House Committee on Government Operations which were not mentioned in the information provided by the agency in response to the OTA request. These include Albendazole which was found carcinogenic in rats and mice by the CVM Cancer Assessment Committee in a July 1984 meeting. In 1979 it had been regarded as a suspect carcinogen and was not approved except for emergency and investigational use. It was ordered off the market in a letter dated Nov. 8, 1984 (Hearings, p. 424). A proposal to withdraw dibutyltin dilaurate was issued Aug. 29, 1978 but, according to an agency memo, products that contain it are still marketed under other NDAs. Dibutyltin dilaurate is a suspect carcinogen because it is related to dibutyltin diacetate which is carcinogenic according to NTP results. (Hearings, p. 187; FDA memorandum ''Re-evaluation of the Status of Certain Marketed Drugs.'') The memo also mentions Ronnel which is potentially contaminated with dioxin which is carcinogenic in rats. According to the memo however, the agency did not have enough evidence to take regulatory action and recommended requesting more data on the level of dioxin contamination in Ronnel (Hearings, pp. 188-189).

It was also noticed that iron dextran complex and several estrogens are regulated according to information in the *Annual Report on Carcinogens* but do not appear on the list provided by the FDA. It also appears from the hearing documents that several substances may have been regulated through informal procedures, such as Albendazole which was regulated by correspondence between the agency and the sponsor so the listing may not be complete.

[b]In 1962 Congress enacted the ''DES Exception to the Delaney Clause'' permitting the use of carcinogenic drugs in animals providing that they could not be detected in edible portions of tissue or foods derived from living animals according to methods determined by the FDA. In 1962, regulations were promulgated permitting the use of DES and establishing detection methods. In the early 1970s however, the USDA found residues at levels below the sensitivity of the prescribed method.

[c]The drug is regulated as a carcinogen under the ''DES Exception'' (see footnote b). Information about the identification of the drug as a carcinogen and the type of evidence relied on was not provided.

[d]Gentian violet is included on the list FDA provided to OTA of compounds being considered for regulation because of concerns about carcinogenicity but no evidence is cited for carcinogenic concern. The FDA held that gentian violet is not ''Generally Recognized as Safe'' (GRAS), however, a jury held that it is GRAS as a mold inhibitor in poultry feed up to 8 ppm (*Fed. Reg.* 47:32480). No earlier citations were provided. According to an agency memo submitted at hearings on ''The Regulation of Animal Drugs by the FDA'' before a subcommittee of the Committee on Government Operations of the House of Representatives, July 24 and 25, 1985 (pp. 180-181) a preliminary review of a long-term feeding study indicated that gentian violet is carcinogenic in mice. Previously, it was only known that the main component of gentian violet, crystal violet, is related to compounds that are known animal carcinogens and two compounds with evidence of human carcinogenicity. The FDA had denied approval for its use Mar. 30, 1979, 44 19035 and denied a hearing on the matter. The decision was overturned on grounds that it was a disputed question as to whether gentian violet was a carcinogen. *Marshall Minerals* v. *FDA* 661 F.2d 410 (11th Cir. 3-28-80).

[e]Ipronidazole was the subject of a causal review in 1980 and was recommended for withdrawal by the Center for Veterinary Medicine in 1983 (a causal review is an agency procedure for reviewing safety and effectiveness data based on problems identified in reports of adverse drug reactions and serves as the basis for requesting regulatory supplemental NADAs). Concern was raised in a petition filed by the Center for Science in the Public Interest (CSPI) because, instead of initiating proceedings to withdraw the drug, in December 1985 it was listed among the ''highest priorities for causal review.'' A decision on the petition is pending.

[f]Zeranol is a suspect carcinogen based on a chronic bioassay of zearalenone which has a similar structure. It is currently under review and chronic bioassays of zeranol are underway.

SOURCES: EPA response to OTA request and cited *Federal Register* notices; hearings on the ''Regulation of Animal Drugs by the FDA'' before the House Committee on Government Operations, 99th Congress, July 24, 25, 1985, p. 66; petition to Withdraw New Animal Drug Applications submitted to the FDA by the Center for Science in the Public Interest.

disease for which there is no safe therapy, it could be regarded as adequately "safe" despite major, even life-threatening, side effects, such as carcinogenicity (233). A drug manufacturer must submit an Investigative New Drug (IND) Application to conduct preliminary investigation of the safety and efficacy of the drug and a New Drug Application (NDA) for marketing approval. The NDA is to include reports of the investigation to show the drug is adequately safe and effective, a list of the drug's composition, samples of the drug, information that might be required for FDA monitoring activity, and proposed labeling.

For approved drugs, the prescription drug labeling must include a precautionary section that states the results of carcinogenicity studies. The usual type of FDA regulatory action in the drug review and approval process is to informally require the drug sponsor to modify the warning information in the physician labeling.

FDA has rarely used the more formal process of publishing a notice in the *Federal Register* to regulate drugs for carcinogenicity, as it has done for certain generic drugs or drug classes. FDA removed or the sponsor recalled from the market as active drug ingredients chloroform (1976), methapyraline (1978), and Phenacetin (1984). Precautionary labeling was instituted by this proc-

ess for estrogenic drugs (1976, 1977), neuroleptic drugs except Reserpine (1978, 1980), and Reserpine (1983) (see table 3-6). FDA relied on human evidence for Phenacetin and estrogens, although positive animal evidence later became available for Phenacetin. FDA relied on animal evidence to evaluate the carcinogenicity of the other drugs.

Questions have been raised about whether FDA has always acted on positive carcinogenicity evidence in this way and whether FDA has always required sufficient information to make appropriate judgments about the safety of potentially carcinogenic drugs. For example, the House Committee on Government Operations has criticized FDA for approving Zomax as a nonsteroidal anti-inflammatory drug even though animal studies indicated a carcinogenic response, and clinical studies did not clearly show that this drug was superior to other available treatments. The committee also expressed concern that a number of other drugs of this class had been approved without adequate evidence on safety (209). FDA officials, on the other hand, argue that Zomax was superior to other treatments and that the animal data did not reveal a "carcinogenic" response, but rather an increase of a benign tumor type that they did not consider to be "a very alarming finding." This increase was to be noted on the drug labeling (69).

Table 3-6.—Human Drugs Regulated as Carcinogens by FDA[a]

Substance	Type of evidence	NPRM	Final	Type of action
Chloroform	Animal	4-6-76	6-24-76	Removed from market
Methapyraline	Animal	—	6-13-78 (order)	Voluntary recall
Phenacetin	Human Animal	8-10-82 — —	10-5-83 amended 2-23-84	Removed from market
Neuroleptic drugs	Animal— rats, mice	—	5-16-78; amended 8-18-78; revised 8-8-80	Cautionary labeling; data reviewed for four drugs, but labeling required for all related drugs (except for reserpine)

[a]This table only includes drugs removed from the market or formally regulated after marketing approval. Carcinogenic drugs which require warning label are discussed in ch. 5.

SOURCE: FDA response to OTA request; cited *Federal Register* notices.

CPSC REGULATORY ACTIONS

Activated in 1973, the Consumer Product Safety Commission (CPSC) is an independent regulatory agency. Its authority to regulate carcinogens is established by both CPSA and the Federal Hazardous Substances Act (FHSA). CPSA authorizes the regulation of most consumer products that pose "unreasonable risks" of injury or illness. FHSA was initially enacted in 1960 as a labeling statute intended to fill gaps in other statutes. FHSA was later amended to permit more drastic action to control hazards and expanded "to cover hazardous substances in general use in the home, and particularly to protect children from hazardous toys and products."

Under CPSA, when a product poses an "unreasonable risk" of injury or illness, CPSC may promulgate a consumer product safety standard, ban the product from commerce when a safety standard would not be adequate to protect the public, bring suit in Federal district court to seize an "imminently hazardous" product or seek an injunction against the distribution of the product, or require certain remedial actions. In 1981 amendments to the act, Congress required CPSC to convene a "Chronic Hazard Advisory Panel" (CHAP) prior to regulating products that present a risk of cancer, mutations, or adverse reproductive effects. Under FHSA, hazardous substances are labeled and may even be banned, but more formal rulemaking procedures are required than under CPSA.

In the 16 years since its creation, CPSC has evaluated and attempted to regulate or begun to regulate 6 individual carcinogens (vinyl chloride, 1974; tris(2,3-dibromopropyl)phosphate (tris), 1977; benzene, 1978; formaldehyde, 1981; diethyl hexyl phthalate (in process); nitrosamines, 1984) and 2 classes of carcinogens—benzidine congener dyes (begun in 1978, not pursued after 1982); and asbestos in various forms (1973, 1977) (353). (See table 3-7 for a summary.)

In 1974, CPSC banned vinyl chloride (used as a propellant in aerosols) as a hazardous substance under the FHSA. The rule was overturned on procedural grounds, but by then the manufacture of aerosols containing vinyl chloride as a propellant had ceased. For this rule, CPSC relied on the risk assessments conducted by other agencies (353). In 1977, CPSC attempted to ban the use of tris in children's sleepwear. In this case, CPSC conducted its own risk assessment. The regulatory action was overturned in court on procedural grounds, but CPSC issued a statement of policy that it was prepared to prove in court that tris products were "banned hazardous products" meriting judicial relief. CPSC brought several suits in 1977 and 1978, and its strategy was upheld (126). In 1978, CPSC proposed to ban benzene as an intentional ingredient or a contaminant in consumer products (except in gasoline and laboratory solvents), but did not finalize its proposed rule because "by 1980, in response to the Commission's action and other factors, the use of benzene in consumer products was virtually nonexistent . . . [thus] . . . The proposed ban was withdrawn in 1980" (353). In 1978, as a result of a petition, CPSC studied the carcinogenic effects of benzidine congener dyes, but concluded by 1982 that their use had virtually ceased in consumer and commercial dye markets, and thus decided no regulatory action was needed (353).

CPSC has regulated asbestos in several different products. In 1973, it banned general use garments containing asbestos (126). In 1977, it banned the use of patching compounds and emberizing materials containing asbestos (the latter was used in artificial fireplace logs), and in 1979, it negotiated voluntary agreements with hairdryer manufacturers to stop using asbestos shields in hairdryers. The agency recently issued an enforcement policy which required labeling of household products that contain intentionally added asbestos that is likely to be released in use (230).

Table 3-7.—CPSC Regulation of Carcinogens

Substance	When identified	Type of evidence	Risk assessment	Petition	NPRM	Final	Type of action	Court action
Asbestos								
—in artificial emberizing materials	—	Human, animal	yes	9-15-77 (EDF)	7-29-77	12-15-77	Banned	—
—in general use garments	—	Human, animal	yes	—	—	9-27-73	Banned	—
—in paper	—	Human, animal	yes	12-14-71	—	—	Petition granted—banned	—
—in patching compounds	—	Human, animal	yes	7-15-76 (NRDC & Consumers Union)	7-29-77	12-15-77	Banned	—
Benzene—in household products except motor gasoline	—	Human	yes	5-5-77 (HRG)	—	Granted	Banned	—
Benzidine—congener dyes—in household packaged dyes & dyed textiles	—	Human, animal	yes	12-18-78 (Art Hazards Project)	—	Denied	Voluntarily removed from consumer household dyes	—
Certain ingredients posing carcinogenic risk	—		—	9-8-78 (EDF)	—	Granted in part	Agreed to investigate room deodorizers and hydrocarbon aerosol propellants for toxicity and consumer exposure, test data rule	—
DEHP—as plasticizer in children's plastic articles (2-di-ethylhexylphthalate)	—	Animal, 2 species—mice, rats	yes	—	—	—	Voluntarily removed from particular children's products; agency action pending decision of adequacy of voluntary action	—
p-Dichlorobenzene	—	Animal, 1 species—mice	yes	—	—	—	Under consideration for regulation	—
Formaldehyde								
—biological specimens for schools	—	—	—	—	—	—	Voluntarily reduced concentration & use	—
—in pressed wood products	—	Animal, 2 species—mice rats	yes	1-5-77 (National Land Corp.)	—	Denied	Banned	—
	—	—	—	8-11-82 (Consumer Federation of America)	—	Denied	Consensus voluntary standard activities to regulate exposure levels	—
—in urea formaldehyde foam insulation	6-80	Animal, 2 species—mice, rats	yes	4-16-81 (Formaldehyde Institute)	6-10-80	—	Withdrawn because inadequate—would have required notice of adverse effects to potential purchasers	—
	—	—	—	10-76 (Denver D.A.'s Consumer Office)	2-5-81	4-2-82	Banned	6-14-82, stay of ban; Gulf South Ins. v. CPSC, 701 F.2d 1137 (5th Cir. 1983)—ban overturned
Methylene chloride—in aerosol paint sprays and paint strippers	—	Animal	yes	9-3-85 (Consumer Federation of America)	—	—	Ongoing activity to determine if hazardous	—
Nitrosamines—in infants rubber pacifiers	—	Animal, several species	—	—	—	—	Less than 60 ppb (proposed voluntary standard 10 ppb per nitrosamine)	—
Perchloroethylene—in coin-operated dry-cleaning facilities	—	Animal, 2 species—mice, rats	yes	—	—	—	Under consideration for regulation	—
Tremolitic talc—in baby powder	—	—	—	2-9-77 (HRG)	—	Granted in part	Banned	—
Tris—flame retardant In children's sleepware	2-4-77 NCI data received	Animal, 2 species—mice, rats	yes	3-24-76 (EDF); 2-9-77 (EDF); 9-27-77 (Hilary Kam)	—	4-8-77	Labeling granted in part; ban granted in part; labeling denied; Tris treated children's clothing banned	Spring Mills v. CPSC, 434 F.Supp. 416 (Dist. of SC 1977)—ban overturned
Vinyl chloride—in aerosols	Noted action by other agencies	Human, animal, 3 species—rats, mice, hamsters	—	2-21-74 (HRG)	5-23-74	8-21-74; 1977	Issued order finding not safe: banned by promulgating order	Pactra Industries Inc. v. CPSC, 555 F.2d 677 (9th Cir. 1977)

SOURCES: CPSC Response to OTA request and cited Federal Register notices.

In 1981 CPSC issued a rule banning urea-formaldehyde foam insulation (UFFI). The Court of Appeals for the Fifth Circuit struck down CPSC's ban. In particular, the court argued that because of uncertainties in the risk assessment based on an animal study, CPSC could not validly conclude that UFFI presented an unreasonable risk (see further discussion in app. A). CPSC is currently engaged in a voluntary effort with the pressed wood industry to develop national consensus standards on formaldehyde emissions from their products and has decided not to convene a CHAP at this time (286).

At present, CPSC is studying diethylhexyl phthalate (DEHP), a plasticizer in polyvinyl chloride products, and nitrosamines found in rubber pacifiers. CPSC convened a Chronic Hazard Advisory Panel in the case of DEHP; the use of DEHP in pacifiers has apparently ceased. CPSC issued a statement indicating it would bring court action under FHSA if nitrosamines in pacifiers exceed 60 ppb (353). CPSC is also working on a voluntary standard to lower the level still further (227).

Finally, CPSC is considering regulating several other carcinogens: methylene chloride, perchloroethylene, and p-dichlorobenzene. In the case of methylene chloride, CPSC is currently engaged in a proceeding to determine if it can be called a hazardous substance under FHSA. No final actions have been taken on any of these chemicals.

In attempting to regulate carcinogens, CPSC has for the most part relied on both human and animal evidence, although for tris and formaldehyde, it relied on animal evidence only (see table 3-7).

EPA REGULATORY ACTIONS UNDER THE CLEAN AIR ACT

Some of the first major environmental statutes enacted in the early 1970s were the Clean Air Act Amendments of 1970. The statute provides an elaborate Federal-State scheme for controlling conventional pollutants, such as sulfur dioxide and carbon monoxide. Because of the emphasis on controlling conventional air pollutants, toxic pollutants were almost ignored. But a provision was added authorizing EPA to set emission standards for "hazardous" air pollutants, which provide "an ample margin of safety." The general scheme is that a pollutant is first listed as hazardous based on pertinent scientific data. Then uniform national standards are to be established for each source category of such pollutants within a specified time.

However, Congress provided no explicit guidance for regulating carcinogens as compared with other hazardous substances under this section. This failure to address carcinogens explicitly has led to considerable controversy in interpreting the statute for application to carcinogens. For a substance with a toxic threshold, that is, a level below which there are no harmful health effects to a group of people, setting a standard would involve determining a "no effects" threshold and providing for a margin of safety. However, for carcinogens, there is no known safe threshold. Thus, providing an ample margin of safety as required by the statute might imply elimination of all exposures by setting an emissions standard of zero, or, possibly, a standard of no detectable concentrations. For these situations, where EPA determines that complete prohibition of emissions would lead to "widespread industry closure" and the costs of that closure would be "grossly disproportionate" compared to the benefits of reduced risk, EPA's strategy has been to require "emission reduction to the lowest level achievable by use of the best available control technology." Recently, however, the D.C. Circuit Court of Appeals ordered EPA to establish a safe level of emissions for vinyl chloride based on health considerations although EPA may consider cost and technological feasibility to establish the actual emission standard (28,142).

In addition, EPA has taken the position that it does not have to regulate exposures that present an "insignificant risk." This policy has been challenged in a case concerning EPA's failure to issue benzene standards. The Natural Resources Defense Council (NRDC) argues that EPA must regulate hazardous air pollutants based exclusively on public health considerations, not technology and costs, and that EPA may not dismiss a health risk by declaring it to be "insignificant" (145).

In the nearly 16 years since enactment of CAA, EPA has listed seven carcinogens as hazardous air pollutants and issued emission standards for six carcinogens (see table 3-8): asbestos, 1973 (five source categories—amended several times); vinyl chloride, 1976 (two source categories); benzene, 1984 (for one source category, others pending); radionuclides, 1985 (four source categories); and arsenic (1986) (two source categories). Beryllium, which is classified in the *Annual Report on Carcinogens* has also been listed and regulated, although not for carcinogenic effects.[9] EPA has listed coke oven emissions (1984), and proposed emissions standards in March 1987, but has not issued them in final form.

Although CAA provides EPA 1 year in which to issue regulations on a pollutant after a substance is "listed," EPA met this deadline only in the case of vinyl chloride. From the date of listing to final action, however, EPA has taken an average of almost 4 years for the six carcinogens for which there are final rules. Four of these carcinogens were regulated or listed under legal pres-

---

[9]Mercury is also regulated under the CAA, but is not classified as a carcinogen.

sure: asbestos, vinyl chloride, radionuclides, and arsenic (see table 3-8).

EPA has indicated an "intent to list" 10 substances: 1,3-butadiene, chromium, carbon tetrachloride, chloroform, ethylene oxide, ethylene dichloride, cadmium, perchloroethylene, trichloroethylene, and methylene chloride. According to EPA, an intent to list a substance as a hazardous pollutant does not legally bind the agency as does a "listing" decision (296). This position was challenged by NRDC in a pending suit (147). NRDC contends that EPA is required to list a substance immediately if it has been determined to cause serious irreversible illness and if EPA has determined that it is a hazardous air pollutant (42).

For the five substances regulated primarily for carcinogenic effects, EPA has relied on human evidence of carcinogenicity (asbestos, vinyl chloride, benzene, radionuclides, and arsenic). For 8 of the 10 substances EPA intends to regulate, it has relied on animal bioassays for evidence of carcinogenicity, and for 2 substances (chromium and cadmium), it has both animal and human evidence of carcinogenicity.

EPA's regulation of potentially hazardous air pollutants has been criticized. A report by the General Accounting Office noted that 4 of 37 hazardous substances identified for possible regulation in 1977 had been regulated by 1983 (198). The report noted both delays in issuing regulations and in obtaining Science Advisory Board approval of EPA's health assessment documents.

EPA REGULATORY ACTIONS UNDER THE CLEAN WATER ACT

First enacted in 1948, CWA has been amended numerous times, most importantly in 1972, 1977, 1981, and 1987. The 1972 act set a goal of achieving "fishable, swimmable" waters by 1983 and for prohibiting the discharge of pollutants and "toxic pollutants in toxic amounts" by 1985, although these deadlines were modified by the 1977 and 1981 amendments.

Toxic substances, including a number of carcinogens, have been regulated under CWA; but

the process has taken a long time, is not yet finished, and has featured considerable litigation. The development of 65 water quality criteria documents for toxic pollutants has been an important part of EPA's risk assessment activities. While these are to be used by the States in developing State water quality standards, few such standards have actually been developed.

Under CWA, pollutors that discharge directly into receiving waters must obtain permits (Na-

Table 3-8.—Carcinogens Considered for Regulation Under the Clean Air Act

Substance	Type of evidence	Intent to list	Listed	NPRM	Final	Industrial category	Petition/court action
Acrylonitrile[a]	—	—	—	—	—	—	—
Arsenic	Human	—	—	6-05-80	7-20-83	8-04-86	High & low arsenic primary copper smelters; glass manufacturing plants—(comment period extended & reopened)	—
Asbestos	Human	—	—	3-31-71	12-07-71	4-06-73	Asbestos mills; selected manufacturing operations; spray-on asbestos materials; demolition operations; surfacing of roadways with tailings	—
Benzene	Human	—	—	6-08-77	4-18-80	—	Maleic anhydride plants	EDF petition 4-14-77
					12-18-80	—	Ethylbenzene/styrene plants	American Petroleum Institute v. EPA (D.D.C.)10-4-83
					12-19-80 1-81	—	Storage vessels	3-6-84, NRDC petition denied
					3-06-84	6-06-84	Fugitive emissions from petroleum refining & chemical manufacturing industries	NRDC v. EPA (D.D.C.) 1-27-84, Court required EPA to publish final rule by 5-23-84
					3-06-84	6-06-84	Withdrawal of prop. stds—maleic anhydride, ethyl-benzene styrene plants & storage vessels	—
					6-6-84		Coke oven byproduct recovery plants	
Beryllium[b]	Insuff. data but concerned about potential carcinogenicity	—	3-31-71	12-7-71	4-73	—	—
1,3-Butadiene	Animal—mice, rats	10-10-85	—	—	—	—	—
Cadmium	Human; animal, 1 species—rats	10-17-85	—	—	—	—	—
Carbon tetrachloride	Animal, 3 species	8-13-85	—	—	—	—	—
Chloroform	Animal	9-27-85	—	—	—	—	—
Chromium	Human, animal	6-10-85	—	—	—	—	—
Coke oven emissions	Human, animal	4-26-82	9-18-84	4-23-87	—	Coke ovens	—
Ethylene dichloride	Animal, 2 species—mice, rats	10-16-85	—	—	—	—	—
Ethylene oxide	Animal	10-2-85	—	—	—	—	—
Methylene chloride (dichloromethane)	Animal, 2 species—mice, rats	10-17-85	—	—	—	—	—
Perchloroethylene (tetrachloroethylene)	Animal, 2 species—mice, rats	12-26-85	—	—	—	—	—
Radionuclides[c]	Human	4-11-79	12-27-79	4-6-83: withdrawn 10-31-84 2-21-85	—	DOE facilities, NRC licensed & non-DOE Federal facilities, elemental phosphorus plants & radon-222 from underground mines	Sierra Club v. Ruckelshaus 84-0656, 2-17-84 (D.C.N.C.A.)
					4-17-85	Radon-222 from underground mines (control technique)	7-25-84 Court order required Agency to take final action within 90 days or find not hazardous
					2-6-85	DOE facilities, NRC licensed & non-DOE Federal facilities, elemental phosphorus plants	12-11-84 EPA held in contempt of Court, final standards required within 30 days, 120 days for radon-222 from underground mines
Trichloroethylene	Animal, 2 species—mice, rats	12-23-85	—	—	—	—	—
Vinyl chloride	Human; animal, 3 species—rats, mice, hamsters	—	12-24-75	12-24-75	4-26-74	Emergency suspension of indoor aerosol pesticides	—
					10-21-76	ethylene dichloride-vinyl chloride plants & polyvinyl chloride plants	—

aTo be regulated through cooperative ventures with State and local governments (pilot project).

bBeryllium was regulated for non-carcinogenic effects, however additional evidence which indicates potential carcinogenicity is under review by EPA.

cThe proposed standards for radionuclides were withdrawn following a court order which required EPA either to promulgate the regulations within a specified period of time or find that the substances are not hazardous air pollutants. EPA determined that "current practice provides an ample margin of safety to protect the public health from hazards associated with exposure to airborne radionuclides" for DOE facilities, NRC licensed & non-DOE Federal facilities & elemental phosphorus plants and issued an advanced notice of proposed rulemaking for control techniques for radon-222 emissions from underground mines. After being held in contempt, EPA issued final standards for the first three categories. These standards were two and a half times higher than the originally proposed standards in order to "accommodate the current level of emissions" (quoted in the Washington Post, Jan. 1, 1985, p. A6).

SOURCES: EPA response to OTA request and cited Federal Register notices.

Table 3-9.—Carcinogens Considered for Regulation Under the Clean Water Act[a]

Substance	Type of evidence	Toxic effluent standards		Number of States that issued water quality standards		Court action[b]
		Proposed	Issued	Aquatic life	Human health	
Acrylonitrile	Human, animal	—	—	—	—	
Aldrin/dieldrin	Animal	12-23-73 6-10-76	12-30-76	10/12	6/10	
Arsenic & compounds	Human	—	—	22	28	
Asbestos	Human	—	—	—	—	
Benzene	Human	—	—	—	—	c
Benzidine	Human	12-23-73 6-30-76	1-12-77	3	4	
Beryllium	Animal	—	—	7	4	
Carbon tetrachloride	Animal	—	—	—	—	c
Chloralkyl ethers	Animal	—	—	—	—	
Chlordane	Animal	—	—	—	—	
Chlorinated ethanes	Animal	—	—	—	—	
Chloroform	Animal	—	—	—	—	
Dichlorobenzidine	Animal	—	—	—	—	
Dichloroethylenes	Animal	—	—	—	—	c
Dinitrotoluene	Animal	—	—	—	—	
Diphenylhydrazine	Animal	—	—	—	—	
DDT	Animal	12-23-73[d] 6-10-76	12-30-76	—	—	
Halomethanes[e]	Animal	—	—	—	—	
Heptachlor	Animal	—	—	5	3	
Hexachlorobutadiene	Animal	—	—	—	—	
Hexachlorocyclohexane	Animal	—	—	7	11	
Nitrosamines	Animal	—	—	—	—	
Polynuclear aromatic hydrocarbons	Animal	—	—	—	—	
PCBs	Animal	12-23-73 7-23-76	2-2-77	11	7	EDF v. EPA, 598 F.2d 62 (D.C. Cir. 1978)
2,3,7,8-Tetrachlorodibenzo-p-dioxin (TCDD)	Animal	—	—	—	—	
Tetrachloroethylene	Animal	—	—	—	—	
Toxaphene	Animal	12-23-73 6-10-76	12-30-76	—	—	Hercules v. EPA, 598 F.2d 91 (DC Cir.1978)
Trichloroethylene	Animal	—	—	—	—	c
Vinyl chloride	Human, animal	—	—	—	—	c

[a]Water Quality Criteria Documents were issued 11-28-80 for all substances listed except TCDD which was issued 2-15-84.
[b]All substances listed are to be regulated with technology-based standards on an industry-by-industry basis as a result of a consent decree in NRDC v. Train 8 ERC. The consent decree was incorporated into the CWA 1977 Amendments, sec. 307.
[c]NRDC v. EPA No. 85-1840 (D.C. Cir. filed 12-26-85); pertained to volatile organic chemicals.
[d]The final rule was never promulgated because EPA determined that there was insufficient evidence to promulgate responsible and defensible standards at the conclusion of a 1974 hearing.
[e]Halomethanes include: chloromethane (methylchloride), bromomethane (methyl bromide), dichloromethane (methylene chloride), bromodichloromethane, tribromomethane (bromoform), dichlorodifluoromethane and trichlorofluoromethane.

SOURCES: EPA response to OTA request for information including list summarizing water quality criteria documents and Federal Register notices cited in above response; 40 C Sec. 129 (1984).

tional Pollution Discharge Elimination System (NPDES) permits) that delineate limitations on the amounts of conventional pollutants (e.g., biological waste material) and toxic substances allowed in discharges. NPDES permits are issued by EPA or by individual States that have an EPA-approved permit program (37 of 54 jurisdictions have been approved). NPDES permits are based on the more stringent of technology-based effluent limitations, State standards, or water quality criteria (166).

Indirect dischargers—industries that discharge into municipal sewers—are not covered under

NPDES permits, but must comply with Federal technology-based effluent limitations or local limits established under federally approved pretreatment programs. The pretreatment standards together with discharge limitations on publicly owned treatment facilities must achieve the same amount of reduction of toxic pollutants as would the use of effluent limitations on a direct discharger.

From 1972 to 1975, EPA issued, under court order, toxic effluent standards under section 307 for six pollutants: Aldrin/Dieldrin, DDT, Endrin, Toxaphene, benzidine, and PCBs. While EPA had thus begun to issue standards for toxic water pollutants on a pollutant-by-pollutant basis, several environmental groups, thinking that EPA was not regulating quickly enough, filed suit against EPA for its failure to regulate toxic pollutants. At the same time industry groups were concerned that a pollutant-by-pollutant approach would require different standards for different pollutants without regard for the availability or compatibility of various strategies of control (79).

Under pressure from both groups, EPA developed a technology-based strategy for regulating pollutants on an industry-by-industry basis. The suits were settled in a consent decree (the "Flannery decree") with EPA agreeing to place specific "numerical limits on the quantities of 65 toxic pollutants in 21 industrial categories" (62,79). The consent decree permitted EPA to regulate toxic substances by means of those sections of CWA designed to control ordinary nontoxic pollutants, and to regulate pollutants on an industry-by-industry basis using effluent limitations (79).

Provisions of this consent decree were incorporated into CWA by Congress in the 1977 amendments, thereby giving congressional sanction to the development of technology-based regulations on toxic pollutants. One result is that under CWA, similar-sounding terms have different meanings. An effluent standard is a control requirement based on the relationship between the discharge of a pollutant and the resulting water quality in a receiving body of water. An effluent limitation, on the other hand, is a technology-based approach. For example, use of BPT, BAT, or "best conventional technology," might be required for direct dischargers.

Effluent limitations are what are used today. EPA has been in the process of issuing technology-based effluent limitations for 65 categories of toxic substances.

The 65 classes of pollutants were chosen in the negotiations leading to the consent decree. For choosing these classes of pollutants, EPA assembled a working group of staff scientists from EPA and other agencies (78). This group conducted a literature search for toxic pollutants using several criteria: 1) evidence that a substance posed potential carcinogenic, mutagenic, teratogenic effects, or adverse effects on any organ system; and 2) evidence of persistence, ability to bioaccumulate in organisms, and synergistic propensities. These general criteria yielded 337 organic compounds, which the committee narrowed to 232. Using the criteria of presence in water effluents and evidence of carcinogenic, mutagenic, or teratogenic effects in animal tests or human epidemiology, or evidence of high toxicity to aquatic organisms or systems, the list was further narrowed to 76. From this list of 76, EPA provided more specific lists of 29, 18, and 18 classes of substances.[10]

List I of 29 classes of substances satisfied three criteria:

1. they were known to occur in point source effluents, in aquatic environments, in fish, or in drinking water;
2. there was substantial evidence of carcinogenicity, mutagenicity, and/or teratogenicity in human or animal studies; and
3. it was likely that point source effluents contributed substantially to human exposure, at least locally.

List II of 18 compounds of second highest priority satisfied the first criterion, but toxicity evidence was based primarily on structural similarity to compounds on list I, mutagenicity tests, or test results that appeared to be incomplete or equivo-

_____

[10]EPA originally developed a fourth list of 12 substances. Although they are present in water effluents, they were judged to present a less substantial direct hazard than the chemicals on lists I-III and were not included in the final list of substances to be regulated.

cal. The 18 compounds on list III all satisfied the first criterion, but there was no substantial evidence that these compounds have "primary carcinogenic, mutagenic, or teratogenic effects" (78). (See table 3-10 for these lists.)

These 65 classes of pollutants, also known as priority pollutants, initially contained more than 129 individual substances, but EPA developed a list of 129 specific substances (177). Three of these were removed from the list, leaving a total of 126 individual substances in 28 industrial classes for which it had to set regulations.[11]

As of today, EPA has issued regulations for 26 of the 28 industry groups. Regulations for the

[11]The exact number of "industries" has varied because definitions were changed.

pesticides industry had been issued, but that regulation was challenged in court, and EPA has remanded the regulation and initiated work to develop new regulations for the pesticide industry. The organic chemicals, plastics, and synthetic fibers industry has yet to be regulated, even though it contributes the largest quantity of organic pollutants of any industry (224).

Moreover, the 28 industry groups do not cover all industries that discharge pollutants. Important industries, such as car washes and other commercial laundries, and paint and ink formulators are excluded completely from effluent guidelines and pretreatment standards. Certain subcategories of other industries, such as adhesives and sealants, are also exempted. Pretreatment standards (for indirect dischargers) were proposed for some in-

Table 3-10.—Classes of Substances Regulated Under the CWA Consent Decree

List I	List II	List III
1. Acenaphthene	1. Chlorinated benzenes (other than dichlorobenzenes)	1. Acrolein
2. Aldrin/dieldrin	2. Chlorinated ethanes	2. Acrylonitrile
3. Arsenic compounds	3. 2-chlorophenol	3. Antimony compounds
4. Asbestos	4. Dichloroethylenes	4. Chlorinated naphthalene
5. Benzene	5. 2,4-Dichlorophenol	5. Chlorophenols (those not on list II)
6. Benzidine	6. 2,4-Dimethylphenol	6. Copper compounds
7. Beryllium compounds	7. Dichloropropane and dichloropropene	7. Cyanides
8. Cadmium compounds	8. Endosulfan and metabolites	8. Dinitrotoluene
9. Carbon tetrachloride	9. Endrin and metabolites	9. Ethylbenzene
10. Chlordane (technical mixture and metabolites)	10. Fluoranthene	10. Hexachlorocyclo-pentadiene
11. Chloroalkyl ethers	11. Haloethers (not on list I)	11. Isophorene
12. Chloroform	12. Halomethanes (not on list I)	12. Nitrobenzene
13. Chromium compounds	13. Hexachlorobutadiene	13. Nitrophenols
14. DDT and metabolites	14. Naphthalene	14. Phenol
15. Dichlorobenzenes (1,2-,1,3-, and 4-dichlorobenzenes)	15. Pentachlorophenol	15. Selenium compounds
16. Dichlorobenzidine	16. Phthalate esters	16. Silver compounds
17. Diphenylhydrazine	17. Tetrachloroethylene	17. Toluene
18. Heptachlor and metabolites	18. Toxaphene	18. Zinc compounds
19. Hexachlorocyclohexane		
20. Lead compounds		
21. Mercury compounds		
22. Nickel compounds		
23. Nitrosamines		
24. Polychlorinated biphenyls (PCBs)		
25. Polynuclear aromatic hydrocarbons		
26. 2,3,7,8-Tetrachlorodibenzo-p-dioxin (TCDD)		
27. Thallium compounds		
28. Trichloroethylene		
29. Vinyl chloride		

SOURCE: Office of Technology Assessment.

dustries, such as textile mills, and plastics molding and forming, but were never issued (224).

The standards that have been issued cover the list of 126 chemicals, although not all 126 chemicals are regulated in each of the 28 industry groups. For each industry, EPA has issued regulations for the pollutants that it judged appropriate to regulate. In a given industry, a particular chemical may not be present, may have no available treatment technique, may be too costly to control, or may be "incidentally" covered by regulations for other pollutants.

While the list of 126 chemicals has been the focal point of the regulation under CWA, many of the chemicals found in industrial effluents are not included in this list. In a nationwide study of wastewater from a wide variety of industries and publicly owned treatment works (POTWs), 4,000 wastewater samples were examined for organic pollutants using gas chromatograph-mass spectrometry. The result is an overall picture of the chemicals discharged by these industries and POTWs. After ranking the top 50 most frequently occurring compounds, the researchers discovered that 16 of them were priority organic pollutants. Thus, 34 of the 50 most frequently occurring compounds are not included in the EPA list of priority pollutants. For the industries studied, the distribution of pollutants differed by industry (i.e., what is important for one industry may not be important in another industry). In general, the priority pollutants make up approximately 25 percent of the most frequently occurring compounds (185).

In addition to setting the technology-based effluent limitations, EPA is also authorized to issue nonbinding water quality criteria documents (under section 304) for substances that might pose hazards to human health or the environment. These are used to guide States in setting water quality standards in their water courses (under section 303) and as guidance for writing NPDES permits. As of 1986, EPA had issued water quality criteria for 65 classes of priority toxic pollutants, including 29 determined to be carcinogenic. These are listed in table 3-10.[12]

The water quality criteria documents present information on protecting human health and aquatic organisms (fish, shellfish, plants, etc.). For human health, the criteria are developed to protect against noncarcinogenic risks as well as carcinogenic risks and are based on two potential exposure pathways—through consumption of drinking water and aquatic organisms. Because safe thresholds for exposure to carcinogens have not been established, the recommended criteria for the maximum protection of human health are water concentrations of zero. But EPA also prepared quantitative risk estimates for carcinogens. The published criteria include the concentrations of the chemical in water that corresponded to calculated lifetime cancer risks of 10^{-5}, 10^{-6}, and 10^{-7}.

States have the option of adopting these numerical criteria, but not all of them have actually done so. Water quality criteria have been prepared for the 65 classes of toxic pollutants covering 126 chemicals—the priority pollutants. For 37 of the 85 organic chemicals from this list, no States have developed standards; for another 32, one State has developed standards. Less than half of the States have developed a water quality standard for any single priority pollutant except arsenic for which standards were established in 28 states (345). Fourteen States have no water quality standards at all for any of the priority pollutants (224). As table 3-10 shows, seven of the water quality criteria that identify chemicals as carcinogenic have been adopted by at least one State.

Although the environmental groups who brought the original suit that resulted in the Flannery decree believed that regulating toxic pollutants by means of technology-based effluent limitations would speed the elimination of these from the Nation's waterways, the regulation has taken considerable time. EPA has missed deadlines and has requested eight separate deadline extensions from the court. EPA's most recent goal of regulating all pollutants for all industries by January 1987 was not achieved. EPA now hopes to complete

---

[12]The list of substances in the table was derived from a list EPA provided that summarizes the water quality criteria documents. OTA included in the table only those substances that were indicated to be carcinogens. Some substances that appear in the *Annual Report on Carcinogens* have been regulated under CWA, but do not appear on this table because the water quality criteria were not based on carcinogenic effects.

the effluent limitations specified in the original 1976 consent decree by September 1987 (166).

In addition, the compliance dates for the industries are usually 3 years after EPA publishes its final rule on the BAT regulations. Thus, even though the consent decree was issued in 1976, not all industries will be in compliance with the regulations until the late 1980s or 1990.

Under the 1987 Amendments to CWA, each State is required to submit to EPA a list of "toxic hot spot waters." These are areas where water quality standards cannot be achieved or maintained because of toxic discharges after the currently required pollution controls have been implemented. The States must also submit a plan to bring these areas into compliance with those standards. The State water quality standards must be based on EPA water quality criteria for toxic pollutants.

EPA REGULATORY ACTIONS UNDER THE SAFE DRINKING WATER ACT

The Safe Drinking Water Act of 1974 regulates the safety of water from public water systems, and it contains several provisions that may be used to regulate hazardous substances, including carcinogens in drinking water. SDWA authorized EPA to regulate contaminants "which . . . may have an adverse effect on the health of persons," and prescribed several steps for EPA to follow.

First, EPA was required to publish national interim primary drinking water standards in 1975. Second, Congress required that EPA commission the National Academy of Sciences (NAS) to "conduct a study to determine . . . the maximum contaminant levels which should be recommended" as national standards. NAS was also required to update this information every 2 years. The NAS study had to consider the impact of contaminants on groups or individuals in the population who are more susceptible to adverse effects than are normal healthy individuals, exposure to contaminants in other media, synergistic effects of contaminants, and body burdens of contaminants in exposed persons. In its 1977 report, *Drinking Water and Health*, (134) NAS provided its first list of contaminants (chosen on the basis of its own criteria) that might have an adverse effect on health and the levels at which those effects are expected based on the best available scientific knowledge. The NAS report, however, did not provide recommended contaminant levels.

Third, within 90 days of the publication of the NAS study, EPA was required to establish "recommended maximum contaminant levels (RMCLs) for each contaminant which . . . may have any adverse effect on the health of persons." Each such RMCL was to be "set at a level at which . . . no known or anticipated adverse effects on the health of persons occur and which allows an adequate margin of safety." RMCLs were nonenforceable health goals which were then used as guidelines for establishing enforceable drinking water standards.

Once EPA established RMCLs for each contaminant, it was required to publish revised national primary drinking water regulations. These regulations were enforceable health standards. The required regulations were to specify a maximum contaminant level (MCL) or require the "use of treatment techniques" for each contaminant for which an RMCL is established. The established MCLs were to be as close to the RMCLs as is "feasible." In determining feasibility, the Administrator could consider "the use of the best technology, treatment techniques and other means . . . [that] are generally available [taking cost into consideration]."

Drinking Water Standards

Under its authority to set drinking water standards, in 1975 EPA promulgated interim drinking water standards for 10 inorganic and 6 organic chemicals and for microbial contaminants (40 CFR 141.11-141.14). The interim drinking water standards issued in 1975 were based on the 1962 recommendations of the U.S. Public Health Service (214).

Among the inorganics, arsenic was specifically cited as being of concern because of carcinogenicity, although EPA decided that the evidence for its carcinogenicity in drinking water was inconclusive. (See table 3-11 for a summary of these data.) Among the organic compounds, endrin, lindane, and toxaphene were regulated based on the effects of acute and chronic exposure, but the substances were later identified as carcinogens by the NAS Drinking Water Committee and by EPA in subsequent proposed drinking water standards. In addition, EPA issued regulations for two

groups of carcinogens: radionuclides in 1976 (40 CFR 141.15) and for total trihalomethanes (four chemicals) in 1979 (40 CFR 141.31).

Thus, the initial approach under SDWA was to issue standards contaminant by contaminant, specifying that every public water system monitor for each contaminant to ensure compliance with the standard. However, it would be both costly and technologically difficult to monitor for relatively small amounts of many synthetic organic compounds and other potentially toxic con-

Table 3-11.—Carcinogens Regulated and Proposed for Regulation Under the Safe Drinking Water Act (SDWA), Actions Before 1980

Substance	Type of evidence[a]	ANPR	NPRM	Final
Inorganics:[b]				
Arsenic	Human, Inconclusive	—	3-14-75	12-24-75
Cadmium	—	—	3-14-75	12-24-75
Chromium	—	—	3-14-75	12-24-75
Lead	—	—	3-14-75	12-24-75
Nitrate/nitrite	—	—	3-14-75	12-24-75
Selenium	—	—	3-14-75	12-24-75
Organics:[c]				
Chlorinated hydrocarbons:				
Aldrin/dieldrin	—	—	d	Not issued
Chlordane	—	—	3-14-75	Not issued[e]
DDT	—	—	d	Not issued
Endrin	—	—	3-14-75	12-24-75
Heptachlor	—	—	3-14-75	Not issued
Heptachlor epoxide	—	—	3-14-75	Not issued
Lindane	—	—	3-14-75	12-24-75
Toxaphene	—	—	3-14-75	12-24-75
Total trihalomethanes[f]				
Bromodichloromethane	Inadequate/ suspected	7-14-76	2-9-78	11-19-79
Dibromochloromethane	Inadequate/ suspected	7-14-76	2-9-78	11-19-79
Tribromomethane (bromoform)	Inadequate/ suspected	7-14-76	2-9-78	11-19-79
Trichloromethane (chloroform)	Animal, 2 species (rats, mice)	7-14-76	2-9-78	11-19-79
Radionuclides:				
Radium 226 & 228	Human, animal	—	8-14-75	7-9-76
Gross alpha particle activation	Human, animal	—	8-14-75	7-9-76
Beta particle & photon radioactivity	Human, animal	—	8-14-75	7-9-76
Treatment standard for all synthetic organic chemicals:[g] [h]		7-14-76	2-9-78	Withdrawn

[a]Type of evidence was derived from the *Federal Register* notices that presented EPA's rationale for regulatory action.
[b]According to the NPRM, the MCLs for inorganics were based on the "possible effects of lifetime exposure." Specific carcinogenic concerns were cited only in response to the comments on arsenic.
[c]Carcinogenic evidence was not cited. The MCLs were based on the "effects of acute and chronic exposure."
[d]Carcinogenic concerns were mentioned but action was delayed pending the results of a nationwide survey to determine the extent of drinking water contamination by these substances.
[e]Carcinogenic concerns were mentioned but final action was delayed pending outcome of FIFRA suspension/cancellation proceedings.
[f]Formed as the result of chlorination of drinking water.
[g]Treatment techniques were proposed for all synthetic organic chemicals (SOCs) as a class rather than individual MCLs because it was not considered feasible to identify and monitor individual substances. Carcinogenic concerns were based on an NAS report, "Drinking Water and Health" which identified 22 SOCs (listed below) as known or suspected carcinogens.
[h]Court challenge: *EDF* v. *Costle*, 11 ERC 1209, No. 752224, 2-10-78.

SOURCE: EPA response to OTA request and cited *Federal Register* notices.

taminants that might occur in drinking water coming from surface water sources. So in 1978 EPA proposed treatment regulations for drinking water systems that used surface waters. The focus was on generic treatment techniques for all synthetic organics without requiring monitoring of numerous individual contaminants.

The proposal for a treatment standard was withdrawn, however, because EPA could not find a clear basis for selecting communities with a synthetic organic chemical contamination problem that would be required to use the treatment techniques. There was also concern about cost and feasibility. Moreover, by 1980 emphasis shifted from surface to ground water contamination. Because of this, EPA began again to focus on setting health standards using a contaminant-by-contaminant approach, rather than pursuing the goal of setting treatment standards for surface water (36).

The contaminants were grouped for the purpose of regulation into volatile synthetic organic chemicals (VOCs), synthetic organic chemicals (SOCs), inorganic chemicals, microbiological contaminants, radionuclides, and disinfectants. This change in approach, from a contaminant-by-contaminant approach to surface water treatment standards and back to the contaminant-by-contaminant approach, delayed the issuance of revised national primary drinking water standards.

In March 1982 EPA issued an Advance Notice of Proposed Rulemaking (ANPRM) for nine VOCs (303). It then held workshops around the country and in June of 1983 issued proposed RMCLs for VOCs (300). It issued final RMCLs for VOCs and proposed MCLs for VOCs in November 1985 (301). Eight final MCLs for VOCs were issued in June 1987 (322).

For 31 SOCs, 16 inorganic compounds (IOCs), radionuclides, and microorganisms, EPA first issued an ANPRM in October 1983 (304), followed by proposed RMCLs in November 1985 (302). Except for fluoride (which is not a carcinogen), EPA has yet to issue final RMCLs, proposed MCLs and final MCLs on these chemicals. In addition, EPA must address a number of the 83 substances for regulation under the requirements of the 1986 amendments.

To determine the RMCLs for carcinogens, EPA classified substances into three categories based on the evidence for carcinogenicity—strong evidence (Category I), equivocal evidence (Category II), and inadequate evidence or lacking evidence (Category III). Classification into these categories for the drinking water standards is based largely on the the classification of the substance in the EPA weight-of-evidence classification (see ch. 2).

Category I chemicals have strong evidence for carcinogenicity from either human or animal studies, i.e., weight-of-evidence group A (sufficient evidence for human carcinogenicity) or group B (probable human carcinogen, based either on limited human evidence or sufficient animal evidence). Category I substances have RMCLs of zero. EPA chose this level for RMCLs based on the legislative history of SDWA (301). MCLs for Category I substances must be set as close to zero as is feasible, taking costs into consideration.

Category II chemicals, with equivocal evidence for carcinogenicity for purposes of drinking water standards, are chemicals including weight-of-evidence group C (possible human carcinogens based on limited evidence in animals). Category III chemicals, those with inadequate evidence or lacking evidence for carcinogenicity, are from weight-of-evidence group D (not classified or inadequate animal evidence) or group E (no evidence for carcinogenicity).

Category II substances have RMCLs set in one of two ways. The first, and preferred approach is to set the RMCLs based on noncarcinogenic chronic toxicity data. Under this approach, EPA calculates an Adjusted Acceptable Daily Intake (AADI). The second method, used when adequate chronic toxicity data are lacking, is to base the RMCLs on the results of quantitative risk assessment using the limited animal carcinogenicity data with the risk level set in the range 10^{-5} to 10^{-6}.

Category III substances have RMCLs based on noncarcinogenic chronic toxicity data. Again EPA calculates an AADI.

For either Category II or Category III chemicals, EPA determines, on the basis of chronic toxicity data, a highest "no observed adverse effect level" (NOAEL) (expressed in mg/kg body weight/day), divides that figure by an appropriate "uncertainty" or "safety" factor (explained below), then multiplies this figure by the assumed weight of an adult (70 kg) and divides by the assumed amount of water consumed by an adult (2 liters/day). The result is an AADI:

$$AADI = (NOAEL/uncertainty\ factor)(70kg/2Liters/day)$$

A safety or uncertainty factor of 10 is "used with valid experimental results on appropriate durations of exposure in humans." A safety factor of 100 is "used when human data are not available and extrapolating from valid results of longterm studies in animals" is involved. A safety factor of 1,000 is "used when human data are not available and extrapolating from studies in animals of less than chronic exposure." Finally, an additional uncertainty factor between 1 and 10 is used when EPA has to use a "lowest observed adverse effect level" (LOAEL) rather than a NOAEL (151,299).

Since Category II substances have some limited if insufficient evidence of carcinogenicity from animal studies, EPA introduces additional safety or uncertainty factors to account for the equivocal evidence of carcinogenicity. Normally, a safety factor of 10 is used, but if data indicate the need for a greater or lesser safety margin, other uncertainty factors can be used (301). In general EPA is more cautious in setting RMCLs and MCLs for Category II substances than for Category III substances.

In its regulatory proceedings so far, five of the nine VOCs were identified as probable carcinogens (Category I). EPA's decision to consider one of them, 1,1-dichloroethylene, as belonging to category II was legally challenged by NRDC and a decision is pending (158). Of 32 SOCs, 10 are probable carcinogens. Of the inorganic substances, two (arsenic and asbestos) were placed in Category II. These substances are carcinogenic when inhaled. EPA has concluded, however, that there is little evidence indicating that asbestos is carcinogenic in drinking water. There is evidence that drinking water exposures to arsenic are associated with skin cancer, although it appears that this is true only for the generally nonfatal forms of skin cancer (36). For the 15 Category I chemicals, 2 were based on human data, and the balance on animal evidence. In addition, all the radionuclide standards were based on human data (see table 3-12). (see Note, page 143.)

Health Advisories

In addition to legally binding regulations, EPA has also provided nonbinding health advisories for contaminants in water. In 1980 the National Academy of Sciences began providing EPA with "suggested no adverse response levels" (SNARLs) for contaminants. These contained acute (24-hour exposure) and short-term (7-day exposure) toxicity information as well as chronic toxicity information. The Office of Drinking Water developed its own SNARLs and issued drafts of them beginning in 1981. Subsequently the term SNARLs was changed to "health advisories" (HAs).

HAs are approximately 10-page dossiers on chemicals that give some indication of their occurrence, their use, short-term toxicity information, chronic toxicity information (including contaminant levels calculated to be associated with different levels of risk, e.g., 1 case per 10,000 people exposed, 1 case per 100,000 people, 1 case per 1 million people), analytical methods of detecting them, and treatment methods that operators of municipal water systems can use. Health advisory concentration numbers are "developed from data based on non-carcinogenic endpoints of toxicity." For suspected carcinogens "non-zero 1-day, 10-day and longer-term health advisories may be derived . . . [but] lifetime exposures may not be recommended." In addition, "projected excess lifetime cancer risks are provided to give an estimate of the concentrations of the contaminant which may pose a carcinogenic risk to humans." These estimates are presented as "upper 95 percent confidence limits derived from the linearized multistage model which is considered to be unlikely to underestimate the probable true risk" (325).

EPA issued its first HAs in 1979, reevaluated them in 1985, and rereleased some 52 for public comment and for evaluation by the EPA Science Advisory Board (36). Health advisories are not legally binding, but are intended to advise public water systems about the health effects of chemicals and their treatment. Health advisories have been widely used in the water industry to determine responses to contamination incidents (36).

Table 3-12.—National Primary Drinking Water Regulations (NPDWR) for Carcinogens, Actions After 1980

Substance	EPA classification & type of evidence	Recommended maximum contaminant level (RMCL)			Maximum contaminant level (MCL)	
		ANPR	NPRM	Final	NPRM	Final
Inorganics (not regulated as carcinogens):						
Arsenic[a]	II	10-5-83	11-13-85	—	—	—
Asbestos	II	10-5-83	11-13-85	—	—	—
Cadmium	III	10-5-83	11-13-85	—	—	—
Chromium	III	10-5-83	11-13-85	—	—	—
Lead	III	10-5-83	11-13-85	—	—	—
Nitrate/nitrite	III	10-5-83	11-13-85	—	—	—
Selenium	III	10-5-83	11-13-85	—	—	—
Synthetic organics:						
Acrylamide	I, animal	10-5-83	11-13-85	—	—	—
Alachlor	I, animal	10-5-83	11-13-85	—	—	—
Chlordane	I, animal	10-5-83	11-13-85	—	—	—
Dibromochloropropane (DBCP)	I, animal	10-5-83	11-13-85	—	—	—
1,2-Dichloropropane	II, animal	—	11-13-85	—	—	—
Epichlorohydrin	I, animal	—	11-13-85	—	—	—
Ethylenedibromide	I, animal	—	11-13-85	—	—	—
Heptachlor	I, animal	—	11-13-85	—	—	—
Heptachlor epoxide	I, animal	—	11-13-85	—	—	—
Lindane	II, animal	10-5-83	11-13-85	—	—	—
Monochlorobenzene	III	—	11-13-85	—	—	—
Polychlorinated biphenyls (PCBs)	I, animal	—	11-13-85	—	—	—
Styrene	II[b]	—	11-13-85	—	—	—
Toxaphene	I, animal	10-5-83	11-13-85	—	—	—
Volatile organic compounds:						
Benzene	I, human	—	6-12-84	11-13-85	11-13-85	—
Carbon tetrachloride	I, animal	3-4-82	6-12-84	11-13-85	11-13-85	—
p-dichlorobenzene	III, animal	—	6-12-84	11-13-85	11-13-85	—
1,2-Dichloroethane	I, animal	3-4-82	6-12-84	11-13-85	11-13-85	—
1,1-Dichloroethylene	II, animal	—	6-12-84	11-13-85	11-13-85	—
Tetrachloroethylene	II, animal	3-4-82	6-12-84	c	—	—
1,1,1-Trichloroethane	III, animal	3-4-82	6-12-84	11-13-85	11-13-85	—
Trichloroethylene	I, animal	3-4-82	6-12-84	11-13-85	11-13-85	—
Vinyl chloride	I, human, animal	3-4-82	6-12-84	11-13-85	11-13-85	—
Radionuclides:						
Radium 226 & 228	Human	10-5-83	—	—	—	—
Gross alpha particle activation	Human	10-5-83	—	—	—	—
Beta particle & photon radioactivity	Human	10-5-83	—	—	—	—
Uranium	Human	10-5-83	—	—	—	—
Radon	Human	10-5-83	—	—	—	—

[a]The proposed regulation was not based on carcinogenic effects because it was considered also to have potential nutrient value.
[b]Proposed regulation not based on carcinogenic effects.
[c]Comment period reopened to consider new data, 11-13-85.

SOURCES: EPA response to OTA request and cited *Federal Register* notices.

Regulation of carcinogens in drinking water has been slow. The SDWA was passed in 1974. As indicated above, in 1975 interim standards based on Public Health Service recommendations of 1962 were issued, and later radionuclides (1976) and trihalomethanes (1979) were regulated. These standards are still in effect today.

As of June 1987, EPA has issued eight final MCLs that constitute national revised primary drinking water standards. Revised HAs designed to provide operators of public water systems with guidance concerning health risks from potential toxic substances are not yet final, although they are under development.

The 99th Congress, concerned that drinking water standards were not being set quickly enough, set regulatory deadlines in the 1986 reauthorization of SDWA.

The 1986 SDWA amendments (Public Law 99-339) required EPA to regulate 83 chemicals in drinking water that the agency had identified as candidates for regulation in 2 ANPRMs in 1982 and 1983 (303,304). The list of 83 are to be regulated in 3 stages by 1989 and include 51 in the process of regulation. EPA must also add 25 chemicals to the list every 3 years after 1989. According to EPA staff, this list of 83 included numerous substances that EPA had not otherwise intended to regulate because of low toxicity or low occurrence in drinking water (36).

The 1986 amendments gave RMCLs a new name: Maximum Contaminant Level Goals (MCLGs). The criteria for these remain the same as for the RMCLs. In addition, MCLs must now be proposed at the same time as the MCLGs rather than in two stages as they were previously. The MCL must be as close to the MCLG as is feasible using the BAT, taking costs into consideration. It must be at least as low, however, as would be achieved using granulated activated carbon (an especially good technique for removing organic contaminants). When an MCL is exceeded, the public must now be notified within 14 days of the detection of the violation.

In addition, the amendments required EPA to develop a list of unregulated contaminants which water utilities must monitor at least once every 5 years. A list of 50 unregulated chemicals for which monitoring will be required is scheduled for publication in June 1987 (36). Congress also authorized stricter enforcement of SDWA and increased fines for violations.

EPA REGULATORY ACTIONS UNDER FIFRA

The Federal Insecticide, Fungicide, and Rodenticide Act was substantially amended in 1972. Under this act, EPA is to screen pesticides through the registration process before they enter commerce and reregister pesticides that were on the market before 1972 to prevent unreasonable adverse health and environmental effects. In both cases, EPA may require manufacturers to submit testing results for product and residue chemistry, environmental fate, and tests in fish, wildlife, and mammals, including long-term bioassays for carcinogenicity.

An applicant for registration of a pesticide must file with EPA certain required information, including a statement of all claims made for the pesticide, directions for its use, a description of tests made upon it, the test results used to support claims made for the substance, and appropriate toxicity data for each pesticide. Specific testing requirements depend on the expected use pattern. EPA now requires carcinogenicity testing in two species for all food use pesticides (40 CFR 158).

In general, EPA must register a pesticide if "it will perform its intended function without unreasonable adverse effects on the environment"—which are defined as "any unreasonable risk to man or the environment, taking into account the economic, social and environmental costs and benefits of the use of any pesticide." The burden of proof to establish the safety of a product lies on the person who wishes to register and market the product.

If EPA finds that a registered pesticide meets or exceeds certain criteria for risk, a special review may be initiated. (Prior to 1983, this was called a "Rebuttable Presumption Against Registration" (RPAR).) This in-depth review of the risks and benefits of the pesticide use determines what regulatory action, if any, is appropriate. The 1975 amendments to FIFRA require that if involuntary regulatory measures are proposed, the EPA recommendations must be submitted to the Department of Agriculture for comment and to the FIFRA Scientific Advisory Panel for review. There is also an opportunity for public comment throughout this process. The outcome of a special review can range from no action, to an immediate emergency suspension, to cancellation of the pesticide. Actions may also consist of modifying the use pattern of the pesticide.

Unless the data on which the special review or RPAR is based are shown to be unreliable or invalid, or the estimated benefits of continued uses outweigh the estimated risks, FIFRA provides for cancellation or suspension of their registration.

In 1972, there were about 50,000 pesticide products and 600 active ingredients already on the market that required reregistration under the new law. Many of the problems of regulating pesticides have arisen with these pesticides. The aim of reregistration was to identify missing information about pesticides, to require registrants to supply it, and to reevaluate the safety of the chemicals in light of the new information. Reregistration involves a number of steps:

1. EPA requests certain pivotal studies on the chemical from the registrant ("data call-in") if studies have not yet been submitted. Data requirements are based on EPA regulations (40 CFR 158). Since 1980, EPA has required carcinogenicity data for reregistration where required (40 CFR 158).
2. EPA reviews all available data on the chemical, including that requested in the data call-in.
3. EPA issues a "regulatory position document" ("registration standard") which summarizes the science position, changes in use necessary to reduce risk, and additional data gaps.
4. Registrant either agrees to fill data gaps or action may be initiated to withdraw the chemical from the market.
5. If "significant risk triggers" are identified in steps 2 through 4, a special review is initiated (21).

The reregistration of active ingredients has taken much longer than originally anticipated. The process was initially to be completed by 1976, but in 1975 Congress extended that deadline to 1977, and in 1978 dropped the deadline completely because of the large number of substance reviews outstanding (203).

The EPA data files for most of the pesticides needing reregistration are still incomplete, although information is currently under development. EPA conducted a data call-in for approximately 600 active ingredients subject to reregistration. As a result of data call-ins, by March 31, 1986, 61 active ingredients had been canceled voluntarily, withdrawn, or suspended, and 124 interim registration standards developed. Of these 124 active

ingredients with registration standards, 6 were voluntarily withdrawn, while 12 were beginning special reviews, and 17 were ready for final regulatory review, while 89 registrants' responses were in progress. One active ingredient is undergoing final standard processing, which would result in final reregistration (203).

Thus, for 72 percent of the chemicals reviewed sufficiently to develop registration standards (89 of 124), EPA is currently waiting to obtain necessary data. Beyond this, over two-thirds of all active ingredients (415 of 600) have not yet been reviewed sufficiently for the agency to issue even registration standards.

Of the 600 active ingredients, about 390 are used on foods. EPA has completed its data call-in for all 390 substances and has issued interim registration standards on 92, with approximately 300 yet to be evaluated. EPA lacked sufficient information to judge the carcinogenic effects of 57 of these 92 ingredients (203).

Actions on Pesticides Already Identified as Carcinogenic

For some 80 active ingredients that have been voluntarily removed from the market or subject to some regulatory activity, carcinogenicity has been an important element. In all, 65 of these 80 substances (81 percent) have had carcinogenic effects.[13]

OTA requested information from EPA on the pesticides it has identified as carcinogenic. As of March 31, 1986, it had identified at least 81 carcinogens. (See tables 3-13 to 3-16 for a summary of these data.)[14] These carcinogens were identified through testing results obtained from a variety of sources, including manufacturers and NTP, and reported in the open literature.

- Of the 81, 18 have been canceled or restricted for some or all uses as a consequence of EPA action (table 3-13).

[13]These percentages are based on the information in "Report on the Status of Chemicals in the Special Review Program, Registration Standards Program, and Data Call-in Program" (331).

[14]This count of chemicals is slightly different from that in other sources of information, including EPA status reports, GAO reports, and correspondence between EPA and Congressman Waxman.

- For one, Daminozide (Alar), EPA published a notice of intent to cancel, but the chemical is still undergoing review (table 3-13).
- Fifteen pesticides have been voluntarily canceled by the registrant (table 3-14).
- At present, 18 of these substances are in special review (SR) (table 3-15). For 10 carcinogens, the SRs have been completed but the chemicals have not been suspended or canceled because EPA decided to allow their continued use based on the balance of risks and benefits (104). For the remaining eight substances, SRs have yet to be completed.
- EPA has identified 29 carcinogens for which it has not started an SR or cancellation proceeding (table 3-16). Of these, 24 are food use pesticides.

Tables 3-15 and 3-16 together indicate that EPA has identified 47 carcinogenic pesticides that have not been canceled. Of these, 18 were made the subjects of SRs (10 completed, 8 still in progress); for others, EPA has decided not to conduct an SR. For example, for 13 of the 47, EPA has determined that either low exposure or risk or the weight of the evidence for carcinogenicity suggest no action need be taken. The remaining identified carcinogenic pesticides are still in use, although EPA suggested to OTA that for many of these regulatory action has been taken to reduce exposure (21). These actions may consist of requiring lowered application rates, enclosed application systems, extension of reentry intervals, a ban on aerial application, and protective equipment and clothing (104).

Moreover, "cancellation" of the 18 chemicals on table 3-13 does not mean that the chemical is no longer in use. Often, a cancellation is for particular uses, while other uses continue. For example, chlordane was canceled in 1978 for food use (table 3-14), but continues in use for termite control. EPA is currently considering whether to cancel that use as well.

For the 18 substances that have been canceled or restricted by EPA (table 3-13), the average time carcinogens were in SR was 44 months, with the shortest time being 13 months and the longest 88 months. Substances in SR earlier took the shortest time with more recent cancellations taking longer. For 10 carcinogens in SR, but not suspended or canceled, the average length of the SR was 63 months, with the shortest SR 36 months and the longest 106 months.

Where EPA has acted on carcinogenic pesticides, the evidence for carcinogenicity has generally been based on positive results from at least two animal species (tables 3-13 to 3-16).

According to figures in a recent NAS report (136), of 289 food use pesticides, 53 or approximately 18 percent have been determined by EPA to be at least potentially carcinogenic. However, for many of these the data are still incomplete or have not been evaluated. Food use pesticides raise special issues with regard to coverage under FDCA.

For raw agricultural commodities, a tolerance for pesticide residues is based on the consideration of risks and benefits (under sec. 408 of FDCA). If processing the food leads to an increase in the concentration of the pesticide residue to a level above that found in the parent raw commodity, then the Delaney clause applies (sec. 409 of FDCA). In this case, the residue is deemed to be a food additive and may not be added to food if it is determined to be carcinogenic. Congress specifically exempted processed food from the Delaney clause if the residue level is no higher than is found in the raw commodity. For commodities that are processed, if no section 409 tolerance may be granted, EPA will not grant a 408 tolerance for the raw commodity either.

While the Delaney clause has been consistently used to deny new tolerances for active ingredients, it has never been used to revoke an existing tolerance. Prior to 1978, section 408 tolerances were generally established without any oncogenicity or residue data. Few of the pesticides approved before 1978 have tolerances for residues in processed foods. Where no tolerance exists for a processed food, EPA simply assumes that residue levels are the same as the level permitted in raw commodities. In the report, the NAS committee concluded that there is no scientific justification for the regulatory distinction between raw and processed foods. The committee also found no scientific justification for the inconsistency between the safety standards applied to old and new pesticides (136).

Table 3-13.—FIFRA[a]
I. Canceled and Restricted Carcinogenic Pesticides

Substance	Registration standard	Type of evidence	Special review initiated	Proposed determination	Notice of intent to cancel/suspend	Voluntary cancellation/ some or all uses	Comments/other regulatory action	Court action or administrative hearing
Aldrin/dieldrin	—	Animal, 2 species — mice, rats	—	—	3-18-71; 6-26-72, susp.; 10-18-74, canc.(most uses)	—	—	Shell Chem. Co. et al., FIFRA Doc. #145; EDF petition filed 12-3-70; EDF v. EPA,465 F2d 528 (DC Cir. 1972) re: suspension pending outcome of hearing
Amitraz (Baam)[b]	9-85	Animal, 2 species	4-6-77	10-7-79	—	—	Conditional registration, for use on pears, not apples	—
Arsenicals (inorganic)[b]	—	Animal, human	10-18-78 wood & non-wood uses	7-13-84 wood use	7-13-84 wood use	—	Labeling requirements	In re: Chapman Co., et al., FIFRA Doc. #529, 7-15-84
copper acetoarsenite	—	—	—	—	—	4-7-77	—	—
copper arsenate	—	—	—	—	—	4-7-77	—	—
arsenic trioxide	—	—	—	—	—	6-19-78	—	—
sodium arsenite	—	—	—	—	—	10-18-78	—	—
Chlordane, heptachlor	—	Animal, 2 species— mice, rats	—	—	7-29-75 susp.; 3-24-78 canc. food uses	8-87 non-food uses	Agreement to discontinue sales unless studies conducted show no residue from outside use	Velsicol Chem. Co. et al. FIFRA Doc. #384, 336; EDF v. EPA, 548 F. 2d 998 (D.C. Cir. 1976) (EDF petition 10-74); NCAMP et al. v. EPA, CIV 87-2089 LFO (D.C. Cir. filed 8-24-87)
Chlordecone	—	Animal, 2 species	3-19-76	4-11-77	4-11-77 all uses	7-27-77	—	Lethelin Products Co. et al., FIFRA Doc. #392, re: cancellation 4-11-77
Chlorobenzilate[b]	12-83	Animal, 2 species	5-26-76	7-11-78	12-13-79	—	Canceled all uses except citrus	—
Coal tar and creosote								
Daminozide (Alar)[b c]	6-1-84	Animal, 2 species	8-18-84	9-12-85	2-11-86	—	Labeling requirement	—
DBCP	—	Animal, 2 species— rats, mice	9-22-77	1-9-85	1-8-77 susp. some uses; 7-18-79 susp. all uses; 11-9-79 susp. all uses— withdrawn 3-31-81; 1-12-84 canc. all uses	3-31-81 voluntary canc. all uses except pineapple	Partial susp. 11-3-77; all uses except pineapple susp. 10-29-79; all uses canc. 1-9-85	Shell Oil Co. et al. FIFRA Doc. #401 & 485
DDT/DDD	—	Animal, 1 species— mice	1969, (USDA review)	—	1969, some uses; 7-7-72, all uses	—	—	Stevens Industries et al., FIFRA Doc. # 63; EDF v. EPA, 439 F.2d 584 (DC Cir. 1971)
Dinitramine	—	Animal	—	—	9-28-82	—	—	—
EDB	—	Animal, 2 species	2-14-77	9-28-83	9-28-83 pending hearing; 2-3-84 susp. grain use	—	Revoked exemption from tolerance; 2-22-84 proposed tolerance grain, citrus papaya, 3-6-84	Vulcan Chem. Co. et al., FIFRA Doc. # 503, re: cancellation: NCAMP et al v. EPA, No. 86-1114 (DC Cir. filed 2-14-86), re: tolerance

continued on next page

Table 3-13.—FIFRA[a]
I. Canceled and Restricted Carcinogenic Pesticides—Continued

Substance	Registration standard	Type of evidence	Special review initiated	Proposed determination	Notice of intent to cancel/suspend	Voluntary cancellation/ some or all uses	Comments/other regulatory action	Court action or administrative hearing
Endrin	—	Animal, 2 species	7-27-76	11-2-78	7-25-79 most uses	10-24-84 other uses	—	—
Goal[e] (product of oxyfluorfen, contaminated with PCE)	—	Animal, 2 species	1-80	6-23-82	6-23-82	—	Limited concentration of PCE	—
Mirex	—	Animal, 2 species	—	—	3-18-71; 9-23-76 canc. all uses	—	5-3-72 labeling restrictions; 6-30-72 reinstated registrations under certain conditions; 4-4-73 modified ban on aquatic application	Allied Chemical Co., FIFRA Doc. #293, 10-20-76, re: cancellation—2-12-76 registrant submitted phaseout plan to settle hearing; McGill v. EPA, 593 F.2d 631 (5th Cir. 1979)
Pentachlorophenol[e] (contaminated with Dioxin)	—	Animal	10-18-78	7-13-84 wood use; 12-12-84 non wood uses	7-13-84 wood uses	—	Labeling & restricted use requirements	In re: Chapman Co. et al., FIFRA Doc. #529, 7-15-84, re: cancellation
Pronamide (Kerb)	4-86	Animal. 1 species—mice	5-20-77	1-15-79	10-26-79	—	Restricted use & labeling required	—
Toxaphene[b c]	—	Animal, 5 species—rats, mice, dogs, monkeys, hamsters	5-25-77	11-29-82	11-29-82	—	Canceled for most uses	—
2,4,5-T/Silvex	—	Animal, 2 species	4-21-78	12-13-79	3-15-79 susp. some uses; 1-2-85 susp. all uses	—	—	In re: Dow Chemical Co., et al., FIFRA Doc. #409/410, re: emergency suspension, FIFRA Doc. #415, re: cancellation, 3-1-79; Dow Chemical Co. v. Blum, 469 F.Supp. 892 (E.D. Mich. 1979)

[a]These tables contain only active pesticide ingredients. In addition to the listed substances, EPA has identified 28 inert ingredients of concern for potential carcinogenicity among 55 inerts of concern for inherent toxicity (as of 6-19-85). The agency is evaluating these inerts to determine the extent of use and exposure, whether they are essential, whether safer alternatives are available, and what regulatory actions would be appropriate. The 28 inert ingredients of carcinogenic concern are—aniline; diethylhexylphthalate (DEHP); asbestos; benzene; betabutyrolactone; cadmium compounds; carbon tetrachloride; chlorobenzene; chloroform; 1,2-dichloropropane; dimethyl 1,1 hydrazine 1,2; dioxane; epichlorohydrin; ethylene dichloride; ethylene thiourea; formaldehyde; hexachlorophene; hydrazine; isophorone; methylene chloride; 2-nitro-propane; perchloroethylene; phenylphenol; propylene oxide; rhodamine; thiourea; 1,1,1-tri-chloroethane (methyl chloroform); trichloroethylene.

[b]Regulated under Sec. 408 FDCA. Registered for food uses—tolerances established for pesticide uses on raw agricultural commodities.

[c]Regulated under Delaney Clause of FDCA Sec. 409(c)(3)(A). A food additive tolerance is required when the pesticide residue level in processed food is greater than the tolerance level for the raw agricultural commodity but may not be established for carcinogenic pesticides.

[d]According to EPA, these compounds do not have oncogenic effects, but they have a metabolite that has shown oncogenic effects in animals.

[e]According to EPA, these compounds do not have oncogenic effects, but they contain a contaminant that has shown oncogenic effects in animals.

SOURCE: EPA response to OTA request and cited Federal Register notices.

Table 3-14.—FIFRA[a]
II. Voluntary Cancellations of Carcinogenic Pesticides

Substance	Registration standard	Type of evidence	Special review initiated	Proposed determination	Notice of intent to cancel/suspend	Voluntary cancellation/ some or all uses	Comments/other regulatory action	Court action or administrative hearing
Acrylonitrile	—	Animal	—	—	—	9-1-76	3 products	—
Benzene	—	Human	—	—	—	7-31-85	All products	—
BHC	—	Animal, 2 species	10-19-76	10-19-76	—	10-19-76, manufacturing only, began to import	7-21-78 amended registration to replace non-gamma isomer content with Lindane	—
Carbon tetrachloride	—	Animal, 2 species	10-15-80	—	—	9-5-83	—	—
Chloranil	—	Animal	—	—	—	1-19-77	—	—
Erbon	—	Animal	—	—	—	10-4-80	—	—
Maleic hydrazide[b c]	—	Animal, 1 species—mice	10-28-77	6-28-82 terminated	—	11-81 some uses	Returned to registration process	—
Monuron	6-83	Animal	—	—	—	8-16-77	Tolerance revoked in 1973	—
Nitrofen (TOK)	—	Animal	—	—	—	9-15-83	—	—
OMPA	—	Animal	—	—	—	5-28-76	—	—
Perthane	—	Animal	—	—	—	6-20-80	—	—
Safrole	—	Animal	—	—	—	2-25-77	—	—
Strobane	—	Animal, 1 species	—	—	—	6-28-76	All products	—
2,4,5-Trichlorophenol	—	Animal	9-15-78	12-31-85	—	All uses, no date	—	—
Trysben	—	Animal	—	—	—	2-9-78	—	—

[a]These tables contain only active pesticide ingredients. In addition to the listed substances, EPA has identified 28 inert ingredients of concern for potential carcinogenicity among 55 inerts of concern for inherent toxicity (as of 6-19-85). The agency is evaluating these inerts to determine the extent of use and exposure, whether they are essential, whether safer alternatives are available, and what regulatory actions would be appropriate. The 28 inert ingredients of carcinogenic concern are—aniline; diethylhexylphthalate (DEHP); asbestos; benzene; betabutyrolactone; cadmium compounds; carbon tetrachloride; chlorobenzene; chloroform; 1,2-dichloropropane; dimethyl 1,1 hydrazine 1,2; dioxane; epichlorohydrin; ethylene dichloride; ethylene thiourea; formaldehyde; hexachlorophene; hydrazine; isophorone; methylene chloride; 2-nitro-propane; perchloroethylene; phenylphenol; propylene oxide; rhodamine; thiourea; 1,1,1-tri-chloroethane (methyl chloroform); trichloroethylene.

[b]Regulated under Sec.408 FDCA. Registered for food uses—tolerances established for pesticide uses on raw agricultural commodities.

[c]Regulated under Delaney Clause of FDCA Sec. 409(c)(3)(A). A food additive tolerance is required when the pesticide residue level in processed food is greater than the tolerance level for the raw agricultural commodity but may not be established for carcinogenic pesticides.

SOURCE: EPA response to OTA Request and cited *Federal Register* notices.

Table 3-15.—FIFRA[a]
III. Carcinogenic Pesticides in Special Review
(A. Presently under review)

Substance	Registration standard	Type of evidence	Special review initiated	Proposed determination	Notice of intent to cancel/suspend	Voluntary cancellation/ some or all uses	Comments/other regulatory action	Court action or administrative hearing
Alachlor[b]	11-21-84	Animal, 2 species	1-9-85	—	—	—	Labeling requirement; restricted use	—
Amitrole	3-30-84	Animal	6-15-84	—	—	—	Labeling requirement; restricted use	—
Cadmium	—	Animal, human	10-26-77	—	—	—	—	—
Captafol[b]	10-84	Animal, 2 species	12-9-84	—	—	—	Labeling requirement; restricted use	—
Captan[b c]	3-86	Animal, 2 species	8-18-80	6-21-85 proposed de-termination	—	—	—	—
Ethylene oxide[b]	—	Animal—insufficient but positive short-term test (in 1978)	1-27-78	—	—	—	—	—
Kelthane (Dicofol)[b] (contains DDT)	12-83	Animal, 1 species—mice (see DDT)	3-21-84	10-10-84, proposed canc.	—	—	—	—
Linuron[b]	7-84	Animal, 2 species	9-26-84	—	—	—	Labeling requirement	—

[a]These tables contain only active pesticide ingredients. In addition to the listed substances, EPA has identified 28 inert ingredients of concern for potential carcinogenicity among 55 inerts of concern for inherent toxicity (as of 6-19-85). The agency is evaluating these inerts to determine the extent of use and exposure, whether they are essential, whether safer alternatives are available, and what regulatory actions would be appropriate. The 28 inert ingredients of carcinogenic concern are—aniline; diethylhexylphthalate (DEHP); asbestos; benzene; betabutyrolactone; cadmium compounds; carbon tetrachloride; chlorobenzene; chloroform; 1,2-dichloropropane; dimethyl 1,1 hydrazine 1,2; dioxane; epichlorohydrin; ethylene dichloride; ethylene thiourea; formaldehyde; hexachlorophene; hydrazine; isophorone; methylene chloride; 2-nitro-propane; perchloroethylene; phenylphenol; propylene oxide; rhodamine; thiourea; 1,1,1-tri-chloroethane (methyl chloroform); trichloroethylene.
[b]Registered under Sec. 408 FDCA. Registered for food uses—tolerances established for pesticide uses on raw agricultural commodities.
[c]Regulated under Delaney Clause of FDCA Sec. 409(c)(3)(A). A food additive tolerance is required when the pesticide residue level in processed food is greater than the tolerance level for the raw agricultural commodity but may not be established for carcinogenic pesticides.

SOURCE: EPA response to OTA request and cited *Federal Register* notices.

Table 3-15.—FIFRA[a]
III. Carcinogenic Pesticides in Special Review
(B. Special review complete but substance not suspended or canceled)

Substance	Registration standard	Type of evidence	Special review initiated	Proposed determination	Notice of intent to cancel/suspend	Voluntary cancellation/ some or all uses	Comments/other regulatory action	Court action or administrative hearing
Benomyl[b][c]	4-86	Animal, 1 species—mice	12-6-77	10-20-82	—	—	Labeling requirements; restrictions on conditions of application	—
Chloroform (trichloromethane)	12-82	Animal	4-6-76	12-82 terminated SR	—	—	More data required; exposures reduced with label changes	—
Diallate[b]	3-83	Animal, 2 species	5-31-77	6-23-82	—	—	Labeling requirements, training & protective clothing	—
Dimethoate	3-31-83	Animal, 2 species	9-22-77	1-19-81	—	—	—	—
EBDCs[b][c]	—	Animal, 2 species	8-10-77	10-27-82	—	—	More data required; labeling requirement	—
Ethalfluralin[b]	—	Animal, 1 species	1-4-84	1-4-84	—	—	Tolerances issued; protective clothing required	—
Lindane[b]	9-85	Animal, 1 species	2-18-77	9-30-83	—	—	More data required; restricted use	In re: Happy Jack Inc. & Continental Chemist Corp. FIFRA Doc. #524 & 526, 10-19-83, re: cancellation
PCNB[b]	—	Animal, insuff.	10-13-77	4-28-82 terminated SR	—	—	4-19-82, negotiated agreement to reduce exposures	—
Thiophonate methyl[b][d]	5-86	Animal, 1 species—mice	12-7-77	10-20-86	—	—	Label precautions	—
Trifluralin[b][c] (contaminated with n-nitrosamine)	—	Animal	8-30-79	8-4-82	—	—	Limit set for amount of n-nitrosamine	—

[a]These tables contain only active pesticide ingredients. In addition to the listed substances, EPA has identified 28 inert ingredients of concern for potential carcinogenicity among 55 inerts of concern for inherent toxicity (as of 6-19-85). The agency is evaluating these inerts to determine the extent of use and exposure, whether they are essential, whether safer alternatives are available, and what regulatory actions would be appropriate. The 28 inert ingredients of carcinogenic concern are—aniline; diethylhexylphthalate (DEHP); asbestos; benzene; betabutyrolactone; cadmium compounds; carbon tetrachloride; chlorobenzene; chloroform; 1,2-dichloropropane; dimethyl 1,1 hydrazine 1,2; dioxane; epichlorohydrin; ethylene dichloride; ethylene thiourea; formaldehyde; hexachlorophene; hydrazine; isophorone; methylene chloride; 2-nitro-propane; perchloroethylene; phenylphenol; propylene oxide; rhodamine; thiourea; 1,1,1-trichloroethane (methyl chloroform); trichloroethylene.

[b]Regulated under Sec.408 FDCA. Registered for food uses—tolerances established for pesticide uses on raw agricultural commodities.

[c]Regulated under Delaney Clause of FDCA Sec. 409(c)(3)(A). A food additive tolerance is required when the pesticide residue level in processed food is greater than the tolerance level for the raw agricultural commodity but may not be established for carcinogenic pesticides.

[d]According to EPA, these compounds do not have oncogenic effects, but they have a metabolite that has shown oncogenic effects in animals.

SOURCE: EPA response to OTA request and cited Federal Register notices.

Table 3-16.—FIFRA[a]
IV. Pesticides Identified as Carcinogenic But Not Reviewed or Canceled

Substance	Registration standard	Type of evidence	Comments/other regulatory action
Acephate[b c]	1-86 (pub. draft)	Animal	—
Acetochlor	—	—	Under review for oncogenic potential
Acifluorfen[b]	—	—	No action taken because EPA determined that exposure/risk from use patterns is below level for concern
Amdro[b]	—	—	No action taken because EPA determined that exposure/risk from use patterns is below level for concern
Asulam[b]...............	Under development	(Potential carcinogen)	—
Azinphos methyl (Guthion)[b]	Under development	(Potential carcinogen)	—
Chlordimeform[b c]	1-86 (pub. draft)	Animal	—
Chlorothalonil[b]	10-84	Animal	More data required; labeling requirement
Cypermethrin[b]	—	—	EPA determined that weight of evidence for carcinogenicity does not support regulatory action
Cyromazine (Larvadex)[b c] ..	—	—	No action taken because EPA determined that exposure/risk from use patterns is below level for concern and that weight of evidence for carcinogenicity does not support regulatory action
Diclofop-methyl (Hoelon)[b] .	—	—	No action taken because EPA determined that exposure/risk from use patterns is below level for concern
Dimethipine (Harvade)[b]	—	—	No action taken because EPA determined that exposure/risk from use patterns is below level for concern and that weight of evidence for carcinogenicity does not support regulatory action
Fenarimol	—	—	Under review for oncogenic potential
Folpet[b]	Under development	(Potential carcinogen)	—
Fosetyl al (Aliette)[b]	6-30-83	Animal	—
Glyphosate[b]	6-86	Animal	—
Methanearsonic acid[b c] ...	—	invalid study	EPA determined that weight of evidence for carcinogenicity does not support regulatory action
Methomyl	9-81	Animal	Manufacturing process restrictions; labeling requirement; restricted use
Metolachlor[b]..............	9-80	Animal	—
Oryzalin[b c]	Pub. draft	Animal	
Oxadiazon[b]	—	—	No action taken because EPA determined that exposure/risk from use patterns is below level for concern
Paraquat[b c]	4-86	Animal	Pre-SR completed 6-29-82, returned to registration process
Parathion[b c]	Under development	(Potential carcinogen)	—
Permethrin[b]	—	—	EPA determined that weight of evidence for carcinogenicity does not support regulatory action
Terbutryn[b]..............	Under development	(Potential carcinogen)	Pre-SR agreement 5-12-82; label changes to reduce applicator exposure
Thiodicarb (Larvin)[b c]	—	—	EPA determined that weight of evidence for carcinogenicity does not support regulatory action

[a]These tables contain only active pesticide ingredients. In addition to the listed substances, EPA has identified 28 inert ingredients of concern for potential carcinogenicity among 55 inerts of concern for inherent toxicity (as of 6-19-85). The agency is evaluating these inerts to determine the extent of use and exposure, whether they are essential, whether safer alternatives are available, and what regulatory actions would be appropriate. The 28 inert ingredients of carcinogenic concern are—aniline; diethylhexylphthalate (DEHP); asbestos; benzene; betabutyrolactone; cadmium compounds; carbon tetrachloride; chlorobenzene; chloroform; 1,2-dichloropropane; dimethyl 1,1 hydrazine 1,2; dioxane; epichlorohydrin; ethylene dichloride; ethylene thiourea; formaldehyde; hexachlorophene; hydrazine; isophorone; methylene chloride; 2-nitropropane; perchloroethylene; phenylphenol; propylene oxide; rhodamine; thiourea; 1,1,1-tri-chloroethane (methyl chloroform); trichloroethylene.
[b]Regulated under Sec.408 FDCA. Registered for food uses—tolerances established for pesticide uses on raw agricultural commodities.
[c]Regulated under Delaney Clause of FDCA Sec. 409(c)(3)(A). A food additive tolerance is required when the pesticide residue level in processed food is greater than the tolerance level for the raw agricultural commodity but may not be established for carcinogenic pesticides.
[d]According to EPA, these compounds do not have oncogenic effects, but they have a metabolite that has shown oncogenic effects in animals.
[e]According to EPA, these compounds do not have oncogenic effects, but they contain a contaminant that has shown oncogenic effects in animals.

SOURCES: EPA response to OTA request for information; EPA correspondence with Representative Henry Waxman (10-2-85); EPA Office of Pesticide Programs, March 1986, "Report on the Status of Chemicals in the Special Review Program, Registration Standards Program, and Data Call-in Program;" GAO, April 1986, "Pesticides—EPA's Formidable Task to Assess to Regulate their Risks" GAO/RCED-86-125; EPA, "Suspended, Cancelled & Restricted Pesticides" Third Revision, January 1985; *Federal Register* notices for each action provided information on the type of evidence used.

Other Pending Issues

In addition to its focus on active ingredients, EPA regards about 55 inert ingredients as "high concern," with 28 of these showing carcinogenic effects. Another 51 substances have "suspected toxicity" while between 800 and 900 other inert ingredients have insufficient health and safety data (130,203). In April 1987, EPA issued a policy statement announcing the intent to encourage the use of the least toxic inert ingredient available, require data necessary to determine the conditions under which it may safely be used, and hold hearings to determine whether the use of certain inert ingredients should continue to be permitted. EPA also intends to reclassify some of them as active ingredients (294).

EPA has inadequate information on a number of nonagricultural pesticides to determine health risks, and these pesticides have not been reassessed by the current standards (202). After reviewing the status of EPA's chronic toxicity data for 50 chemicals, selected because they are used in large quantities, the General Accounting Office found that "EPA had done preliminary assessments for 18 of the 50 nonagricultural chemicals [pesticides not used on crops] and found that it did not have enough chronic toxicity data on 17 of the 18 chemicals to complete the assessments." A tiered approach (explained below in the section on TSCA) in obtaining chronic test data on nonagricultural pesticides is currently taking place through the data call-in (21).

Finally, EPA has invalid or fraudulent health data on 36 pesticides, including 35 food use pesticides. Some or all of these products may have been tested by Industrial Biotest Laboratories, which submitted invalid, and in some cases fraudulent, test data in the mid-1970s (203,206). These data are being replaced through the data call-in and reregistration process.

EPA REGULATORY ACTIONS UNDER TSCA

The Toxic Substances Control Act was enacted in 1976. With TSCA Congress established the policy that chemical manufacturers are responsible for developing data about the health and environmental effects of their chemicals, that the government regulates chemical substances that pose unreasonable risks of injury to health or the environment, and that regulatory efforts should not unduly impede industrial innovation. Singled out for special concern were substances that present or will present significant risks of cancer, gene mutations, or birth defects.

EPA actions under TSCA cover both new and existing chemicals. For new chemicals, the principal focus is a premanufacturing review. After review of available information, EPA can request or require additional toxicity testing, can require certain workplace practices and controls, and can require that the manufacturer notify EPA before putting the chemical to a "significant new use." For existing chemicals, EPA can require testing for toxicity and environmental effects, designate the chemical for priority review, require that "significant new uses" be reported, require manufacturers, importers, and processors to report production and use information, or require that they submit copies and lists of unpublished health and safety studies to EPA. EPA can also issue regulations restricting or banning the production of a chemical or limiting its uses.

New Chemicals

Under section 5 of TSCA, chemical manufacturers must submit a premanufacturing notice (PMN) for any "new chemical." "New chemicals" are those not found in the TSCA inventory of chemicals in commerce. At the end of EPA's review, any one of four actions are possible:

1. The substance described in the PMN can be manufactured without restriction.
2. The substance can be manufactured for the uses described in the PMN, but the Agency can require that it be notified if manufacture for a significant new use is considered. If EPA decides that a potential new use of the substance might be associated with an unreasonable health or environmental risk, it

can by a separate rulemaking procedure issue a Significant New Use Rule (SNUR) to restrict the manufacture or distribution of the substance (section 5(a)(2)).

3. The manufacture, processing, distribution, use or disposal of the new substance can be regulated pending the development of additional information about the substance (section 5(e)).

4. The manufacture, processing, distribution, use, or disposal of the new substance can be regulated because it presents or will present an unreasonable risk (section 5(f)) (222).

From July 1979, when the program started, until September 1986, EPA received 7,356 valid PMNs (see table 3-17). EPA decided that no further action was necessary for 5,761 of these, or about 80 percent. Of the remaining chemicals that raised EPA concerns, 523 were subject to some kind of action. EPA concerns led to the manufacturers' agreeing to voluntary testing in 64 cases, voluntary control actions such as the use of personal protective equipment in 33 cases, and the complete withdrawal of the PMN in 139 cases. Thus, in 236 cases or about 45 percent, the threat of EPA action lead to informal, "voluntary" responses by the manufacturers. The other half of the time the actions were more formal: 271 chemicals subject to consent orders under which the manufacturers agree to controls for worker exposure, or restrictions on production use or disposal until testing is done (section 5(e)); 12 unilateral orders under which EPA imposes restrictions or bans pending until testing is completed (section 5(e)); and 4 chemicals subject to immediately effective proposed rules setting permanent requirements for the production of these chemicals (sections 5(f) and 6(a)). EPA has received notices of commencement of manufacture for 3,678 of the 7,356 chemicals.

In 1983, OTA prepared a background paper, *The Information Content of Premanufacture Notices*, describing the nature and extent of information reported on PMNs submitted during a 2-year period from 1979 to 1981, and on PMNs submitted for June of 1982. That study found that about half the submitted PMNs reported no toxicity information and "only 17 percent of PMNs have any test information about the likelihood of

Table 3-17.—TSCA: New Chemicals, Section 5 PMN Reviews

	Total (1979-86)
Valid PMNs received	7,356
PMNs requiring no further action	5,761
Some action	523
Voluntary testing	64
Voluntary control actions	33
PMNs voluntarily withdrawn	139
Section 5(e) consent orders	271
Unilateral 5(e) orders	12
Section 5(f) rules	4
New chemicals subject to SNUR:	
Proposed	55
Final	15

SOURCE: EPA response to OTA request for information.

the substance's causing cancer, birth defects or mutations—three biological effects that were singled out for special concern in TSCA" (222). Because long-term experiments in whole animals are expensive, the tests conducted for all these PMNs were short-term mutagenicity tests.

The submitters of nine of the PMNs examined by OTA did not begin manufacture because of EPA actions. Six of the substances were phthalates, of special concern because of a recent National Cancer Institute study showing some phthalates to be carcinogenic (222). Two of the remaining three were benzidine dyes, which have long been associated with human carcinogenicity. The submission of these PMNs shows that substances of demonstrated toxicity, even substances closely related to known carcinogens, are still considered for possible manufacture and use.

Because most PMNs do not contain any toxicity test information, EPA is forced to rely on structure-activity relationships in attempting to predict, from the chemical structure of the substance, the hazards it poses. In this case, EPA uses computerized databases on chemical structures to identify related chemicals or analogs. EPA then searches toxicity databases for information on the analogs. EPA staff admit that the identification of analogs is a "rather subjective process," and that the final assessments concerning a new chemical rely on the "knowledge and professional judgments" of the staff performing the evaluation (11).

Existing Chemicals—Obtaining Additional Data

Section 4(e) of TSCA created an Interagency Testing Committee (ITC) to review existing chemicals and make recommendations to EPA about testing chemicals for health and environmental effects. EPA is to review these recommendations and then decide whether testing should or should not be required. TSCA provides that ITC shall give "priority attention" to chemicals that cause or are suspected of causing cancer, mutations, or birth defects. ITC can simply commend a substance or mixture to EPA's attention, or ITC can "designate" it, in which case EPA has 12 months to initiate a proceeding to require testing or publish reasons for deciding that testing is unnecessary.

The recommendations are in the form of a list of chemical substances and mixtures. This "Priority List" of "designated" chemicals awaiting EPA action at any one time cannot exceed 50.

By statute, ITC consists of representatives from EPA, OSHA, NIOSH, the National Institute of Environmental Health Sciences, the National Cancer Institute, the National Science Foundation, the Council on Environmental Quality, and the Department of Commerce. In addition, seven other agencies belong to ITC as "liaison members": CPSC, FDA, NTP, the U.S. Department of Agriculture, the U.S. Department of Defense, the U.S. Department of the Interior, and the National Library of Medicine. They participate in the reviews of chemicals and the meetings of ITC, although they do not vote to select the chair of ITC or on the contents of the final reports.

TSCA specifies a number of factors for ITC to consider in designating substances and mixtures:

- the quantity that is or will be manufactured,
- the quantity that enters or will enter the environment,
- the extent of occupational exposures (both numbers of workers and durations of exposure),
- the number of people exposed (presumably including people exposed outside the workplace),
- the similarity to other substances known to present unreasonable risks to health or the environment,
- the existence of data on health and environmental effects,
- the extent to which testing may result in data useful for predicting effects on health or the environment, and
- the availability of testing facilities and personnel (sec. 4 (e)).

Using a periodic process called a "scoring exercise," ITC narrows down the universe of chemicals in commerce (over 50,000) in several steps to select a manageable number (40-50) for detailed review by ITC members. The most recent scoring exercise—the sixth, completed in January 1987—began with a list of approximately 20,000 organic chemicals. The first step was to remove chemicals previously reviewed by ITC, as well as common metabolites, chemicals "generally recognized as safe (GRAS)," food additives, drugs, pesticides, certain regulated chemicals, and widely occurring natural products. The remaining chemicals on the master list were then arrayed in different lists based on potential health effects, potential ecological effects, presence in the environment, potential high workplace exposures, and potential high consumer exposure. Each chemical then received a "score" based on the frequency of occurrence on these lists. Chemicals not produced or imported in significant quantities were removed, and the remaining substances were scored for exposure potential, health effects, and ecological effects. From these scores, chemicals were selected for more detailed review by ITC. More detailed information was prepared for these chemicals, and then ITC made its recommendations to EPA (335). A major problem in this process is the difficulty of obtaining current data on the volumes of domestic production and imports (16).

ITC transmits an updated list of recommended substances to EPA at least every 6 months. These reports are published in the *Federal Register* for public comment. The first ITC report was published in October 1977; report number 19 came out in November 1986.

EPA's responses to the ITC lists have generated concern, both for the length of time they have taken to develop responses and for the procedures chosen for obtaining the necessary test data. During the first years of the program, from the first ITC report in 1977 until the sixth in November 1980, EPA never responded within the statutory deadline of 12 months. Because of this failure to act, NRDC sued EPA. In a decision in early 1980, a New York District Court ruled that EPA had failed to fulfill its responsibilities to act on ITC-designated chemicals. As a result, for reports 7 through 16, EPA has responded within the deadline for all designated chemicals (195).

At the time of the initial lawsuit, EPA also developed a new procedure to negotiate agreements with chemical manufacturers concerning the conduct of toxicity tests. These negotiated testing agreements were designed to avoid what EPA viewed as the difficulties of the rulemaking procedures mandated by TSCA. Six hundred studies were produced under 22 of these agreements (341).

In 1983, EPA was sued a second time by NRDC, on the grounds that TSCA did not provide for these voluntary negotiated testing agreements and that such agreements did not trigger certain other provisions of the law. The court agreed and ordered EPA to reconsider its decisions on several chemicals. In 1985, NRDC and the Chemical Manufacturers Association (CMA) approached EPA concerning these issues. The result was a new set of EPA procedures under which EPA attempts to negotiate a consensus among all interested parties on testing needs. If negotiation fails, EPA will develop a test rule. In either case, EPA appears committed to the prompt resolution of these issues, i.e. meeting the statutory deadline (338).

In 1985, ITC created a "recommended with intent-to-designate" category for chemicals it intends to designate in the future. This new category enables EPA to begin gathering production and use data, and information on unpublished health and safety studies (see discussion below), before the statutory 1-year clock starts ticking. EPA regards this category as essential for using the negotiated consent agreement process since negotiations require an additional 10 weeks (341). Any information obtained in this way can also be reviewed by ITC before making the final decision to "designate" the chemical.

Counting the total number of chemicals recommended by ITC and considered by EPA is difficult, because a single recommendation may cover a single chemical or a group of chemicals. In the 19 reports published through September 1986, ITC has recommended testing for 101 chemicals or groups of chemicals. Of these, 94 were "designated."[15] Counting all of the discrete chemicals in the various groups of chemicals under consideration yields a total of 389 chemicals (17,18).

In all, EPA has issued 20 final rules on testing for various health and environmental effects (not necessarily carcinogenicity). Another 24 proposals for test rules are pending, and 5 chemicals or chemical groups have been the subjects of ANPRMs. Finally, for 51 chemicals or chemical groups, EPA has decided that testing is not needed ("decisions not to test"), and for 1 category, EPA returned the category to ITC (195).

Table 3-18 presents the 17 chemicals from the ITC reports for which EPA has required, proposed, or negotiated carcinogenicity testing. These include 11 chemicals to be tested in carcinogenicity bioassays and 6 tested in "tiered testing," which involves conducting a battery of short-term tests and then evaluating the results before deciding whether to require a long-term bioassay.

EPA can require testing under section 4 for chemicals in addition to the ones recommended by ITC, but this has been less frequent than consideration of ITC-recommended chemicals. Recent proposals concerning 1,1-dichloroethylene, diethylene glycolbutyl ether, 2-ethylhexanol, and a group of 73 substances found at hazardous waste sites are the exceptions. The latter two were nominated by other program offices at EPA, the Air Office and the Office of Solid Waste and Emergency Response, respectively. This may mark the beginning of a new trend, but thus far the testing agenda under TSCA has been set by ITC and its recommendations. According to EPA this is because, under section 4(e), ITC designations have top priority, and, also because of resource limitations (341).

---

[15]In the 19th report, two additional chemicals were recommended with "intent-to-designate." The 94 "designated" chemicals include three that had been similarly recommended with "intent-to-designate" in reports 17 and 18.

Table 3-18.—TSCA: Existing Chemicals;
Testing Required To Determine Carcinogenicity

Name	ITC nomination date of Federal publication	Negotiated testing agreement	NPR	Final	Type*
Alkyl phthalates	10-12-77	1-5-82	a	—	C
Benzyl butyl phthalate	11-25-80	1-5-82	9-6-85	Pending[b]	C
Antimony trioxide	6-1-79	9-2-83	c	—	C
Aryl phosphates	4-19-78	—	12-19-83 (ANPR)		T
2-(Butoxyethyl) ethyl acetate	12-14-83	—	8-4-86	Pending	C
Chlorinated benzenes	10-12-77	—	1-13-84	7-8-86	C
4-Chlorobenzotrifluoride	2-5-82	7-18-83	d	—	T
2-Chlorotoluene	5-22-81	4-28-82	d	—	T
Cresols	10-12-77	—	7-11-83	4-28-86	C
Cumeme	11-29-84	—	11-6-85	Pending	C
Diethylene triamine	5-22-81	—	4-29-82	5-23-85	C
Ethyl toluenes	5-25-82	—	5-23-83	5-17-85	T
Fluoroalkenes	11-25-80	6-4-84	11-6-85	Pending	C
Glycidol & derivatives	10-30-78	—	12-30-83 (ANPR)		C
Mesityl oxide	6-1-79	—	7-5-83	12-20-85	T
Oleylamine	12-14-83	—	11-19-84	Pending	T
Phenylenediamines	5-28-80	—	1-6-86	Pending**	C

*Type C = carcinogenicity testing in 2-year bioassays; type T = tiered testing.
**Decision not to test certain phenylenediamines. Others still under review.
[a]Phase I negotiated Testing Program completed. Data under review to determine further testing needs.
[b]Adequate environmental data submitted. Health testing needs under review. See footnote a.
[c]Industry currently performing carcinogenicity bioassay under Negotiated Testing Agreement.
[d]Data adequate. Carcinogenicity testing not triggered.

SOURCE: EPA response to OTA request for information.

EPA may issue rules, under section 8(a), to require manufacturers, importers, and processors to provide information on production and uses of a chemical, plant characteristics, process characteristics, environmental releases, and worker exposures. Rules adopted under section 8(d) require that manufacturers, importers, and processors submit to EPA copies and lists of unpublished health and safety studies. (This is different from the general obligation under section 8(e) to notify EPA of studies revealing any "substantial risks." A section 8(d) rule requires the submission of all studies: positive and negative, "substantial risks" or not.)

Again, ITC recommendations have dominated EPA activity under these sections of TSCA. EPA has issued 8(a) and 8(d) rules for all the substances recommended by ITC, and in fact now routinely adds the chemicals from each new ITC report to the 8(a) and 8(d) lists within a few days after receipt of the report (108). Until 1986 there had been relatively few chemicals included under 8(a) and 8(d) that were not included in ITC reports. This may now be changing. For example, in the past year, 33 chemicals nominated by the Office of

Solid Waste were added to the 8(d) list (320), and in May 1987 EPA published a final rule under section 8(d) that listed 107 chemicals nominated by the Offices of Toxic Substances, Water, and Solid Waste within EPA as well as by the CPSC (344). For section 8(a) rules in general, EPA has proposed a Comprehensive Assessment Information Rule, to establish uniform reporting requirements for chemicals that are subject to 8(a) rules. The proposal includes a list of 47 chemicals nominated by EPA's Air Office and Office of Toxic Substances, CPSC, OSHA, and NIOSH (308).

Under section 8(e) of TSCA, manufacturers, processors, and distributors of chemicals are required to notify EPA of information supporting the conclusion that a chemical "presents a substantial risk of injury to health or the environment." Since January 1977 when this requirement became effective, EPA has received over 600 such notifications. In addition, EPA also receives "for your information (FYI)" notifications which do not fit the statutory requirements of section 8(e). The section 8(e) and FYI notifications are evaluated and EPA staff identify which chemicals should receive further attention, such as the preparation of Chemical Hazard Information Profiles (CHIPs). The information provided in the 8(e) notices is made available to the interested public and is reported through the use of bulletins, published chemical status reports, and a computer database (341). Finally, EPA promulgated an 8(a) rule to update the information in the TSCA inventory of chemicals in commerce by requiring manufacturers and importers to report current data on production volume and plant site. The data in the original inventory was reported in 1977 and is out of date. Under the new rule the data must be reported every 4 years (335).

The various reporting requirements of TSCA have undoubtedly stimulated increased awareness in the chemical industry. Companies must evaluate information they obtain to determine whether it meets the notification requirements. Even if it does not, the information may be reported to EPA. To some extent, companies will voluntarily reduce exposures, conduct additional studies, change product labels and Material Safety Data Sheets (which provide information on product hazards and safe handling), and notify exposed workers, consumers, or others (341).

Existing Chemicals— Identified Hazards

For some existing chemicals, there may be enough information to determine that they cause cancer. For these chemicals, the issues are determining whether this risk of cancer is "unreasonable" and whether actions are needed to reduce or eliminate these risks. Table 3-19 lists 40 substances or groups of substances that EPA's Office of Toxic Substances has identified as carcinogenic, based on information provided to OTA. EPA has prepared hazard analyses and/or risk assessments for 25 of those substances.

This list does not cover all the substances for which TSCA evaluations, such as CHIPs, have been prepared. A CHIP provides a concise summary of available information on a chemical, including physical and chemical properties, estimates of exposure and environmental fate, health effects, environmental effects, and existing standards and recommendations. As of March 1987, CHIPs have been prepared for 93 different chemicals with carcinogenic concerns (336). CHIPs are initiated when EPA receives information on potential hazard from NTP bioassays, published scientific papers, or submissions to EPA under section 8(e). Using the information in a CHIP, EPA's Office of Toxic Substances may decide to develop more information on exposures, refer the chemical to other agencies or EPA programs, or begin action under TSCA.

EPA provided OTA with limited information about the kinds of evidence used for these identifications and risk assessments; most have relied on animal data. From the list in table 3-19, only asbestos and certain aromatic amines are known to be carcinogenic from human studies. The remainder are based on animal data, including 15 tested in NCI/NTP bioassays.

Regulatory actions for this group of existing chemicals consist of 4 designations under section 4(f) for an expedited review (4,4'-methylene dianiline, 1,3-butadiene, formaldehyde, and methylene chloride), 10 proposed or final SNURs for existing chemicals, and proposed section 6 regulatory actions for 4 substances.

The first section 4(f) designation, for 4,4'-methylene dianiline, occurred in 1983. Formaldehyde, designated in May 1984, had been considered in 1982 for designation. John Todhunter, the first director of the Office of Pesticides and Toxic Substances in the Reagan Administration, however, decided against designation—a controversial decision at the time. Following a lawsuit by NRDC, EPA formally acknowledged that section 4(f) applied to formaldehyde and announced a regulatory investigation in an ANPRM (158).

EPA response to its evaluation of chemicals under section 4(f) has been to refer chemical exposures limited to the workplace to OSHA under a provision of TSCA (sec. 9) that provides for referrals if EPA believes that OSHA, may be able to address the hazard. This has happened for acetaldehyde, acrylamide, chloromethane, 4,4-methylene bis (2-chloroaniline), toluenediamine, 4,4'-MDA, 1,3-butadiene, and for the occupational exposures associated with formaldehyde (see table 3-19). EPA is still investigating risks associated with nonoccupational exposure to formaldehyde in pressed wood products. The section 4(f) find-

Table 3-19.—TSCA: Existing Chemicals Identified as Carcinogens

Name	Hazard/risk assessment	Action
Acetaldehyde	Yes	Section 9 referral to OSHA
Acrylamide	Yes	Section 9 referral to OSHA; section 6 designation
Acrylates		SNUR for new acrylates under consideration
II-Aminoundecanoic acid		SNUR issued
Aniline	Yes	ANPR for testing
Aromatic amines		
Asbestos	Yes	Section 6: rules on asbestos removal; proposed ban/ phase-down
1,3-Butadiene[a]	Yes	Section 4(f) designation; section 9 referral to OSHA
Clarified slurry oil		
Chlorinated paraffins[a]	Yes	
Chloromethane	Yes	Section 9 referral to OSHA
3-Chloro-2-methylpropene (CMP)[a]	Yes	
C.I. Disperse Yellow 3[a]		
D & C Red 9[a]		
1,4-Dichloro-2-butene (DCB)	Yes	EPA has taken no action because use is in closed systems
1,4-Dichlorobenzene[a]	Yes	
Dihydro safrole		
Epichlorohydrin		Proposed SNUR
Ethanolamines & metal working fluids	Yes	Advisory warnings; section 6 proposal
Etheylenediaminetetra methylenephosphonic acid		
Formaldehyde	Yes	Section 4(f) designation; termination of investigation & section 9 referral to OSHA for occupational exposures
Glycol ethers	Yes	Section 9 referral to OSHA
Hexachlorobenzene (HCB)	Yes	SNUR & 8(a) rules issued
Hexachloronorbornadiene (HexBCH)	Yes	SNUR & 8(a) rules issued
Hexafluoropropylene oxide		Proposed SNUR
Hexamethyl phosphoramide (HMPA)		SNUR issued
4,4-Methylenebis (2-chloroaniline)	Yes	SNUR proposed, section 8(a) rule issued. Section 9 referral to OSHA
Methylene chloride[a]	Yes	Section 4(f) designation
4,4'-Methylenedianiline[a]	Yes	Section 4(f) designation; section 9 referral to OSHA
Naptha solvent		
2-Nitropropane	Yes	
Paradichlorobenzene	Yes	Advisory warning
Pentachloroethane (PCE)[a]		SNUR issued; 8(d) rule
Perchloroethylene (perc)[a]	Yes	Decision not to designate under 4(f)
Phthalates		
DEHP[a]	Yes	
Polychlorinated biphenyls[b]		Section 6 rules
Propylene oxide[a]	Yes	EPA's quantitative risk assessment indicates risk is low
Toluenediamine	Yes	Advisory warning, section 9 referral to OSHA
Urethane		SNUR issued
Vinylcyclohexene[a]		

[a]Tested by NCI/NTP.
[b]Not in "Part 2 Summary," but regulated by EPA under TSCA.

SOURCE: EPA, OTS, "Part 2: Summary Narratives of Chemical Dispositions," in EPA response to OTA request for information on substances identified as carcinogenic.

ing for methylene chloride initiated an interagency regulatory investigation of hazards associated with methylene chloride and five other chlorinated solvents. All four designations under section 4(f) are based on results of animal studies.

After EPA issues a SNUR, a manufacturer must notify EPA before beginning production that falls under the terms of the SNUR. If EPA receives such notice, it can take action under section 5(e) or 5(f) to impose controls or prohibit production.

SNURs can be based on concern for any of the health and environmental effects regulated under TSCA, including carcinogenicity. Table 3-21 lists the existing chemicals for which carcinogenicity

Table 3-20.—TSCA: 4(f) Reviews

	4(f) Notice of accelerated review	ANPR summarizing evidence	Final action
4,4'-Methylene dianiline	4/27/83	9/20/83	7/5/85—Section 9(a) referral to OSHA
1,3-Butadiene	1/5/84	5/15/84	10/10/85—Section 9(a) referral to OSHA
Formaldehyde	11/18/83	5/23/84	3/19/86—Announced termination of investigation concerning occupational exposures; non-occupational exposures still being investigated
Methylene chloride	5/14/85	10/17/85	Pending

SOURCE: EPA response to OTA request for information.

was one reason for the SNURs. The table shows that for existing chemicals considered to be carcinogenic, EPA began proposing SNURs in 1984, nearly 7 years after enactment of TSCA.[16] The table shows that EPA has proposed six SNURs (for eight chemicals) and issued four final SNURs on chemical substances of carcinogenic concern. For Hex-BCH it also issued a section 8(a) rule requiring reporting of production volumes. EPA issued an 8(a) rule for MBOCA instead of an SNUR. (See table 3-21.)

Section 6 provides wide-ranging authority to limit production and uses, which includes banning substances. Table 3-22 presents EPA actions concerning carcinogens under this section. Congress prohibited the manufacture of PCBs in the act itself, and required that EPA issue rules concerning their use and disposal.

EPA also issued rules in 1982 concerning identification and notification of asbestos in school buildings. In 1987, EPA proposed to require removal of this asbestos in certain circumstances. EPA has also issued rules to regulate the exposures of asbestos-removal workers who are not covered by OSHA standards. In 1986, EPA issued a proposal concerning asbestos, including proposed bans of certain asbestos applications (for which EPA has concluded there are available substitutes) and to "phase down," over 10 years the amount of asbestos that may be mined or imported. That proposal has not been issued in final form.

One other group of substances—chlorofluorocarbons (CFCs)—have been acted on under section 6. EPA banned use of CFCs as aerosol propellants and in 1980 issued an ANPRM concerning a possible limit on production and consumption of these substances for other uses, which may harm the ozone layer. That harm might have large consequences on the earth's climate, as well as increasing the amount of ultraviolet radiation reaching the earth's surface. Increased ultraviolet exposure would lead to an increase in the rate of skin cancer; thus CFCs in the atmosphere might, indirectly, be considered carcinogens. If further regulations or control of CFCs are proposed, such actions will not be proposed under TSCA, but rather, under section 157(b) of the CAA (47).

Another section 6 action that has not been issued in final form is a 1984 proposal concerning certain potentially carcinogenic compounds that may form in metalworking fluids.

Table 3-21.—TSCA: Existing Chemicals, Significant New Use Rules for Carcinogens

Name	Proposed	Final
Epichlorohydrin	1-2-87	—
Hexachloronorbornadiene (HEX-BCH)	2-22-85	11-19-85[a]
Hexafluoropropylene oxide	1-2-87	—
Hexa methylphosphoramide (HMPA)	10-10-84	3-19-86
4,4'-Methylenebis (2-chloroaniline)	4-26-85	4-18-86[b]
Pentachloroethane	3-24-86	9-9-86
Trichlorobutylene oxide	1-2-87	—
Urethane	10-10-84	3-19-86

[a]SNUR and section 8(a) rule.
[b]Issued as section 8(a) rule.

SOURCE: EPA response to OTA request for information.

---

[16]The total number of SNURs issued for carcinogenic and other concerns is 9 proposed (for 11 chemicals) and 8 final.

Table 3-22.—TSCA: Existing Chemicals;
Section 6 Actions

Substance	Regulation	ANPRM	NPRM	Final
Asbestos	Statement of policy on coordination of regulatory activities	10-17-79		
Asbestos	Asbestos in schools: identification and notification		9-17-80	5-27-82
Asbestos	Asbestos abatement projects/worker protection		7-12-85	4-25-86 2-25-87 (revised final)
Asbestos	Mining and import restrictions and manufacturing; importation and processing prohibitions (asbestos ban & phase out)		1-29-86	
CFCs*	Prohibition of several uses		5-13-77	3-17-78
CFCs*	Production restriction	10-7-80		
Metalworking fluids	Prohibition of nitrites in		1-23-84	pending
PCBs	Ban rule			5-31-79
PCBs	Exclusions, exemptions, and use authorizations for PCBs under 50 ppm		12-8-83	7-10-84
PCBs	Electrical equipment		4-22-82	8-25-82
PCBs	Use in closed and controlled waste manufacturing processes		6-8-82	10-21-82
PCBs	Amendment to use authorization for PCB railroad transformers		11-18-81	1-3-83
PCBs	Approval for PCB disposal facilities (procedural amendment)	3-30-83	11-17-83	3-30-83 7-10-84
PCBs	Use in microscopy and R&D proposed rule			
PCBs	Manufacture, processing, distribution in commerce exemptions		11-1-83	7-10-84
PCBs	Policy for compliance and enforcement of PCB storage for disposal regulation	11-17-83		11-17-83

*Not a carcinogen, but possible effects of CFCs on atmospheric ozone might increase skin cancer rates.

SOURCE: EPA response to OTA request for information.

EPA REGULATORY ACTIONS UNDER RCRA

The Resource Conservation and Recovery Act of 1976 was enacted to protect health and the environment from chemical wastes and to conserve material resources. Subtitle C of RCRA establishes a hazardous waste management system. EPA has issued regulations on the identification and listing of hazardous wastes, established recordkeeping and reporting requirements for generators, transporters, storers, and disposers of hazardous wastes, and permit requirements for treatment, storage, and disposal of hazardous waste. EPA requirements also establish a manifest system for tracking the movement of wastes from generation to disposal.

Wastes are subject to regulation under RCRA if they have the characteristics of hazardous waste, if they are listed as hazardous wastes, or if they are mixtures containing listed hazardous wastes. EPA did not include carcinogenicity and other acutely toxic effects among the characteristics of a hazardous waste, but instead regulates these substances through the listing mechanism. According to EPA, the test protocols for such characteristics are either insufficiently developed or too complex and dependent on the use of highly skilled personnel and specialized equipment to place the burden on generators of the waste (288). Among the characteristics, however, is "Extraction Procedure (EP) Toxicity." This provision specifies the maximum amount of 14 particular contaminants that may be found in a waste using a specific detection method. Most of the 14 contaminants are carcinogens.

A solid waste may be listed as a hazardous waste if it exhibits the characteristics of hazardous waste, if it is acutely hazardous, or if it contains toxic constituents listed in 40 CFR Appendix VIII. Carcinogenicity is one of the criteria for listing a chemical in Appendix VIII (107,221). In May 1980, EPA published three generic lists of wastes considered to be hazardous and subject to the RCRA Subtitle C hazardous waste management regulations (40 CFR 261.31-40 CFR 261.33).

EPA's lists contain 361 commercial chemicals and 85 industrial waste processes, with others proposed as additions (not including Appendix VIII). Where possible, EPA emphasized waste streams from commercial processes, rather than specific hazardous substances to relieve waste generators of testing burdens and uncertainties in "relating a waste containing many substances to a list of specific substances" (290). Appendix VIII includes 391 constituents of the listed commercial compounds; many of them are carcinogens. The distinction between carcinogenicity and acute toxicity or other chronic health effects was not of particular concern to EPA when the list was compiled, since all of these criteria were used as the basis for the Appendix VIII listing.

Any listed waste is subject to RCRA "proper handling" regulations unless it is "delisted." Delisting a substance requires a petition for a regulatory amendment and is subject to requirements for public notice and comment. Although the Appendix VIII list of hazardous constituents is nonregulatory, inclusion of a substance on it may provide the basis for listing a commercial chemical product on the regulatory list. Waste generators may also be required to monitor groundwater for the constituents as a condition of their permits.

RCRA was passed in 1976. Congress gave EPA 18 months from the date of passage to provide criteria for characteristics of hazardous wastes and to list hazardous wastes. EPA issued its proposed rules on these topics in December 1978. In May 1980, it issued its first final rule concerning lists and characteristics of hazardous wastes (some other regulations are to follow). Also in 1980, EPA issued final rules for "proper handling" of hazardous wastes.

EPA has had considerable difficulty in adding to the list of hazardous wastes. Since issuing its list of 361 commercial chemicals and 85 industrial waste processes as well as the 4 generic waste characteristics in 1980, EPA has added 5 additional wastes, and no new characteristics. Moreover, EPA does not know whether the existing lists cover 90 percent of potentially hazardous wastes or 10 percent. Some of the as yet unlisted wastes are highly toxic, such as certain pesticides and known carcinogens (200).

Congress, in the 1984 RCRA amendments, addressed a number of aspects of the solid waste program. Relevant to this discussion of regulating chemicals for carcinogenicity were a series of congressional deadlines for EPA action and automatic bans on land disposal (known as "hammers") for particular, specified wastes if EPA fails to act by the deadlines for issuing treatment standards. The 1984 amendments also required that EPA review its waste list in three stages (ending in 1990) and decide whether to ban land disposal of these wastes (334).

EPA REGULATORY ACTIONS UNDER CERCLA

While RCRA was prospective—designed to prevent problems from hazardous wastes in the future—the Comprehensive Environmental Response, Compensation, and Liability Act of 1980, also known as "Superfund," was designed to address the cleanup of inactive hazardous waste sites, manage emergency response to the release of hazardous substances into the environment, and provide for liability and compensation. CERCLA addresses problems ranging from spills requiring immediate responses, to hazardous

waste dumps leaking into the environment and posing long-term health and environmental hazards.

The list of hazardous substances is established under section 101(14) of CERCLA. Section 102 of CERCLA authorizes EPA to list additional hazardous substances—those substances which, "when released into the environment may present substantial danger to the public health or welfare or the environment." Also, EPA is to set "reportable quantities" (RQs) for all of these substances. The RQs are set by statute at 1 pound except when different reportable quantities have been set under section 311(b)(4) of the Clean Water Act. EPA is authorized to adjust RQs by regulation (42 U.S.C. 9602). Section 103 sets requirements notifying appropriate government officials in the event that a hazardous substance is released in amounts greater than the relevant RQ.

"Hazardous substances" under section 101(14) of CERCLA include substances specified by sections 307 and 311 of the Clean Water Act, section 3001 of the Resource Conservation and Recovery Act, section 112 of the Clean Air Act, and section 7 of the Toxic Substances Control Act, and any substance designated as hazardous under section 102 of CERCLA (42 U.S.C. 9602). Of the CERCLA hazardous substances, 191 were identified as "potential carcinogens" (table 3-23).[17]

In 1985, EPA issued a final rule that clarified reporting procedures and set final RQ adjustments for 340 substances from its list of 717 hazardous substances (307). In 1986, EPA finalized RQ adjustments for an additional 102 hazardous substances (316). These adjustments did not cover potential carcinogens. In March 1987, EPA proposed RQ adjustments for the 191 substances identified as potential carcinogens on the CERCLA list (314).

To identify potential carcinogens on that list, EPA used IARC monographs, the *Annual Report on Carcinogens* (see ch. 5), final EPA determinations on carcinogenicity for other regulatory programs, and determinations by EPA's Carcinogen Assessment Group. EPA then developed a hazard ranking for the CERCLA substances that appeared on these various lists of carcinogens. The hazard ranking is a method used to sort a list of potential carcinogens into levels of relative carcinogenicity, which may then be equated to RQ levels for notification purposes (314).

The hazard ranking consisted of both qualitative and quantitative evaluations. For the qualitative portion, the chemicals were grouped using the CAG weight-of-evidence classification scheme

(see ch. 2 and below). For the quantitative assessment, a potency factor was estimated by EPA's Carcinogen Assessment Group. For this hazard ranking, each potential carcinogen was placed in one of three potency groups, based on the estimated dose required to induce cancer in 10 percent of a population exposed for 70 years. In the final step in the ranking, the qualitative weight-of-evidence and quantitative potency grouping are combined to yield a relative hazard ranking—high, medium, or low—for each chemical.

For this ranking, 191 chemicals were placed in the weight-of-evidence categories. Fourteen chemicals were placed in group A (sufficient human evidence), 8 in group B1 (limited human evidence, sufficient animal evidence), 102 in group B2 (sufficient animal evidence only), 20 in group C (limited animal evidence only), and 7 in group D (no evidence for carcinogenicity). Two more chemicals were given a range: one in groups B1 and B2, and one in groups B2 and C. Finally, for 37 chemicals, there were no data for directly determining their carcinogenicity, but these chemicals were either compounds of known carcinogenic metals (for example, compounds of arsenic, beryllium, cadmium, chromium, and nickel) or members of chemical families of known carcinogens (for example, PCBs) (56). Thus, most of the chemicals were classified based on animal evidence.

The final grouping, after combining weight-of-evidence and potency estimates, had 62 chemicals in the "high," 77 in the "medium," and 45 in the "low" hazard groups; 7 chemicals were not ranked because of lack of evidence for carcinogenicity.

CERCLA was enacted in 1980. Congress, perhaps aware regulatory agencies are sometimes slow in issuing regulations for toxic substances, put requirements in the statute itself; it specified toxic substances that were to be listed for CERCLA regulation by incorporating previously established lists, and it set reportable quantities for many of these substances at 1 pound until EPA issued more appropriate reportable quantities. As of today, EPA has not modified the reportable quantities for CERCLA carcinogens, although, as indicated above, these regulations have been proposed.

---

[17]The carcinogens on the CERCLA list were identified in a technical background document prepared by Environmental Monitoring Services Inc. (56) in support of the proposed rule to adjust the reportable quantities for carcinogens issued in March 1987 (314).

Table 3-23.—Substances Listed in CERCLA That Were Identified as Potential Carcinogens

2-Acetylaminofluorene (Acetamide, N-9H-fluoren-2-yl)	Sodium chromate	Methylthiouracil
Acrylonitrile	Strontium chromate	Mitomycin C
Aldrin amitrole	Chrysene	1-Naphthylamine
Arsenic	Coke Oven Emissions	2-Naphthylamine
Arsenic acid	Creosote	Nickel
Arsenic disulfide	Cyclophosphamide	Nickel ammonium sulfate
Arsenic pentoxide	Daunomycin	Nickel carbonyl
Arsenic trichloride	DDD	Nickel chloride
Arsenic trioxide	DDE	Nickel cyanide
Arsenic trisulfide	DDT	Nickel hydroxide
Cacodylic acid	Diallate	Nickel nitrate
Calcium arsenate	Diaminotoluene (mixed)	Nickel sulfate
Calcium arsenite	Dibenz (a,h) anthracene	2-Nitropropane
Cupric acetoarsenite	1,2:7,8-Dibenzopyrene	n-Nitrosodi-n-butylamine
Dichlorophenylarsine	1,2-Dibromo-3-chloropropane	n-Nitrosodiethanolamine
Diethylarsine	3,3-Dichlorobenzidine	n-Nitrosodiethylamine
Lead arsenate	1,2-Dichloroethane	n-Nitrosodimethylamine
Potassium arsenate	1,1-Dichloroethylene	n-Nitrosodi-n-propylamine
Potassium arsenite	Dieldrin	n-Nitroso-n-ethylurea
Sodium arsenate	1,2:3,4-Diepoxybutane	n-Nitroso-n-methylurea
Sodium arsenite	1,2-Diethylhydrazine	n-Nitroso-n-methylurethane
Asbestos	Diethylstilbestrol	n-Nitrosomethylvinylamine
Auramine	Dihydrosafrole	n-Nitrosopiperidine
Azaserine	3,3-Dimethoxybenzidine	n-Nitrosopyrrolidine
Aziridine	Dimethyl sulfate	5-Nitro-o-toluidine
Benz (c) acridine	Dimethylaminoazobenzene	Pentachloroethane
Benz (a) anthracene	7,12-Dimethylbenz (a) anthracene	Pentachloronitrobenzene
Benzene	3,3-Dimethylbenzidine	Pentachlorophenol
Benzidine and its salts	Dimethylcarbamoyl chloride	Phenacetin IARC (H)
Benzo (b) fluoranthene	1,1-Dimethylhydrazine	Polychlorinated biphenyls (PCBs)
Benzo (k) fluoranthene	1,2-Dimethylhydrazine	Aroclor 1016
Benzo (a) pyrene	Dinitrotoluene (mixed)	Aroclor 1221
Benzotrichloride	2,4-Dinitrotoluene	Aroclor 1232
Benzyl chloride	2,6-Dinitrotoluene	Aroclor 1242
Beryllium	1,4-Dioxane	Aroclor 1248
Beryllium chloride	1,2-Diphenylhydrazine	Aroclor 1254
Beryllium fluoride	Epichlorohydrin	Aroclor 1260
Beryllium nitrate	Ethyl carbamate (urethane)	1,3-Propane sultone
alpha-BHC	Ethyl 4,4-Dichlorobenzilate	1,2-Propylenimine
beta-BHC	Ethylene dibromide	Saccharin
gamma-BHC (Lindane)	Ethylene oxide	Safrole
Bis (2-chloroethyl) ether	Ethylenethiourea	Selenium sulfide
Bis (chloromethyl) ether	Ethyl methanesulfonate	Streptozotocin
Bis (2-ethylhexyl) phthalate	Formaldehyde	2,3,7,8-Tetrachlorodibenzo-p-dioxin (TCDD)
Cadmium	Glycidylaldehyde	1,1,1,2-Tetrachloroethane
Cadmium acetate	Heptachlor	1,1,2,2-Tetrachloroethane
Cadmium bromide	Heptachlor epoxide	Tetrachloroethylene
Cadmium chloride	Hexachlorobenzene	Thioacetamide
Carbon tetrachloride	Hexachlorobutadiene	Thiourea
Chlorambucil	Hexachloroethane	o-Toluidine
Chlordane	Hydrazine	p-Toluidine
Chlornaphazine	Indeno (1,2,3-cd) pyrene	o-Toluidine hydrochloride
Chloroform	Isosafrole	Toxaphene
Chloromethyl methyl ether (technical grade)	Kepone	1,1,2-Trichloroethane
4-Chloro-o-toluidine, hydrochloride	Lasiocarpine	Trichloroethylene
Chromium	Lead acetate	Trichlorophenol (mixed)
Ammonium bichromate	Lead phosphate	2,4,5-Trichlorophenol
Ammonium chromate	Lead subacetate	2,4,6-Trichlorophenol
Calcium chromate	Melphalan	Tris (2,3-dibromopropyl) phosphate
Chromic acid	Methyl chloride	Trypan blue
Lithium chromate	3-Methylcholanthrene	Uracil mustard
Potassium bichromate	4,4-Methylenebis (2-chloroaniline)	Vinyl chloride
Potassium chromate	Methyl iodide	
Sodium bichromate	n-Methyl-n-nitro-n-nitrosoquanidine	

SOURCE: Notice of proposed rulemaking issued 3-16-87. *Federal Register* 52:8140. 1987

Moreover, most of EPA's activity has focused on adjusting the RQs of substances already on the list rather than adding to the list. The original list was issued in April 1985 and contained 698 chemicals (307). Since then, EPA has added 19 chemicals to the list which had been added to RCRA in January 1985 (291) bringing the total to 717.

Several sections of the 1986 Superfund Amendments and Reauthorization Act (SARA) pertain to carcinogen testing and regulation. Section 121 requires that cleanup at Superfund sites "assures protection of human health and the environment" and achieves compliance with standards established under other Federal and State environmental laws (including the MCLGs of the new SDWA amendments, formerly called RMCLs, which are zero for carcinogens).

Section 110 requires the Agency for Toxic Substances and Disease Registry (ATSDR) of the Centers for Disease Control to compile a list of substances commonly found at Superfund sites and to prepare toxicological profiles of those substances. The first such list was issued on April 17, 1987 (306), and was drawn from the 717 substances already listed as hazardous under CERCLA.

ATSDR must also initiate a testing program in cooperation with NTP to determine the health effects of substances on the list for which there are not adequate data. The type of testing required is within the discretion of ATSDR, but the agency must consider recommendations of the Interagency Testing Committee established under TSCA. EPA is required to issue regulations under TSCA to recover the costs of testing from the responsible parties. ATSDR must also prepare health assessments for Superfund sites included on the National Priorities List established by EPA and other sites in response to petitions from affected citizens.

Section 313 of Title III of SARA which is the Emergency Planning and Right to Know Act requires companies to report annually to EPA on the amounts of certain substances used and discharged into various media such as air and water. These substances are listed in Committee Print 99-169 of the Senate Committee on Environment and Public Works. EPA may add a substance to this list if it causes or can reasonably be anticipated to cause ". . . various chronic human health effects," including cancer.

EPA'S CARCINOGEN ASSESSMENT GROUP

At EPA, risk assessment and risk management activities are separated more clearly than they are at several other agencies. The Carcinogen Assessment Group (CAG) was established in 1976 to centralize the conduct of risk assessments at EPA. CAG is organizationally independent from EPA's program offices (such as the air and water programs). Its personnel are responsible only for risk assessment, not risk management. CAG develops most risk assessments of carcinogens at EPA, although the Office of Pesticides and Toxic Substances (administering FIFRA and TSCA) usually conduct their own.

The first chemical given a quantitative risk assessment at EPA was vinyl chloride, which was assessed in 1975 for the Air Program. In its early years, CAG worked on a number of pesticides, including Chlordane/Heptachlor, Toxaphene, Lindane, and Endrin. In 1977, work began on

assessment of airborne carcinogens, and in 1978 the development of the risk assessments for the water quality criteria documents began. The latter were published in draft form in three installments in 1979, and in final form in November 1980. Between the draft and final versions, CAG abandoned the one-hit model for extrapolating from high doses to low doses in favor of the linearized multistage model.

A full CAG risk assessment represents a thorough review and evaluation of the carcinogenic risks of a particular substance and averages 200 pages in length. These assessments include those prepared for the water quality criteria documents and the health assessment documents for EPA listing of hazardous pollutants under CAA. Other risk assessments are much shorter, such as the Health and Environmental Effects Profiles (HEEPs) prepared to support decisions to list substances under RCRA (typically, 30 to 40 pages) or to develop reportable quantities for hazardous substance releases under CERCLA (10 to 20 pages) (164).

Table 3-24 lists the substances for which the full assessments were performed. These assessments are all extensively reviewed. For hazard identification, CAG indicates the level of evidence in humans and animals, as well as the overall grouping in EPA's classification scheme (see ch. 2). CAG also performs dose-response assessments to calculate the estimated carcinogenic potency. The slope of the dose-response line is also presented in table 3-24. When using data from animal studies, EPA estimates the upper confidence limit of the slope of the dose-response curve as derived from the multistage model (see ch. 2). Calculations based on human epidemiologic data are best estimates using a linear nonthreshold model. The slope represents the degree of carcinogenic response associated with a given exposure. The greater the slope, the more potent the carcinogen. Thus exposure to a relatively potent carcinogen (such as tetrachlorodibenzo-p-dioxin (TCDD))

will lead to a much higher probability of cancer than exposure to the same amount of a less potent carcinogen (for example, epichlorohydrin).

To date, CAG includes 57 chemicals in the list of full assessments complete with calculated potencies. Most of these have relied on animal data. Of the CAG list, nine were judged to have sufficient evidence in humans and eight of these nine (all except arsenic) also had sufficient evidence in animals. Three more were sufficient in animals with limited evidence in humans, while 37 were sufficient in animals with inadequate evidence in humans. The remaining seven were grouped in EPA classification C because the evidence was inadequate in humans and limited in animals.[18] Additional substances have been evaluated by CAG at the request of EPA program offices, but have not received the level of review of the 57 listed in table 3-24.

---

[18]The total is only 56 because technical grade hexachlorocyclohexane is not classified, although 3 isomers of this chemical are.

OFFICE OF MANAGEMENT AND BUDGET

During the 1970s, a series of executive orders gave groups in the Executive Office of the President a role in reviewing regulatory proposals. Over time this centralized review has greatly increased, culminating in extensive involvement in regulation by the Office of Management and Budget (OMB) under the Reagan Administration (219). OMB's role has been quite controversial.

President Reagan made "regulatory relief" an important goal for his administration early in 1981, and has issued two Executive orders on this subject. Executive Order 12291, issued February 17, 1981, requires that agencies prepare "regulatory impact analyses" on all major regulations, and requires that "regulatory action shall not be taken unless the potential benefits to society for the regulation outweigh the potential costs." It further specified that "regulatory objectives shall be chosen to maximize the net benefits to society" and required, to the extent possible, that all ben-

efits and costs be quantified in monetary terms. This Executive order also centralized regulatory review in OMB and required agencies to submit proposed and final regulations and regulatory impact analyses prior to publication. Although the legal authority to propose and issue regulations remains with the heads of regulatory agencies, in practice Executive Order 12291 has required agencies to receive approval from OMB prior to publication (219).

Executive Order 12498 (issued January 4, 1985) primarily requires each agency subject to Executive Order 12291 to submit an annual agenda of "significant regulatory actions" to OMB. OMB also reviews agency regulations and research proposals for compliance with the Paperwork Reduction Act (Public Law 96-511).

Proponents of OMB review argue that presidents have always taken steps they thought nec-

Table 3-24.—Substances Evaluated by the EPA Carcinogen Assessment Group for Carcinogenic Potency as of Aug. 1, 1986

Compounds	Level of evidence[a] Humans	Level of evidence[a] Animals	Grouping based on EPA criteria	Slope[b] (mg/kg/day)$^{-1}$
Acrylonitrile	L	S	B1	0.24(W)
Aldrin	I	S	B2	16
Allyl chloride	I	S	B2	1.19×10^{-2}
Arsenic	S	I	A	15(H)
B[a]P	I	S	B2	11.5
Benzene	S	S	A	2.9×10^{-2}(W)
Benzidene	S	S	A	234(W)
Beryllium	I	S	B2	8.4(W)
1,3-Butadiene	I	S	B2	1.8(I)
Cadmium	L	S	B1	6.1(W)
Carbon tetrachloride	I	S	B2	1.30×10^{-1}
Chlordane	I	S	B2	1.3
Chlorinated ethanes				
1,2-Dichloroethane (Ethylene dichloride)	I	S	B2	9.1×10^{-2}
Hexachloroethane	I	L	C	1.42×10^{-2}
1,1,2,2-Tetrachloroethane	I	L	C	0.20
1,1,2-Trichloroethane	I	L	C	5.73×10^{-2}
Chloroform	I	S	B2	8.1×10^{-2}
Chromium VI	S	S	A	41(W)
Coke oven emissions	S	S	A	2.16(W)
DDT	I	S	B2	0.34
3,3-Dichlorobenzidine	I	S	B2	1.69
1,1-Dichloroethylene (Vinylidene chloride)	I	L	C	1.16(I)
Dichloromethane (Methylene chloride)	I	S	B2	1.4×10^{-2}(I)
Dieldrin	I	S	B2	20
2,4-Dinitrotoluene	I	S	B2	0.31
Diphenylhydrazine	I	S	B2	0.77
Epichlorohydrin	I	S	B2	9.9×10^{-3}
Bis (2-chloroethyl) ether	I	S	B2	1.14
Bis (chloromethyl) ether	S	S	A	9300(I)
Ethylene dibromide (EDB)	I	S	B2	41
Ethylene oxide	L	S	B1	3.5×10^{-1}(I)
Heptachlor	I	S	B2	4.5
Heptachlor expoxide	I	S	B2	9.1
Hexachlorobenzene	I	S	B2	1.67
Hexachlorobutadiene	I	L	C	7.75×10^{-2}
Hexachlorcyclohexane technical grade				2.0
alpha isomer	I	S	B2	2.7
beta isomer	I	L	C	1.5
gamma isomer	I	S-L	B2-C	1.1
Hexachlorodibenzodioxin	I	S	B2	$6.2 \times 10^{+3}$
Nickel refinery dust	S	S	A	0.84(W)
Nickel subsulfide	S	S	A	1.7(W)
Nitrosamines				
Dimethylnitrosamine	I	S	B2	25.9(not by q_1)
Diethylnitrosamine	I	S	B2	43.5(not by q_1)
Dibutylnitrosamine	I	S	B2	5.43
N-nitrosopyrrolidine	I	S	B2	2.13
N-nitroso-N-ethylurea	I	S	B2	32.9
N-nitroso-N-methylurea	I	S	B2	302.6
N-nitroso-diphenylamine	I	S	B2	4.92×10^{-3}
PCBs	I	S	B2	4.34
Tetrachlorodibenzo-p-dioxin (TCDD)				$1.56 \times 10^{+5}$
Tetrachloroethylene (Perchloroethylene)	I	S	B2	5.1×10^{-2}
2,4,6-Trichlorophenol	I	S	B2	1.99×10^{-2}
Toxaphene	I	S	B2	1.13
Trichloroethylene	I	S	B2	1.1×10^{-2}
Unleaded gasoline vapor	I	S	B2	3.5×10^{-3}
Vinyl chloride	S	S	A	1.75×10^{-2}(I)

[a]S = Sufficient evidence; L = Limited evidence; I = Inadequate evidence.

[b]Animal slopes are 95% upper-bound slopes based on the linearized multistage model. They are calculated based on animal oral studies, except for those indicated by I (animal inhalation), W (human occupational exposure), and H (human drinking water exposure). Human slopes are point estimates based on the linear nonthreshold model. Not all of the carcinogenic potencies presented in this table represent the same degree of certainty. All are subject to change as new evidence becomes available. The slope value is an upper bound in the sense that the true value (which is unknown) is not likely to exceed the upper bound and may be much lower, with a lower bound approaching zero. Thus, the use of the slope estimate in risk evaluations requires an appreciation for the implication of the upper bound concept as well as the "weight of evidence" for the likelihood that the substance is a human carcinogen.

SOURCE: Environmental Protection Agency.

essary to control executive branch agencies (38). Further, even in a technological age, not all decisions are technical ones, and regulatory decisions involving both technical and policy decisions should be subject to the President's oversight. OMB's role has also been supported as necessary to achieve reform of the regulatory process, ensure "good regulation," reduce the costs of regulation, curb overzealous agencies, and gain control over a potentially expensive process. By reviewing regulations and requiring cost-benefit analysis, OMB forces agencies to confront problems of "covert redistribution and overzealous pursuit of agency goals," thus making agencies accountable to the President (38).

But OMB's review of agency actions has generated criticism (206,208,225). These criticisms raise constitutional issues, other legal issues, and more public policy issues about OMB's role in reviewing agency decisions.

Even with presidentially delegated authority, it may be unconstitutional for OMB to control decisions delegated by Congress to executive branch agencies (152). Furthermore, if OMB, in reviewing and approving or disapproving regulations, uses considerations not authorized by statute, its actions may not be permissible, especially when this conflicts with expressed congressional intention (152). In addition, since congressional committees have documented some evidence of *ex parte* and secret contacts between OMB and regulated industries (206,225), several commentators suggest this undermines the Administrative Procedure Act's public participation requirements for informal rulemaking (132,153).

Critics have also argued that delays imposed by OMB

... are paid for through the decreased health and safety of the American public ...; [that OMB review] ... places the ultimate rulemaking decisions in the hands of OMB personnel who are neither competent in the substantive areas of regulation, nor accountable to Congress or the electorate in any meaningful sense ... [and that Executive Order 12498] ... allows OMB to cut off investigations before they even begin, making it nearly impossible to attack OMB's decision that a potential rule is "unnecessary" (132).

One problem with the regulation of carcinogens discussed in this chapter involves delays in regulating problematic substances, delays between the time a statute is passed and the time an agency is authorized to regulate toxic substances, and delays between the time an agency has information that a substance is a carcinogen and the issuance of regulations. In recent years there have been additional delays because of OMB's review of major regulations.

EPA has compiled data on the average number of days rules are extended past the time limits specified by Executive Order 12291, for example, 30 days for a minor rule and 60 days for a major rule (one with an impact on the economy of $100 million or more).

The average extension in April 1985 was slightly under 50 days; previous peaks were near 100 days ... [thus] ... OMB holds minor rules for an average of over two months [rather than the 10 days Executive Order 12291 specifies for minor rules] and major rules for over four months [rather than the 60 days specified by Executive Order 12291] (225).

This average conceals some much longer delays involving rules that generated considerable dispute between OMB and some of the regulatory agencies:

- EPA's proposed ban on certain uses of asbestos and its phase-out over time of most other uses of asbestos were delayed more than 1 year.
- OMB delayed by at least 5 months EPA's recommended maximum contaminant levels "for approximately forty organic and inorganic chemicals under the authority of the Safe Drinking Water Act ...," leading Senator Durenberger to introduce an amendment to require OMB to complete its review of the RMCLs by a certain date.
- For EPA's proposed National Priorities List under Superfund, OMB forced EPA to choose between "delaying an entire executive action or sacrificing a part of it to gain OMB's approval of the major portion."
- Eleven proposed New Source Performance Standards for new and modified stationary air pollution sources, all submitted to OMB 3 to 13 months ahead of a statutory dead-

line, were delayed beyond that time. Fourteen months of the delay in publishing these rules were due to OMB's review under Executive Order 12291.

- High-level radioactive waste storage rules proposed by EPA were delayed by 1 year, leading to a suit by the Environmental Defense Fund against both OMB and EPA for missing a statutory deadline. Several congressmen filed an amicus brief on the side of the Environmental Defense Fund (225).

Shortly after the report from the Senate Committee on Environment and Public Works was completed, the District Court for the District of Columbia ruled that OMB has no authority to delay the issuance of the final rule which was the subject of suit or to delay issuance of any other rules "subject to statutory or judicial deadlines

under the Hazardous and Solid Waste Amendments of 1984" (225).

Under the Paperwork Reduction Act, OMB also reviews research proposals that involve government survey research. One example of OMB's involvement in a study concerning a carcinogen was a NIOSH proposal to evaluate the risk to human beings from MBOCA. This study was delayed by OMB for 6 months. This was not a regulation, but a proposed study of human beings, when NIOSH already had evidence that MBOCA was carcinogenic in three animal species (207). The Paperwork Reduction Act mandates that agencies obtain approval from OMB concerning these recordkeeping requirements. In some cases, regulations will require that records be kept by industry.

POSSIBILITIES FOR IMPROVING AGENCY TIMELINESS

This chapter has described the activities of the Federal agencies in assessing and regulating carcinogenic chemicals. A number of chemicals have been regulated for carcinogenicity and exposures have been reduced or eliminated. In some cases, the agencies have determined that the risks posed by particular chemicals are low and that there was no need to regulate. In other cases, the agencies are still obtaining toxicity and other information needed to regulate or are developing the analyses required by their statutes and by OMB. Finally, there probably are cases in which the necessary data have been collected, the analyses have been performed, and agency staff are simply waiting for decisions whether to regulate.

A constant in this chapter's overview of Federal activities is that the regulatory process is often a lengthy one.[19] To force regulatory action, Congress has legislated a variety of statutory mechanisms. These include statutory deadlines, congressionally mandated regulations, and institutional review and response mechanisms.

The most common of these have been statutory deadlines; although they have led to regulatory action, they are also frequently missed by the agencies. A report on the statutory deadlines in 15 environmental protection statutes affecting EPA found approximately 328 deadlines for setting regulations, issuing reports and studies, achieving compliance, setting guidelines, and accomplishing other tasks. The report estimated that 14 percent of these deadlines were actually met. However, while the original deadlines were missed, an estimated 41 percent of the EPA actions that were required had been completed by September 1985. The report concluded that, while statutory deadlines have brought about action by EPA and secured advances in environmental protection, deadlines are not sufficient to speed action. Moreover, Congress sets more deadlines for EPA than the Agency can meet, thus diluting the effectiveness of any one deadline, and it often sets unrealistic deadlines (48).

Congress has also mandated particular regulatory requirements. For example, in section 6 of the Toxic Substances Control Act, Congress ordered EPA to regulate the disposal of PCBs within 6 months and to prohibit further manufacture, except in closed systems, within 1 year of enactment.

---

[19]Concern about regulatory delay is not new. See, for example, a 1977 paper by Sidney Wolfe (364).

Another form of mandate is to require an agency to regulate a specified list of chemicals. For instance, in the Occupational Safety and Health Act, Congress ordered OSHA to adopt within 2 years established Federal occupational health and safety standards and the standards adopted by consensus standards organizations. In CERCLA, EPA was required to adopt a list of chemicals that had already been regulated under other environmental laws. In the 1977 Clean Water Act amendments, Congress codified a list of 65 classes of pollutants that had been developed for a consent decree settling a lawsuit directed against EPA for failing to regulate water pollutants. For the deadlines set in the 1984 RCRA amendments, Congress added "hammers"—statutory bans that take effect if EPA misses deadlines; depending on one's perspective, this combines either the best and worst of these two approaches to congressional mandates.

One final mechanism is requiring agencies to consider or, stronger still, respond to recommendations of another agency or organization. For example, OSHA is to consider the recommendations of NIOSH when developing new occupational health and safety standards. Stronger is the requirement that the Mine Safety and Health Administration must respond within 60 days to NIOSH recommendations. However, NIOSH has not yet sent any recommendations to MSHA that trigger this requirement. Under TSCA, EPA must respond to ITC's nominations of chemicals for toxicity testing, although one possible response is a decision not to require testing. In fact, the lists of chemicals recommended by ITC have dominated EPA activities on the testing of existing chemicals.

In the Safe Drinking Water Act of 1974, Congress required EPA to commission a study on drinking water by NAS. The study was to identify potentially harmful contaminants in drinking water and make recommendations concerning national standards for maximum contaminant levels. It was expected that EPA would use those recommendations. However, the NAS study decided that it was not appropriate to recommend contaminant levels, concentrating instead on developing information on potential toxic effects. Later amendments to the SDWA removed the requirement for EPA response to the NAS reports.

None of these mechanisms is a panacea. Congressional deadlines and mandated lists may force action, but also may divert regulatory agencies from chemicals and regulations more in need of regulation. Lists, in particular, may immerse Congress in extensive detail on particular chemicals. An institutional mechanism allows a group of experts or a scientific agency to sort through the lists of particular chemicals and recommend the ones of highest priority. On the other hand, establishing regulatory implications may mean that the recommending group should have a formal process for decisionmaking, including response to public comments. Establishing such a process might slow the whole operation down. In addition, establishing a regulatory linkage might dissuade scientists from participating who do not want to be involved in decisions with regulatory implications. Finally, such recommendations may inappropriately redirect agency priorities and create pressure for regulatory action. Such regulatory action may not always be necessary and may impose costs on regulated industries.

PUBLISHER'S NOTE:

When no current regulations or concentration limits exist, the NSF standards for direct and indirect drinking water additives use standardized methods for calculating risk for both carcinogens and non-carcinogens. These methods follow those supported by EPA (51 FR 33992, September 24, 1986). If the contaminant is a non-carcinogen, the NOAEL from the most sensitive observed effect is used with an appropriate safety factor to calculate a Maximum Drinking Water Level (MDWL). The MDWL also takes into account assumed body weight and drinking water consumption for the referenced consumer. If the contaminant is a carcinogen, the MDWL is calculated using a 10^{-6} risk factor and a dose-response value obtained by linear extrapolation of existing dose-response information to low-exposure ranges at the 95% confidence level. These approaches are used by both EPA and FDA in risk estimation. The MAL for unregulated contaminants cannot exceed 10% of the MDWL.

Chapter 4

The National Toxicology Program

CONTENTS

Figures

Tables

The National Toxicology Program

BACKGROUND

In the 1960s, government agencies, especially the National Cancer Institute (NCI), used animal tests to predict carcinogenicity, though at first to learn more about the relation between chemical structure and carcinogenicity and not for regulatory purposes. In November of 1978, the Secretary of the then Department of Health, Education, and Welfare (DHEW) established the National Toxicology Program (NTP), aware of the need to test chemicals for carcinogenicity (and other toxic end points), the limited ability of existing programs to keep up with the demands of new legislation, and the lack of coordinated testing.

Cancers often develop more quickly in animals than in humans, although not in relation to lifespan. Still, animal tests take time, 2 years of exposure for rodents, for example, and the tests are costly. In the 1970s, based primarily on the work of Bruce Ames, a high correlation was found between tests for mutagenicity of chemicals in microorganisms and carcinogenicity in animals (217). These genetic toxicology tests, and a second generation of short-term tests that followed, can be performed in days rather than years, and are much less costly than animal tests. The hope was expressed in DHEW that "by 1985 . . . better test systems will begin to replace the tedious and costly animal assay now required" (60). This

optimism has proved unfounded; the new tests have not proven superior to the original Ames test, which itself is an imperfect predictor of animal carcinogenicity.

Most animal carcinogenicity testing was transferred from NCI to NTP in 1981. In the first part of this chapter, the origins, support, and organization of NTP are described and the NTP process of selecting chemicals for testing is analyzed. A discussion follows of the relation of the results of the short-term tests to those of animal carcinogenicity studies, and finally the predictability of human carcinogenesis from animal tests. Some of these issues were discussed in chapter 2; they will be examined here only with regard to NTP.

Since this background paper focuses on the relation of carcinogen studies to regulatory decisions and the research activities of NTP, many of them conducted with the National Institute of Environmental Health Sciences (NIEHS) and the National Center for Toxicology Research (NCTR), receive scant consideration. NTP's goals include understanding the mechanisms by which cancers are initiated and propagated and developing better and quicker methods of determining chemicals' carcinogenicity.

THE NEED FOR TESTING

In 1980, NTP contracted with the National Research Council (NRC) to conduct a study, with a charge "to characterize the toxicity-testing needs for substances to which there is known or anticipated human exposure" (140). From approximately 5 million chemicals the Study committee compiled a list of 53,500 chemicals in 7 categories of human exposure. By systematic sampling of chemicals in each category, 675 chemicals were selected from this list. Multiple sources were examined to determine whether toxicity testing had

been conducted on these 675 chemicals. Extrapolating to the entire list, the committee estimated that there was no toxicity information on 38 percent of pesticides, 56 percent of cosmetic ingredients, 25 percent of drugs and excipients used in drug formulations, 46 percent of food additives, 78 percent of chemicals in commerce of which over 1 million pounds were produced in 1977, 76 percent of chemicals of which under 1 million pounds were produced, and 82 percent of chemicals whose production status was unknown or indeterminable.[1] Tests for chronic toxicity were performed most frequently on drugs (39 percent) and least frequently on chemicals in commerce (3 to 4 percent). From the list of 675 chemicals, the committee selected 100 on which some toxicity information was available; 10, 15, or 20 were selected from each of the 7 categories to determine the type and adequacy of testing. The report concluded, "Only about 8 percent of the tests met the standards of the reference protocol guidelines, and about another 19 percent were judged to be adequate." In discussions with OTA in 1986, Dr. Ernest McConnell, Director of the Toxicology Research and Testing Program, which is the principal NIEHS component of NTP, estimated that approximately 1,000 chemicals with high human exposure potential should be tested (120).

[1]Chemicals that were environmental decomposition products, manufacturing contaminants, or natural substances were not systematically included.

HISTORY OF MAJOR FEDERAL EFFORTS IN CARCINOGENICITY TESTING

NCI Testing Activities

NCI began animal testing of chemicals for carcinogenicity in 1961. Elizabeth Weisburger, one of NCI's charter researchers, described the aims of the project:

> There was no mention of a program for large-scale bioassay of industrial or environmental materials. To quote, "first priority should be given to chemicals most likely to make a contribution to our knowledge of the etiology of cancer and deepen our understanding of their mode of action" (357).

In the late 1960s, NCI responded to demands for testing chemicals in the environment. In 1970, for instance, it initiated contracts for studies of 40 pesticides approved for use in the United States. Appropriations under the National Cancer Act of 1971 provided sufficient funds to initiate a greater number of long-term animal studies, which reached a peak of 200 in 1972. The increase was so rapid that the consequences were not fully appreciated. "Neither NCI nor the prime contractor had enough assistance in pathology to examine all the microscope slides which resulted" (357). The backlog of chemical studies was not eliminated until 1979.

Government laboratories could not accommodate the volume of testing. Moreover, these laboratories were designated primarily for basic research, not for the routine testing of chemicals for toxicity.[2] Consequently, most of the animal tests were performed contractually by nongovernment laboratories. In 1973, NCI contracted with Tracor Jitco, Inc., to oversee the bioassay operations of the other contractors. Tracor Jitco also supplied data on chemicals being considered for testing to the Chemical Selection Working Group (CSWG) in NCI, which was responsible for the actual selection. A General Accounting Office report (201) found fault with this system. As a result, NCI instituted stricter monitoring of Tracor Jitco and its other contractors.

Despite concern over carcinogens in the environment, which contributed to the flood of testing in the 1970s, the process of notifying regulators was neither easy nor uncontroversial. Publications in the scientific literature indicating the carcinogenicity of 1,2-dibromoethane and 1,2-dibromo-3-chloropropane (in 1973) "led to no consternation or notice among regulatory agents [sic]" (357). To overcome this, NCI issued a "memorandum of alert" in 1975, when it became evident that trichloroethylene (TCE) was causing an increase of liver tumors with lung metastases in some animals. According to Weisburger:

[2]In a congressional hearing in 1981, Dr. Vincent DeVita, director of NCI, commented that NCI "never developed this [animal testing] program to be a source of information for the regulatory agencies. Therefore, when suddenly there was pressure for us to provide routine information, we were not able nor properly constructed to do that" (213).

Criticism of the "memo of alert" was so great that this mechanism was not used again. Instead, the complete record of any bioassay was compiled in a Carcinogenesis Technical Report; draft versions of the reports were sent to the regulatory agencies for their information prior to release to the public (357).

The furor over the preliminary publication arose because the Food and Drug Administration (FDA) could no longer consider TCE as an acceptable solvent for decaffeination under the Delaney clause. According to Weisburger, NCI staff were unaware that it was used for that purpose.

Problems of communication generally intensified as more agencies became involved in regulating carcinogens. These agencies had the capability to perform tests for carcinogenicity or the authority to require industry to do so. No channels existed for agencies to communicate about the chemical tests that were in progress or recommended; duplicate testing sometimes resulted.

Sometimes testing was beyond the capability of an agency, yet information on the toxicity of a chemical would have been helpful in making regulatory decisions. No formal mechanism existed for regulatory agencies to request NCI testing. "The entire process [of test selection] was quite informal with discussion among [NCI] staff only" (357).

Establishment of the National Toxicology Program

NTP was established in November 1978 by the Secretary of the then Department of Health, Education, and Welfare "to strengthen the Department's activities in the testing of chemicals of public health concern, as well as in the development and validation of new and better integrated test methods" (268). To accomplish its goals, NTP was "comprised of the relevant [Public Health Service] activities" within FDA (namely, NCTR), NCI, Centers for Disease Control (CDC) (namely, the National Institute for Occupational Safety and Health (NIOSH)), and the NIEHS. Dr. David Rall, Director of NIEHS, was named Director of NTP, reporting to the Assistant Secretary for Health. The organizational structure of NTP is shown in figure 4-1.

A 1981 paper prepared by NIEHS as background for a congressional hearing on NTP (90), commented that NTP was established as an interim measure because there was disagreement within DHEW as to how testing should be organized. The NTP Director was expected to coordinate the activities of various departmental components, but he could not "allocate resources, either funds or personnel, to areas of greater need and priority, except through agreement with the other agency heads." At the hearing, Dr. Rall and Dr. Ronald Hart, director of NCTR, indicated that they were in frequent communication and that there was "a minimal amount of confusion" (213). Dr. Rall also delineated NTP's role from that of the regulatory agencies. NTP's responsibility was in risk identification and quantification, primarily in animals; the agencies' responsibilities were determining human exposures and evaluating human risks and benefits. In response to a question from Congressman Albert Gore, Dr. Rall indicated that the allocation of so much of NTP's budget on testing was not optimal and that more should be devoted to developing better methods. Dr. Hart, Dr. Vincent P. DeVita, Director of NCI, and Dr. Millar, Director of NIOSH, emphasized NTP's role in testing chemicals.

In October 1981, the Secretary of the Department of Health and Human Services (DHHS) granted NTP permanent status. The funding arrangements remain voluntary. As stipulated in the original announcement, memoranda of understanding are signed by the head of each cooperating agency and the NTP Director, specifying the resources to be devoted to NTP, and identifying by organizational title the supporting elements of the participating agencies and their responsibilities (e.g., specific studies to be undertaken). With the transfer of the NCI Carcinogenesis Testing Program to NIEHS in July 1981, the vast majority of funds (87 percent of the NTP budget at the time) come from NIEHS. Dr. David Rall determines how much NIEHS will contribute to NTP, as the heads of NCTR and NIOSH determine their agencies' contributions. At the present time, most

Figure 4-1.—National Toxicology Program (NTP)

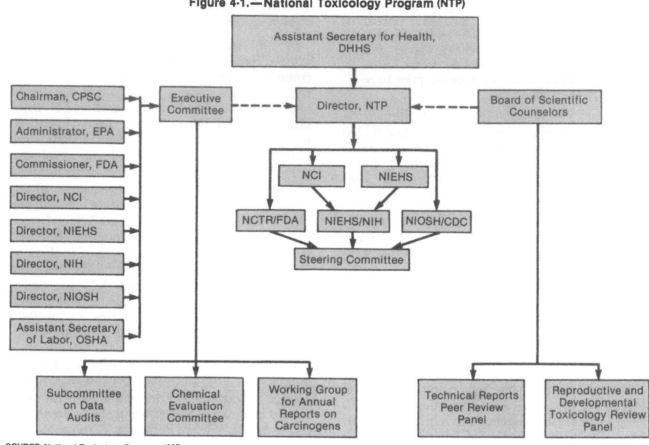

SOURCE: National Toxicology Program, 1987.

of the staff assigned to NTP comes from NIEHS, consistent with the contribution NIEHS makes to NTP's budget. The memorandum of understanding between NIEHS and NTP indicates that the NIEHS Toxicology Research and Testing Program is "dedicated to the National Toxicology Pro-

gram." It lists the NIEHS program elements and key scientists committed to NTP, accepts responsibility for administration, and indicates its contributions to NTP in person-years, budget, and space (234).

ORGANIZATION OF NTP

Structure

The NTP Steering Committee was formed in 1980 to promote "cooperative working relationships" among the contributing DHHS agencies. The committee consists of the NTP Director and the heads of NIEHS, NCTR, and NIOSH. It meets three to four times yearly to review programs and

projects, resolve interagency problems, and make agency allocations for approved chemical toxicological studies (257).

The documentation of NTP activities and plans is accomplished through an annual plan, as stipulated by the Secretary of DHEW (268), who also specified the plan's contents: information on current toxicology testing capacity and capacity in the coming year on plans for test development and validation, on the compounds to be tested, and on the regulatory and scientific opportunities that were considered in developing the plan. The Secretary established an Executive Committee to approve and monitor the annual plan. This committee consists of the heads of four regulatory agencies—the Environmental Protection Agency (EPA), FDA, the Occupational Safety and Health Administration (OSHA), and the Consumer Product Safety Commission (CPSC)—and the heads of the National Institute of Health (NIH), NCI, NIOSH, and NIEHS. On April 1, 1987, the Executive Committee voted to add the Agency for Toxic Substances and Diseases Registry to its membership. In addition, the Assistant Secretary for Health of DHHS is a nonvoting member and the Director of NCTR is a nonvoting consultant.

In testimony before Congress in 1980, Dr. Rall indicated that he had proposed that the Departments of Energy and Agriculture join the NTP Executive Committee (205). This has not happened. The composition of the Executive Committee provided the regulatory agencies outside of DHHS input to the planning and operation of NTP. These agencies, partly through the Executive Committee, have an important role in NTP activities, particularly in setting priorities for chemicals studies and in coordinating testing. In 1980, NTP reported that several chemicals recommended for industry testing by the Interagency Testing Committee (ITC) under the Toxic Substances Control Act (TSCA) were under test or scheduled for test by NTP (253). Since 1980, NTP has had a liaison representative with ITC to avoid redundancy of testing.

Resources

The number of chemicals tested depends primarily on the resources available. The budget for NTP activities (including NCI's contribution for testing) increased approximately 40 percent between 1979 and 1981. From fiscal year 1981 to 1987 the total NTP budget rose (including contributions from NCTR and NIOSH) from $70.5 to $77.9 million. After adjustment for inflation, this represents a small decline. The budget percentage devoted to testing fluctuated between 66

and 74 percent. Remaining funds were used for developing and validating testing methods and for management expenses. After NTP was established in 1978, participating agencies expected that they would receive larger appropriations to allocate to testing under NTP.

Since 1981, the inclusion of additional short-term tests, more detailed prechronic testing, and the use of three experimental doses and controls instead of two in chronic studies, which entails a greater number of animals, has increased the costs of testing a single substance. To resemble human exposures more closely, inhalation studies are being used more frequently than in the past. These studies entail special equipment and are the most costly of the chronic studies. Expenditures for analytical chemistry, a chemical repository, and auditing of data and laboratory practices have also increased.

The costs of various types of tests in fiscal year 1986 are shown in table 4-1. The prechronic study (to identify target organ toxicities and determine the doses to be used in the chronic phase) and the chronic study of a single chemical often cost over $2 million. (The cost of the Salmonella assay or, "Ames" test, is about $3,300.) In fiscal year 1987 an estimated 43 chemicals will be in the prechronic phase of testing, including beginning studies ("starts") on an estimated 30, and 137 will be in the chronic phase, including starts on an estimated 7.

NCTR's budget allocated to NTP activities fell from $6.1 million in fiscal year 1981 to $4.3 million in 1986, with only $1 million estimated for 1987. As a result of the budget reductions recently necessitated by the Gramm-Rudman-Hollings Act, NCTR discontinued long-term animal tests under NTP on one antihistamine and continued two others only when NTP through NIEHS agreed to fund their completion. NCTR is testing other chemicals through NTP. NIOSH's allocations to NTP have fluctuated considerably between 1981 and 1986: $4.1 million in 1981 (the highest allocation), $1.8 million in 1985 (the lowest), and $3.7 million estimated for 1987. Not all of these funds are used for testing. Staff scientists who serve as chemical managers at NCTR and NIOSH continue to design protocols for test-

Table 4-1.—Costs of NTP Studies of Fiscal Year 1986

Type of study	Cost per study[a]	
Mutagenicity:		
Drosophila	$ 11,083	
Salmonella	3,328	
Cytogenetics	12,932	
Mouse lymphoma	6,500	
Fertility & Reproduction:		
Fertility assessment	80,300	
Sperm morphology	5,300	
Teratology:		
Conventional	68,000	
Inhalation	350,000	
Prechronic Studies:[b]	Low range[c]	High range[c]
Dosed feed/dosed water	440,000	730,000
Gavage	505,000	785,000
Skin paint	520,000	785,000
Inhalation	655,000	1,285,000
Chronic:[d]		
Dosed feed/dosed water	1,210,000	1,860,000
Gavage	1,460,000	1,860,000
Skin paint	1,460,000	1,960,000
Inhalation	1,960,000	2,460,000

[a]Costs include actual contract award, support contracts, plus in-house operating costs.
[b]Includes studies such as: 14 day, 90 day, sperm morphology, vaginal cytology, clinical chemistry, urinalysis, hematology, chemical disposition, in-vivo short-term characterization.
[c]Costs based on range of awards made in fiscal year 1986. Where awards were not made in some of the categories, estimates were prepared.
[d]Three dose levels, one interim sacrifice, clinical chemistry and possibly hematology.

SOURCE: National Toxicology Program.

ing selected chemicals, but NIEHS provides most of the funds and staff for NTP testing.

Nominating Chemicals for Testing

NTP invites the nomination of chemicals for testing from any source. In fiscal year 1982, the NTP Executive Committee agreed to an FDA request that each participating agency could nominate one "priority" chemical per year for carcinogenicity testing. FDA wanted to ensure that more chemicals of concern be tested. The "agency priority" chemicals would be placed on a "fast track" for selection, as will be discussed further. If an agency fails to nominate a chemical in one year, it can still only nominate one the following year. Only five chemicals have been nominated by this fast-track route. They are D&C Yellow No. 11 (FDA), gallium arsenide (NIOSH), 2-butoxyethanol (CPSC), t-butylhydroquinone (FDA), and styrene (NIOSH). Nominating a chemical in this way does not preclude an agency from nominating other chemicals by the normal process in the same fiscal year. Any organization or individual can nominate as many chemicals as desired. The NTP Chemical Selection Coordinator and an Assistant to the Director of NTP remove from further processing chemicals that have already been or are being tested, or have been previously rejected for testing (although renominations that are submitted with additional information may be considered). "Draft executive summaries" are then prepared on the remaining chemicals, except those that have been nominated for genetic toxicology testing only. For these, the nominations are presented directly to the Chemical Evaluation Committee (CEC) with summary data on production levels.

Until 1984, the summaries were prepared by NCTR. In 1984, NTP contracted with Dynamac Corporation (Rockville, Maryland) to prepare the draft executive summaries. After selecting Dynamac as the successful bidder, one of the unsuccessful bidders objected to the award, claiming that Dynamac did not qualify as a small business. This delayed Dynamac's start and resulted in a backlog of nominations. The summaries prepared by Dynamac include information on chemical and physical properties, production, use, exposure, toxicology, and related regulatory activity (26). They are presented to the CEC, which makes the initial recommendation in selecting chemicals for testing. The nomination and selection process is diagramed in figure 4-2.

Selection

In congressional hearings for fiscal year 1981 appropriations, the directors of NCI and NTP proposed that funding be sufficient to begin 100 animal tests on chemicals in 1981. Dr. Rall proposed that this level of starts be maintained each year until 1984. At that time he said, "implementation of the Toxic Substances Control Act will allow us to begin to decrease the number of chemicals tested" (205). In commenting further on industry's role at the hearings, Dr. Donald Fredrickson, then NIH Director, said that it was impossible to attach commercial interests to chemicals that had been in the common domain for many years. He anticipated that when the Toxic Substances Control Act was "fully implemented, however, the NTP should be able to scale down"

Figure 4-2.—NTP Chemical Nomination and Selection Process

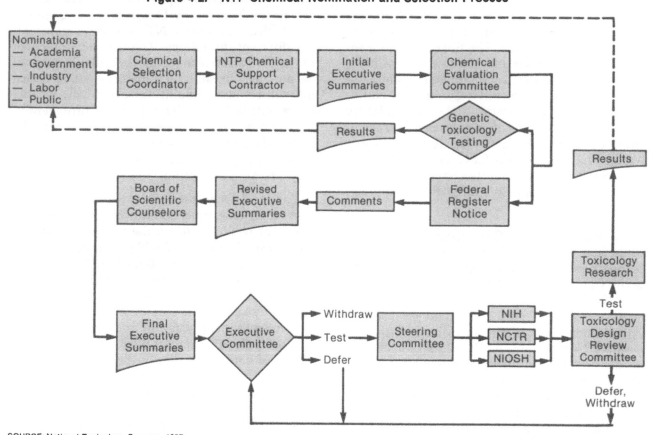

SOURCE: National Toxicology Program, 1987.

(205). The budget increase that would have supported 100 test starts annually never materialized and NTP had more nominated chemicals than it could study at that time (26).[3] Under such circumstances, nominations of chemicals had to be considered carefully, and priorities for testing established.

The Executive Committee, composed of representatives from CPSC, EPA, FDA, OSHA, NCI, NIEHS, NIOSH, NCTR, and NTP, meets about

four times a year to evaluate the drafts and recommend the types of testing, if any, to be performed along with their priorities. By having each member serve as a primary or secondary reviewer of each chemical nomination, NTP is assured of better participation from the other agencies. The CEC makes the final decision about nominated chemicals only for genetic toxicology testing. Approximately 2 months after a chemical is considered by the CEC, it is listed in the *Federal Register*, together with the Executive Committee recommendation. Its decisions on chemicals nominated only for genetic toxicology testing are not published in the *Federal Register*. Thirty days are given for responses but all responses are considered regardless of the date of receipt.

The executive summaries are revised to include public comments, and then submitted to NTP's Board of Scientific Counselors. The Board is com-

---

[3]After it became evident that funds would not be available to support as much testing as was anticipated, the participating agencies were each asked to designate the 10 highest priority chemicals from among the 140 chemicals that had already been recommended for testing by CEC by March 1980. Thirty-nine were chosen and approved by the NTP Executive Committee; for scientific reasons, not all were tested. Approximately 50 additional chemicals among the 140 were also approved by the NTP Executive Committee for testing by mid-1982.

posed of eight nongovernmental scientists appointed by the Assistant Secretary for Health for staggered 4-year terms. It meets two or three times a year in public sessions. The Board's recommendations and suggested testing priorities, are incorporated into the executive summaries and submitted to the NTP Executive Committee. Notice of the one priority chemical that each regulatory agency is permitted to nominate each year is not sent to the Board, but goes directly to the Executive Committee from the CEC. Neither the CEC nor the Executive Committee have rejected any of the five agency priority nominations submitted to date.

The NTP Executive Committee makes the final decision on prechronic and chronic testing and testing priorities for those chemicals recommended by the CEC or Board. It has done this by selecting "priority chemicals" for testing each year commensurate with NTP resources. It does not set priorities among these chemicals at the time of selection. Notifications on chemicals CEC recommends for chemical disposition, genetic toxicology, or reproductive studies are not sent to the Executive Committee but to the program leaders in NTP-NIEHS responsible for corresponding areas; they make the decisions about testing. They also can select other chemicals for testing within their program subject to budgetary limitations.

Once chemicals are approved by the Executive Committee, the NTP Steering Committee refers them to one or more of the three constituent agencies of NTP (NIEHS, NCTR, and NIOSH) where, in turn, they are assigned to chemical managers.

These scientists develop testing protocols to submit to the Toxicology Design Review Committee (TDRC), a group of NTP scientists representing different disciplines. The chemical manager of the TDRC can also recommend that testing not be pursued, because, for instance, of technical difficulties, unavailability of chemicals, or adequate outside testing. This happens infrequently. Based on the studies under their supervision, the chemical managers can nominate additional chemicals for study or additional studies on chemicals they are already testing.

The *NTP Technical Bulletin* provided information on the Executive Committee's selections and plans for testing, and on the results of mutagenicity tests. More than 7,000 people received this bulletin, which was discontinued in 1983. According to NTP staff, a similar publication would be useful, to present the results of prechronic studies and plans for chronic studies and other information for public information and comment. NTP is considering a plan to publish the experimental design of chronic studies in the *Federal Register* to permit responses from interested readers. Beginning in 1986, NTP publicly named chemicals on which short-term toxicology studies had been completed, specifying the administration route, species, and duration for proposed prechronic studies on these chemicals; the names of the responsible chemical managers were also specified. Comments were invited on chemicals' current production, uses, exposure levels, and toxicology data, to help NTP decide whether additional studies, including long-term toxicology and carcinogenicity studies, are needed (254).

NTP CARCINOGENICITY TESTING

Methods of Study and Analysis

Before a bioassay for carcinogenicity can be performed, preliminary information is needed. This is obtained by gathering data on chemical exposures and from studies done elsewhere. When there is some question about biological availability, chemical disposition and pharmacokinetic studies are conducted prior to prechronic and chronic studies. Based on information gathered before the prechronic studies, and also on budgetary constraints, chemical testing may be deferred or dropped. Such decisions have been made for 37 chemicals since 1982.

Chemical Disposition

Information must be obtained on how a chemical selected for testing is absorbed through various administration routes, for example through gastrointestinal and respiratory tracts and skin. The route of administration usually selected for bioassay is the route through which humans will most likely be exposed, unless the compound cannot be absorbed by that route. Such a finding may also lead to a decision not to test.

Pharmacokinetic Studies

Determining the rates of absorption and conversion to other compounds at various doses helps in selecting the doses for prechronic and chronic studies. When several related compounds in a class are tested it is also important to know whether they are converted to a common metabolite; it may then be possible to test only one member of the class.

Prechronic Studies

In prechronic studies, animals (usually mice and rats) are administered various doses of the chemical first for 14 days and then for 13 weeks, to evaluate organ-specific pathological changes, body and organ weight changes, clinical signs, and other indicators of toxicity. From the evaluation, an estimated maximum tolerated dose (EMTD) is determined. The EMTD is usually the highest of three doses administered in the chronic studies that follow. The Ad Hoc Committee of the NTP Board of Scientific Counselors recommended that when results are nonlinear, additional intermediate doses should be used, and that consideration should be given to having the lowest dose in a chronic study in the range of human exposure (258). Such a dose is likely to yield a significant number of tumors only when larger doses cause tumors in a very high percentage of animals, as occurred in the original chronic inhalation studies on methylene chloride. Over half the male and female mice developed lung tumors at 2,000 ppm, and over half of them developed liver tumors at 4,000 ppm. NTP has decided to conduct additional studies in female mice, probably using doses of 2,000, 1,000, and 500 ppm to elucidate the chemical's mechanism of action. The OSHA maximum peak dose for 5 minutes in any 2-hour period is 2,000 ppm, and for an 8-hour time-weighted average is 4,500 ppm; its accept-able ceiling is 1,000 ppm. NIOSH-recommended exposure limits are lower (131).[4]

Chronic Studies

In the chronic studies, the chemical is usually administered to both sexes of mice and rats for 2 years, at which time the surviving animals are sacrificed. Usually 60 animals of each sex and species receive each dose of the chemical for the duration of the study; an additional 60 of each sex and species serve as controls, receiving no chemical. Usually 10 of each experimental group (defined by species, sex, and dose) are sacrificed at 15 or 18 months to determine whether any tumors have already developed.

Evaluation of Results

The incidence of tumors in each group is determined, as are nontumorigenic effects, through necropsy and histopathologic examination. In the study of a single chemical, about 40,000 tissue selections may have to be examined (93). Omitting examination of certain sections succeeded in reducing costs. However, the calendar time required for testing was longer because of added review steps. Therefore, the attempt to reduce pathological studies is no longer made in standard studies.

A number of different statistical techniques are used to determine whether there is a significant increase in tumors associated with exposure to the chemical and, if so, whether there is a dose-response relationship (83). Data are also compared with those on tumor incidences in the NTP historical control database. The large number of

---

[4]The very high frequency of lung tumors (and also liver tumors in female mice) in methylene chloride tests, compared with controls, raised the possibility that the cancers reflected an acute toxic event causing high cell turnover. If so, the number of mutations per cell division need not be increased, but simply the number of cell divisions. Light microscopy provided no evidence for acute toxic effects in the lung or of increased cell turnover; more sensitive techniques will be used in the second study. It would be unlikely that increased cell turnover would occur at the lower doses, so that an excess of tumors would suggest a significant carcinogenic effect. The high frequency of chemically related liver tumors in female mice also affords an opportunity to determine whether these tumors have the same oncogene pattern as the background liver tumors. Such oncogene studies may provide added information on the significance of liver tumors.

these historical controls improves statistical power in determining whether rare tumors are in fact related to a chemical under study.

Classification of Carcinogenicity

The study in one sex of one species constitutes an "experiment." The NTP classifies carcinogenicity for each individual experiment. Based on statistical and biological significance, the results of each experiment are classified into one of five levels of evidence for carcinogenic activity:

1. clear evidence—a dose-related increase of malignant neoplasms or—a combination of benign and malignant neoplasms, or a marked increase of benign neoplasms that may progress to malignancy;
2. some evidence—the strength of the evidence for carcinogenicity is less than for the first category;
3. equivocal evidence—a marginal increase of neoplasms that may be chemically related;
4. no evidence; and
5. inadequate study—a major quantitative or qualitative limitation prevents interpretation (93).

Quality Assurance

Prechronic and chronic testing, necropsy, and histopathologic examination are performed primarily in contract laboratories. In May 1982, after testing had been transferred from NCI to NTP, Tracor Jitco ceased to provide oversight of contractor testing and NTP assumed greater responsibility for monitoring the tests. In 1983, NTP withdrew a draft report of carcinogenicity studies on methylene chloride administered by gavage because of a contractor's poor testing practices. As a result, NTP developed stringent quality assurance procedures (19,93). These include retrospective data audits of each step in testing, from analysis of the chemical to review of the histopathological sections. Contract laboratories are also visited at least once annually and must submit monthly progress reports. A report by the General Accounting Office in 1984 concluded that NTP's new auditing procedures greatly strengthened quality control of testing (201).

Histopathological sections are examined by two independent groups of pathologists. Although nei-ther is blinded with regard to the dose the animal received or the gross lesions, the Chairman of the NTP Pathology Working Group (PWG) selects sections for additional examination by PWG members, who are not told whether the sections come from exposed or control animals.

Review and Publication

Following a retrospective audit of all study data and resolution of any discrepancies, a technical report is prepared by the chemical manager. It is first reviewed by NTP staff, and then submitted for peer review to the Technical Reports Review Subcommittee of the Board of Scientific Counselors. The public is informed when the results will be considered by the Peer Review Panel in open meeting. Since industry, labor, and academia are represented on the subcommittee, the classifications receive a full and candid critique from the principal parties concerned. Draft reports are made available to anyone on request.

Following approval by the Peer Review Panel, a final technical report is printed and distributed, usually within 9 months. When the preliminary histopathological evidence suggests that a chemical is highly carcinogenic, the agencies represented on the NTP Executive Committee are notified before the technical report is completed, as are manufacturers, trade associations, labor unions, public interest groups, and other groups monitoring carcinogenicity, such as the International Agency for Research on Cancer (IARC). Such a procedure was followed, for example, for the inhalation studies of methylene chloride and 1,3-butadiene. The notifications stated that the findings were preliminary.

NTP has an agreement with the National Library of Medicine to enter the results of NTP studies in TOXLINE, a computerized database to which the public has access, before printing and distribution of studies' technical reports. This will result in wider and earlier availability of the Summary results.

The results of genetic toxicity tests are published in peer-reviewed journals, for which there are frequently long delays between submission and publication. Results had been published in the *NTP Technical Bulletin* before its publication was terminated in 1983.

OBSERVATIONS ON THE NTP NOMINATION AND SELECTION PROCESS

Criteria for Nomination and Selection

Nominating sources are asked to submit a description of the chemical and its properties and any available information on:

1. production, uses, occurrences, and analysis;
2. toxicology;
3. disposition and structure-activity relationships;
4. ongoing toxicological and environmental studies; and
5. a rationale for the recommendation and suggested studies (257).

The Ad Hoc Panel charged by NTP's Board of Scientific Counselors to examine NTP's testing and evaluation program criticized the lack of emphasis on human exposure, either its magnitude or frequency, in the process of selection (258). In its reply to the Ad Hoc Panel, NTP maintained that exposure was considered in selecting chemicals for tests, and that it would obtain current information from manufacturers on production volumes and exposures during production and use (259). NTP now communicates both with manufacturers and trade associations. It also uses information on potential worker exposures from the National Occupational Hazard Survey and the National Occupational Exposure Survey conducted by NIOSH. The NRC report earlier cited also emphasized the importance of considering exposure along with "suspicion of toxic activity," in planning toxicity tests (138).

There are a number of problems, however, in emphasizing exposure in this way. The first is inadequacy of information. The NRC committee found that, of all types of information needed for health hazard assessments, the least information was available on exposure. For 36 chemicals in its subsample of 100 tested chemicals "no data were available from which the committee could determine the extent of exposure, and, for 75 of the substances in the subsample, no information

was available from which trends in exposure could be estimated."[5]

A second problem relates to the second point that the NRC committee suggested should provide a basis for selecting chemicals, "suspicion of toxic activity." Despite strides in understanding the chemical substituents that may cause toxicity, great ignorance remains. Again, the original intent of the NCI program was to learn more about the structure-activity relationships of carcinogenicity, and this focus was carried over into the NTP. What has become increasingly apparent, however, is the inability to predict with certainty the carcinogenicity of a chemical from its structure. This unpredictability creates a dilemma in setting policy for testing. On the one hand, priority could be given to chemicals whose testing could reveal more about the relation of chemical properties to carcinogenicity. On the other, it could be given to chemicals for which potential or actual human exposure (or exposure of other components of the ecosystem) is great, or which is suspected of being a human carcinogen. Not all chemicals fit into both categories. (Nor is it always possible to predict or determine exposure.) A toxicologist with the Environmental Defense Fund commented to OTA:

It is much more likely for a substance with strong structural resemblance to a known carcinogen or mutagen to be nominated for testing, rather than a compound for which there is quasi-epidemiologic reason to suspect carcinogenicity (186).

_____

[5]The NRC report gave several reasons for the lack of exposure data: 1) There are few reporting requirements. "Even data on production volumes of substances and numbers of people involved in manufacture, distribution, use, and waste disposal are limited." 2) There is little incentive for voluntary reporting. 3) Monitoring for compliance of standards focuses on specific substances, few of which were included in its subsample. "Furthermore, data collected for compliance monitoring may be of limited value in evaluating population exposures." 4) "Little is known about physical processes and procedures that affect the exposure potential for uses other than those intended. For example, the intensity of occupational exposure is strongly influenced by the choice of process and control equipment, and the intensity of environmental exposure is strongly influenced by the selection of waste-disposal technique, chemical reactivity, and degree of biodegradability"(138).

A third problem in emphasizing exposure is duplication of NTP efforts with efforts by manufacturers or processors required by regulatory agencies. Section 4 of TSCA established ITC to "designate" or "recommend" chemicals in commerce, as defined by TSCA, to be tested for certain health or environmental effects. EPA can then require manufacturers and processors to test these chemicals. (See ch. 3.) As mentioned earlier, several chemicals recommended for industry testing under TSCA were under test or scheduled for test by NTP in 1980 (117).

At the suggestion of EPA officials, who were concerned about NTP's activities overlapping with those of TSCA, NTP established eight "chemical selection principles," drafted in 1979 by EPA staff members (table 4-2). The introduction to the principles recognizes that industry has responsibility for testing chemicals under the authority of agencies created by Congress, although principle 8 indicates that there may be special situations in which NTP would test chemicals that "have potential for large-scale and/or intense human exposure," even if industry could be required to test them. Principle 3 recognizes improving the understanding of structure-activity relationships as a criterion for selection. Principle 5, permitting testing of previously tested chemicals "to cross-compare testing methods," follows from NTP's goal of developing and validating new tests. The remaining principles implicitly recognize the importance of human exposure as a basis for selection, but for those chemicals that industry cannot be required to test. These include chemicals in the environment not associated with commercial activities (principle 1), old chemicals whose manufacturers derive "too little revenue to support an adequate testing program" (principle 6), and groups of chemicals manufactured by different companies for which the companies "probably cannot be required" to test (principle 7).

Duplication of NTP and regulatory agency testing is also avoided through liaison between NTP and ITC; the NIEHS is a voting member of ITC, and EPA, under whose authority ITC operates, is represented on CEC. Before conducting a detailed review, ITC asks NTP for information. At the present time, ITC and NTP use the same contractor, Dynamac, to prepare the documen-

Table 4-2.—NTP Chemical Selection Principles[a]

1. Chemicals found in the environment that are not closely associated with commercial activities (11);
2. Desirable substitutes for existing chemicals, particularly therapeutic agents, that might not be developed or tested without Federal involvement (1);
3. Chemicals that should be tested to improve scientific understanding of structure-activity relationships and thereby assist in defining groups of commercial chemicals that should be tested by industry (91);
4. Certain chemicals tested by industry, or by others, the additional testing of which by the Federal Government is justified to verify the results (27);
5. Previously tested chemicals for which other testing is desirable to cross-compare testing methods (8);
6. "Old chemicals" with the potential for significant human exposure which are of social importance but which generate too little revenue to support an adequate testing program (some of these may be "grandfathered" under FDA laws) (15);
7. Two or more chemicals together, when combined human exposure occurs (such testing probably cannot be required of industry if the products of different companies are involved) (1); and
8. In special situations, as determined by the Executive Committee, marketed chemicals which have potential for large-scale and/or intense human exposure, even if it may be possible to require industry to perform the testing (39).

[a]Numbers in parentheses indicate the number of times the principle was used to support a CEC recommendation for testing in animals.

SOURCE: National Toxicology Program.

tation for their reviews (18). Liaison reduces the likelihood of duplicate testing under the NTP principle 8. ITC has nominated chemicals for NTP testing to decide whether to recommend chemicals for more extensive industry testing (principle 3). Most chemicals nominated by ITC have been for short-term genetic toxicology testing (see table 4-3).

In table 4-2, the number of times CEC cited each principle to support a recommendation for extensive animal testing between fiscal year 1981 and 1986 is shown in parentheses following each principle for 123 chemicals; more than 1 or even 2 principles were used to justify the testing of some chemicals. Of the 193 citations of principles, the most frequent was of principle 3 (referring to structure-activity relationships), a total of 91 times. Although principle 7, focusing on combinations of chemicals, has only been invoked once, NTP has other initiatives under way on mixtures that are not part of the chemical nomination and selection process. These include a contract with the National Academy of Sciences to develop

Table 4-3.—Source of Nomination of Chemicals for Mutagenicity and Bioassay Testing by NTP by Year

Source	1979 Mut	1979 Bio	1980 Mut	1980 Bio	1981 Mut	1981 Bio	1982 Mut	1982 Bio	1983 Mut	1983 Bio	1984 Mut	1984 Bio	1985 Mut	1985 Bio	1986 Mut	1986 Bio
Government:																
CPSC												1[a]				
EPA								1	34[b]		19	28	1			
FDA										1		1		7		5[c]
ITC							2		5	1	1			1		1
NCI	176		56		48		14			6		4		15[d]	5	24[e]
NIEHS					136		195		62		30	15	121	3		6
NIOSH					92			2								4
OSHA										1				1		
State agencies				1		1	5							1		
Nongoverment:																
Individuals								1	7	3		4	1	2	1	3
Industry						1										
National Academy of Sciences										2						
Professional associations														1		
Unions								9				5		1		1
Totals	NA	303[f]	300[g]	60[h]	228	50	197	32	108	14	50	58	122	33	6	44

[a]2-butoxyethanol also nominated by UAW International Union; only shown under CPSC.
[b]Alkyl epoxides nominated jointly by EPA and NIEHS; only shown under EPA.
[c]Includes 4 benzodiazepines nominated by NIEHS in 1984.
[d]Includes 8 submitted for reconsideration and one (oxymetholone) also nominated by NIEHS (shown only under NCI).
[e]Includes 6 resubmissions by NCI and one nominated by NIEHS in 1984.
[f]NTP could not provide information on the nominators for bioassay testing for fiscal years 1979, 1980, and 1981 other than from NCI.
[g]Nominations came from EPA, NIOSH and FDA (NCTR). Breakdown not available. The EPA nominations were submittted in response to ITC designations of chemical classes for possible industry-required testing and to aid in the pre-manufacture notification program.
[h]Added in.
KEY: Mut—mutagenicity tests only; Bio—bioassay (extensive testing in animals); NA—not available; EPA—Environmental Protection Agency; FDA—Food and Drug Administration; ITC—Interagency Testing Committee; NCI—National Cancer Institute; NIEHS—National Institute of Environmental Health Sciences; NIOSH—National Institute for Occupational Safety and Health; OSHA—Occupational Safety and Health Administration.
NOTE: Data on sources of nominations for mutagenicity tests is complete only after 1981. In prior years, blank spaces mean that data was not available, not that nominations from the particular source were not made. For the same reason, the same applies to blank spaces for bioassay nominations prior to 1982.
SOURCE: National Toxicology Program.

guidelines for studying complex mixtures; chemical disposition studies of well-defined mixtures; and a Superfund-sponsored effort to characterize the toxic potential of chemicals and mixtures found at waste dump sites.[6]

OTA obtained from NTP a list of the chemicals for which principle 3 was invoked to recommend either genetic toxicology or animal studies. Eighty-eight chemicals on the list were unequivocally positive in one or more short-term genetic toxicology tests, but have not been tested further by NTP. OTA asked ITC and the EPA Office of Toxic Substances whether they had required, or considered requiring, industry to test any of these chemicals in accord with principle 3. Of the 77 chemicals that were in the TSCA inventory, ITC has considered 61 chemicals and conducted detailed reviews on 19 that might present the greatest human health hazards. Of the 19, it has deferred further consideration of 12 and recommended 7 for testing to EPA. Deferral is based on low production volumes, low exposure potential, or adequate knowledge of closely related chemicals. Three of the seven recommended chemicals were tested before NTP was created. In 1983 EPA issued an Advance Notice of Proposed Rule Making on one of these, bisphenol A diglycidyl ether (320). Although EPA noted the mutagenicity of several glycidyl compounds, it did not cite NTP studies among its sources. EPA has also issued a Notice of Proposed Rule Making on two of the other chemicals recommended by ITC after NTP was established, namely, meta- and ortho–phenylenediamine (321). The mutagenicity of these compounds was reported before the creation of NTP, and neither the ITC recommendation nor the pro-

[6]In general, NTP strives to test single, pure chemicals. Seldom, however, are humans exposed to isolated pure chemicals.

posed rule cites the NTP genetic toxicology results. The other two reports on chemicals ITC recommended for testing do not mention a positive NTP mutagenicity result. Of the 12 chemicals whose consideration ITC deferred, 4 were reviewed before NTP was organized. Five of the remaining eight (including four xylidines) mention a positive NTP genetic toxicology result (18).

Thus, a small proportion of the chemicals studied under principle 3 that have positive genetic toxicology results have been recommended or proposed for industry testing. Many of the chemicals studied under principle 3 were of little interest to ITC or EPA, however, usually because of low production or low human exposure potential. An NTP official told OTA that many of the chemicals selected under principle 3 represented chemical classes nominated solely for Salmonella testing, to examine the effects of structural modifications on the genotoxic potential of the class and to ascertain the usefulness and predictivity of the assay for the different classes of chemicals. In these cases, the principle was not used to propose chemical candidates for further industry testing (26). A rewording of principle 3, or perhaps dividing it into two principles, one focusing on structure-activity relationships and the other on defining groups of chemicals for industry testing, might clarify the situation.

Number, Source, and Disposition of Nominated Chemicals

Table 4-3 indicates the sources of nominations submitted each year to NTP for mutagenicity testing and bioassays (animal testing). From 1980 to 1986, 1,011 chemicals have been nominated only for mutagenicity testing. There has been a steady downward trend in the number nominated for mutagenicity tests, except in 1985. In that year, NIEHS nominated 121 dump site chemicals. Since 1981, 54 percent of all nominations for mutagenicity tests have been made by NIEHS; its nomination of fewer chemicals accounts for most of the decline. In 1981 and 1982, most of the NIEHS nominations requested examining structure-activity relations, which has not been the case more recently.

A total of 594 chemicals have been nominated for bioassays from 1979 to 1986. In January 1979, shortly after NTP was created, all participating agencies were asked for their nominations and several hundred were received. Thereafter the numbers were considerably smaller. Since 1980, the number of chemicals nominated for animal testing has fluctuated between 14 and 60 per year without any discernible trend.

Of the 942 nominations for all types of testing between 1981 and 1986, 42 came from nongovernment sources; unions nominated chemicals most frequently, a total of 16.

As a result of positive mutagenicity tests, some chemicals were nominated for more extensive testing. New information about a chemical can also lead to renomination and recommendations for additional testing. For instance, methyl isocyanate was originally selected only for genetic toxicology tests, because extended human exposure was considered unlikely (26). After the disaster at the Union Carbide plant in Bhopal, India, in which methyl isocyanate was accidentally released, studies on the delayed effects of a single exposure were undertaken (257).

Of 186 chemicals nominated for more than mutagenicity testing and reviewed by CEC between 1981 and 1986, 114 were recommended for testing (61 percent). CEC recommended animal testing of 59 percent of chemicals nominated by government agencies and of 72 percent of those nominated by nongovernment sources. It did not recommend testing of more than half of the chemicals nominated by FDA and about half of the chemicals nominated by NIEHS. In 1986, CEC did not recommend 53 percent of chemicals for testing, a greater proportion than ever before.

Examination of the data provided by NTP indicated the initial response of the Board of Scientific Counselors to 164 chemicals acted on by CEC (table 4-4). Regarding 130 chemicals (79 percent), CEC and the Board agreed in their recommendations. For eight (5 percent), CEC recommended testing and the Board recommended that testing not be done. For four (2.4 percent), CEC did not recommend testing while the Board recommended testing. These four chemicals were all reconsidered by the Board after 1984.

The number of chemicals selected for testing is consistent with the NTP budget. Because of budgetary cutbacks in fiscal year 1986, priorities were recently set for chemicals selected for testing, but for which studies had not yet begun. Chemicals given low priority may not be tested unless new information about their toxicities or exposures raises their rank or additional funding becomes available.

Duration of Testing Process Stages

OTA obtained information from NTP on the time from nomination to NTP Executive Committee action for every chemical nominated for bioassay reviewed by CEC in fiscal year 1981 or 1982 (table 4-5). For those chemicals in this "cohort" whose bioassays the Executive Committee approved, OTA obtained the time from Executive Committee action to testing status as of January 1987 (table 4-6). For most chemicals, it took over 2 years from the time of nomination to action by the Executive Committee. It took approximately half of this time before CEC acted. There times do not appear to have decreased appreciably in more recent years. Action by the Board of Scientific Counselors added another 3 to 8 months. For 7 chemicals for which CEC recommended no testing and for 13 for which it recommended testing (see table 4-4) the Board deferred action; it also deferred action on one chemical as had CEC. These deferrals add further time. Of these 21 chemicals, 6 have been rereviewed by the Board. In each case, on rereview the Board agreed with CEC's original recommendation. It should be recalled that Board review is omitted for any regulatory agency's priority chemical.

No chemical approved for animal testing in either fiscal year 1981 or 1982 has yet passed through the entire process. Of the 30 chemicals approved for testing in those 2 years, only four have reached the stage of chronic testing. Twenty-two are being tested in the prechronic phase. Testing of four has been deferred or withdrawn because of budget constraints. Of these four chemicals, one presented unusual technical difficulties (benzoyl chloride), one was no longer produced (m-chloroaniline), one had been tested for carcinogenicity by industry and found by NTP disposition studies not to be absorbed from the gas-

Table 4-4.—Initial Response of Board of Scientific Counselors to Actions of the Chemical Evaluation Committee (CEC) on 164 Chemicals, Fiscal Year 1982 - Fiscal Year 1986

CEC action	Board's recommendation (number of chemicals)		
	No test	Defer	Test
No test	35	7	4
Defer	0	1	1
Test	8	13	95

SOURCE: National Toxicology Program.

trointestinal tract (CI Vat Blue No. 1), and one was found not to metabolize to the suspected carcinogen (1,3-dichloro-5,5-dimethylhydantoin). Considering the time to develop protocols, announce, accept, and negotiate contracts for testing, perform chemical disposition studies to determine the extent of absorption through various administration routes, and conduct prechronic studies, it becomes clear that the 2 years of the chronic phase itself greatly underrepresents the total time needed for carcinogenicity bioassays.

Table 4-7 lists the five priority chemicals nominated by the regulatory agencies for "fast track" analysis. This process took no longer than 5 months for any of the chemicals from the time of nomination as an agency priority chemical to NTP Executive Committee selection, a time shorter than for any nonpriority chemical by 3 months. The shorter time is due to rapid preparation of the executive summaries by NTP, rather than by a contractor, and to omitting consideration by the Board of Scientific Counselors. Although none of the agency priority chemicals have reached the stage of chronic testing yet, the earliest was selected in 1984. Thus, while the early processing for the chemicals has been prompt, too little time has elapsed to conclude whether their testing will be completed more rapidly.

Case Study in Nomination and Selection: the Benzodiazepines

From documents obtained from NTP and interviews with some of those involved, OTA tracked the nomination and selection of five widely used benzodiazepine (BDZ) drugs that were recently approved conditionally for NTP testing. OTA cannot say whether the process has been similar

Table 4-5.—Time (in months) From Nomination of Chemicals Considered by the Chemical Evaluation Committee (CEC) in Fiscal Year 1981 and Fiscal Year 1982 to Intermediate and Final Points in the Process of Approval for Testing by NTP

| | From time of nomination to: | | | | | | | | | | |
| | CEC recommendation (months) | | | Action by Board of Scientific Counselors (months) | | | | Action by Executive Committee (months) | | | |
Year of CEC review	<13	13-18	19-24	<13	13-18	19-24	>24	<19	19-24	25-30	>30
Fiscal year 1981: Number of chemicals[a]	33	12	1	9	23	14	0	1	3	19	0
Fiscal year 1982: Number of chemicals[b]	28	18	1	0	16	30	1	0	0	4	8

[a]23 nominated chemicals were not recommended for prechronic and/or carcinogenicity testing and were not, therefore, referred to the Executive Committee.
[b]35 nominated chemicals were not recommended for prechronic and/or carcinogenicity testing and were not, therefore, referred to the Executive Committee.

SOURCE: National Toxicology Program.

Table 4-6.—Status of Chemicals Approved for Testing by the NTP Executive Committee in Fiscal Year 1981 or Fiscal Year 1982 (status as of January 1987)

| | | Prechronic phase | | | | Chronic phase | |
Year of executive committee approval	Deferred or withdrawn	TDRC[a] approval	Contract awarded	Testing initiated	TDRC[a] review	Contract awarded	Necropsy completed
Fiscal year 1981: Number of chemicals.	4[b]	2	0	8	0	3[c]	1
Fiscal year 1982: Number of chemicals.	0	2	1	8	1	0	0

[a]Toxicology Design Review Committee.
[b]Three chemicals were deferred in July, 1986 because of budget constraints. One chemical was withdrawn by the Executive Committee in October, 1984.
[c]Chronic testing initiated for one chemical in December 1986. Contract awards anticipated for the other two by March, 1987.

SOURCE: National Toxicology Program.

Table 4-7.—History of Agency Priority Chemicals (status as of April 1987)

Chemical	Nominating agency	Date of priority nomination	NTP Executive Committee selection date	Status, April 1987
D&C Yellow No. 11	FDA	9/27/83	1/27/84	Prechronic testing
Gallium arsenide[a]	NIOSH	7/16/84	8/31/84	Contracted for prechronic testing
2-Butoxyethanol	CPSC	9/27/84	3/07/85	Out for bid (prechronic)
t-Butyl-hydroquinone . . .	FDA	7/29/85	12/19/85	Out for bid (prechronic)
Styrene	NIOSH	12/13/85	2/13/86	Protocol for prechonic in preparation

[a]Originally nominated Dec. 8, 1983.

SOURCE: National Toxicology Program.

for other chemicals. Questions on studies of classes of chemicals and on industry's role in testing were also raised in NTP's consideration of these drugs for testing.

Four of the BDZs that were nominated are frequently used to relieve anxiety: diazepam (Valium), chlordiazepoxide (Librium) and clorazepate (Tranxene), and oxazepam (Serax). The fifth, flurazepam (Dalmane), is used as a hypnotic (sleep inducer). The first three are metabolized to oxazepam. These drugs were marketed prior to 1968, when FDA began to require carcinogenicity test-

ing. They are still extensively used, with over 2.5 million prescriptions written for the least frequently prescribed (oxazepam), and over 25 million for the most frequently prescribed (diazepam) in 1983. Diazepam ranked third in "new prescriptions" in 1985 and fourth for "new and refilled" prescriptions.[7]

---

[7]This and the following information on toxicity studies was obtained from the draft executive summaries submitted by Dynamac to NTP on October 28, 1986. Citations are to the original sources.

In 1980, an NTP senior toxicologist, Dr. James Huff, sent a memo to the Deputy Director of NTP, in which he suggested that diazepam, chlordiazepoxide, and oxazepam "be tested first in the Genetic Toxicology Component and at the same time be nominated for long-term carcinogenesis bioassay." His concern was based on a report of liver cell adenomas in mice receiving oxazepam. Huff considered the study inadequate because of the small number of animals used and the short duration (94). No nomination was submitted to the NTP office responsible for processing nominations. In 1984 Huff sent to the Director of Toxicology Research and Testing Program and the assistant to the director of NTP, who are responsible for processing nominations, a new report of liver tumors in male mice administered large doses of another BDZ, ripazepam, together with a copy of his earlier memo. Shortly thereafter, the three, as well as clorazepate, were nominated for study by NIEHS (25).

The FDA joined in the nomination on March 31, 1986, after NTP requested that it provide unpublished data on toxicologic testing of the drugs that could be included in the executive summaries (26). At that time, FDA also nominated flurazepam for study. FDA based its nominations on

- the extensive use of these drugs,
- the inadequacy or absence of carcinogenicity studies on them,
- an increased incidence of liver tumors in mice and benign thyroid tumors in rats given some newer BDZs that were required to undergo extensive carcinogenicity testing prior to marketing,[8] and
- the need to determine whether the types of tumors observed in the newer drugs were characteristic effects of the class of BDZs or specific for each chemical (361).

In fact, FDA has not approved new drug applications for some of the newer BDZs because of the finding of tumors in animals (193).

The nominations were not reviewed by CEC until September 1986, after their nomination by FDA. The delay between the time of initial nomination and CEC review was due largely to a backlog that had accumulated while NTP was attempting to get a new contractor to prepare the summaries for the CEC. CEC recommended the drugs for bioassay (prechronic and chronic testing two with high priority (diazepam and flurazepam) and one (oxazepam) with moderately high priority. (The recommendations were based on principles 3 and 4; see table 4-2.) It did not recommend clorazepate and chlordiazepoxide for testing because the three related BDZs were recommended. The summaries prepared for CEC do not make clear, however, that oxazepam is the major metabolite of both drugs.[9] Additional reasons for not recommending these drugs were that previous chronic studies indicated minimal toxicity of clorazepate, although the reviewer noted that "no carcinogenicity studies were done," and that the use of chlordiazepoxide "was declining" (261).

The Board of Scientific Counselors reviewed the five BDZs in November 1986. One of the questions raised by a counselor at the meeting was whether the results for one drug could be extrapolated to others that have the same metabolite. The answer given by NTP staff was "no." In fact, one of the problems of doing "class" studies is that when one or more members of a class is found to be carcinogenic, a new member of the class, which may differ by only a single substituent, must be tested in order to establish its carcinogenicity. It is for this reason that NTP has tested several derivatives of benzidine and several phthalates. The difficulty arises because of the imperfect ability to predict carcinogenicity from structure.

---

[8]The 1968 FDA guidelines for carcinogenicity testing called for an 18-month study in rats only. Manufacturers now submit data on both sexes of rats and mice in their new drug applications, although there is no formal requirement for them to do so.

[9]Data in the summary prepared for CEC does not indicate oxazepam in the blood or urine following chlordiazepoxide administration to human volunteers. The summary on clorazepate does not indicate the urinary metabolites following administration of the drug. Desmethyldiazepam, an immediate precursor of oxazepam, "is the major metabolite of clorazepate in the blood." Diazepam is also converted to desmethyldiazepam. In one study, only 16 percent of the administered dose was excreted as a conjugate of oxazepam, the major form of oxazepam excretion.

The Board voted unanimously to recommend that NTP test all five BDZs subject to an extensive review of existing studies "to determine sex/species combination in which to test individual chemicals" (260). The NTP Executive Committee accepted the recommendation on December 18, 1986, giving highest priority to the three drugs originally recommended for testing by the CEC. Because of budgetary constraints, it is not yet clear how many of the five drugs will be tested in animals.

The only BDZ selected by NTP for genetic toxicology testing prior to review by CEC was diazepam, and this occurred in December 1985. Following action of the CEC in September 1986, NTP genetic toxicology staff added flurazepam and oxazepam. To OTA's knowledge these tests have not yet been conducted. When the CEC was considering the BDZs the question was asked why FDA did not require industry to test the compounds. Although the patents on the older BDZs had expired, FDA could still have required companies to test drugs that they are marketing. (They did get Wyeth and Hoffmann-LaRoche to perform

carcinogenicity studies in rats on oxazepam and diazepam, respectively, after they had been marketed but while they were still under patent. In view of the finding on the newer BDZs, FDA is now interested in mouse studies.) FDA apparently chose not to require such testing in 1986 for at least two reasons. First, they would have to get several companies to collaborate in carrying out a suitable protocol; since the drugs were no longer under patent, several companies were marketing each BDZ. (There is precedent, however, for FDA's requiring manufacturers to agree on a common protocol and to contribute to testing of pharmaceutical agents (193).) Second, FDA had limited leverage on the companies. Without evidence of these drugs' human carcinogenicity, regardless of their widespread use, it is not likely that FDA would prevail if it sought to remove the drugs from the market. FDA could require that the drug label state that the drug has not been tested adequately for carcinogenicity (if that was the case), but an FDA spokesperson doubted that most clinicians would be affected greatly by such a statement (193).

MUTAGENICITY TESTING: CORRELATION WITH ANIMAL CARCINOGENICITY

The initiation of cancer may involve the mutation of particular nucleotides that form the backbone of DNA. Evidence that chemicals cause mutation or combine with DNA or affect its function can be obtained rapidly by a variety of in vitro and in vivo methods. The recent discovery that certain genes (proto-oncogenes) can be converted to oncogenes by known chemical carcinogens (chemicals associated with the presence of tumors) extends this work and promises better short-term methods to determine carcinogenicity (115). Work on developing such methods is being conducted at NTP and NCTR.

At the present time, the "Ames" test, which measures mutant frequencies in one or more strains of Salmonella bacteria, is the most extensively used test for genetic toxicity. At NTP, such results are considered in deciding how to proceed

with bioassays for carcinogenicity in rats and mice. NTP results of Salmonella mutagenesis tests have been published on 775 chemicals. These tests were performed in one or more of three contract laboratories. The laboratories did not know the identity of the chemicals. Multiple doses were each tested in triplicate. Positive and negative control chemicals (that is, known mutagens and known nonmutagens) were also used in each experiment. Reproducibility within and between laboratories was documented in most cases. Of the 775 chemicals tested, 194 (25 percent) were clearly mutagenic, and 49 (6 percent) gave questionable results (in one laboratory) or different results in different laboratories. The remaining 532 (69 percent) were negative (87,133,367).

At the request of the Board of Scientific Counselors, NTP compared the results of short-term tests for genetic toxicity with those of animal tests for carcinogenicity (194). It performed Salmonella mutagenicity tests on 44 chemicals that were carcinogenic in NTP tests in at least 1 animal experiment (an experiment is defined as the test in 1 sex of 1 species), on 20 chemicals that were negative in all animal experiments (both sexes of 2 species, usually mice and rats), and on 9 chemicals that gave equivocal results in the animal studies. Except for 10 chemicals that could not be tested for technical reasons, these 73 chemicals were all of those tested in NTP long-term studies of carcinogenicity in which the animals were sacrificed in 1977 or later and on which the conclusions had been approved by peer review before 1985.[10] Counting the equivocal animal tests as negative, 20 of the 24 chemicals that were mutagenic in the Salmonella test had proved to be carcinogenic in animals (83 percent, predictive value positive (PVP)), and 25 of the 49 chemicals that were nonmutagenic had proved to be noncarcinogenic (51 percent, predictive value negative (PVN)). The chance that an animal carcinogen would give a positive Salmonella test (the test's sensitivity) was 45 percent, and the chance that a noncarcinogen would give a negative test (the test's specificity) was 86 percent. There is some increase in the PVN when the following factors are taken into consideration: carcinogenic potency (the lowest dose producing tumors in animals),[11] malignancy of animal tumors, number of animal experiments that gave positive results, number of organ sites with tumors, and exclusion of liver tumors. But there are decreases in the PVP when each of these factors is considered. With 224 chemicals in the entire NCI-NTP database on which Salmonella and carcinogenicity testing had been done, the PVP was 69 percent; the PVN, 45 percent; the sensitivity, 54 percent; and the specificity, 70 percent. The range of concordance—agreement between Salmonella and animal test results (positive and positive, negative and negative)—when each of these factors is examined separately varies between 62 and 74 percent.

NTP performs additional short-term tests on most chemicals in the genetic toxicology program. These include the mouse lymphoma (ML) mutagenesis assay, tests for chromosome aberrations (CA) in Chinese hamster ovary (CHO) cells, and sister chromatid exchanges (SCE) in CHO cells. These tests are all performed in vitro. NTP also examined the predictivity of these tests for animal carcinogenicity alone and in combination with each other or the Salmonella test. The PVP of each was less than the Salmonella test (ML 66 percent, CA 73 percent, and SCE 67 percent), and the PVN was about the same (50 to 52 percent). There was no combination of two, three or all of these tests that gave much higher concordance with animal test results than the Salmonella test alone (66 percent v. 62 percent). Nor were chemicals that caused positive responses at lower doses in any of these genetic tests more likely to be carcinogenic in animal bioassays than those for which higher doses were needed. Nor did a sequential approach—in which chemicals negative in one genetic test were then subject to the other three short-term tests—significantly improve the ability to distinguish animal carcinogens from noncarcinogens. NTP found no combination of tests or other test to represent a substantial improvement over the Salmonella test.

NTP performed 2 other short-term genetic tests on some of the 73 chemicals: the test for unscheduled DNA synthesis in rat primary hepatocytes on 44, and the sex-linked recessive lethal assay in Drosophila on 27. The specificity and PVP were both high (with a specificity of 0.93 and 1.0 respectively, and a PVP of 0.86 and 1.0 respectively), but sensitivity and PVN were much lower for these than for the other tests.

Some short-term tests involving mammalian cells can be conducted in vivo. The chemical is administered to the animal, and tests are performed on cells obtained either from liver or bone marrow. In vivo tests for unscheduled DNA synthesis in rodent liver cells were generally negative for animal carcinogens and noncarcinogens, in tests of 16 chemicals. Two short-term tests, SCE

---

[10]This and the following information was obtained at the NTP Board of Counselors meeting, November 25, 1986.

[11]This was not true for all chemicals. For instance, very low doses of 2,3,7,8,-tetrachlorodibenzo-p-dioxin ("dioxin") and polybrominated biphenyl (PBB) mixtures were positive in all animal experiments, as was reserpine in three of four animal experiments. Yet all three chemicals gave negative results in the Salmonella and other tests for genetic toxicity (described below). Dioxin and PBBs are known to act as tumor promoters under certain circumstances.

and CA, were performed on the same 16 chemicals in vivo on mouse bone marrow cells. Of the eight that were known animal carcinogens, all gave positive SCE results in vitro; only six gave positive results in vivo. Seven of the animal noncarcinogens gave positive SCEs in vitro; only four gave positive results in vivo. The carcinogens also gave fewer positive in vivo CA responses than positive responses in vitro (four compared to six); the noncarcinogens gave fewer positive in vivo responses (one compared to four in vitro). When in vivo tests on mouse bone marrow were performed on seven carcinogen-noncarcinogen pairs of structural analogs, the CA test correctly identified the carcinogenic member of five of the chemicals without giving any false positives. In the SCE test, four of the carcinogens were positive, but five of the noncarcinogens were also positive. NTP will perform in vivo assays of additional chemicals. It will also incorporate them into prechronic tests. It is possible that nonmutagenic chemicals do not act as initiators, that is, do not cause mutations in DNA, but instead act as promoters, whose mechanism of action is poorly understood. Animals administered promoters sometimes show regression of preneoplastic or neoplastic lesions after the administration of the chemical is stopped. For some chemicals selected for carcinogenicity testing, but which are negative in short-term tests, NTP conducts "stop" studies, in which administration is discontinued in some animals but not in others (115); regression or absence of tumors in the first set, but not in the second would be consistent with the observations made so far for several promoters.

Although the results of these studies show a fair degree of consistency among the different tests, suggesting that not all studies need be performed, they fail to show very good agreement with the results of animal testing. In discussing the results, NTP staff concluded that the short-term tests could not be used as surrogates for long-term rodent studies, but could be helpful in assessing carcinogenic potential (260).

RELATION OF CHEMICAL CARCINOGENICITY IN ANIMALS TO CANCER IN HUMANS

IARC investigators recently compiled data on the ability of animal tests to predict that a chemical is a human carcinogen.

Of 30 exposures (to chemicals) for which there is sufficient evidence of carcinogenicity to humans, the animal data provide sufficient evidence for 19 . . . limited evidence for seven . . . and inadequate evidence or no data for four . . . Of the 14 exposures for which there is limited evidence of carcinogenicity to humans, the experimental data provide sufficient evidence for eight . . . limited evidence for three . . . and inadequate evidence or no data for three.
. . . The four exposures for which there is sufficient evidence of carcinogenicity to humans that have not been adequately tested in experimental animals are: certain combined chemotherapy regimens including MOPP (mechlorethamine [nitrogen mustard] oncovine [vincristine], procarbazine, prednisone), conjugated oestrogens, smokeless tobacco products and treosulphan. However, for some individual components of MOPP—nitrogen mustard and procarbazine—there is sufficient evidence of carcinogenicity in experimental animals . . . Further, it is reasonable to believe that conjugated oestrogens would react similarly to other oestrogens in experimental animals . . .; for some oestrogens there is sufficient evidence of carcinogenicity to animals (360).

For 37 of the 44 chemicals considered to be carcinogenic in humans, there was evidence of carcinogenicity in animals; in these cases, the evidence was "sufficient" for 27 and "limited" for 10.[12] (The remaining seven chemicals were not studied adequately.) Further strengthening this association was the finding that for every chemi-

---

[12]IARC's definition of "sufficient evidence" in animals includes increased incidence of malignant tumors in multiple species or strains. The definition of "limited evidence" includes studies involving a single species or strain, inadequate dosage levels or period of followup, too few animals, and high rate of spontaneous tumors.

cal supported by sufficient evidence of carcinogenicity for both humans and animals, the same organ was involved in both; the types of tumor were often identical or similar. Some chemicals that subsequently proved to be human carcinogens were first demonstrated to be carcinogens in experimental animals (4-aminobiphenyl, diethylstilbesterol, melphalan, methoxsalen with ultraviolet A, mustard gas, and vinyl chloride).

It does not follow from these studies that all chemicals carcinogenic in animals will prove to be carcinogenic in humans. It would be helpful to demonstrate that chemicals that are not carcinogenic in humans are also noncarcinogenic in animals. This would be an expensive undertaking. Moreover, no universally accepted list of human noncarcinogens exists. Although FDA requires that all new drugs for long-term or widespread use be tested, most other chemicals are selected for testing because it is suspected they cause cancer in humans.

SUMMARY

The establishment of NTP has improved coordination of testing within the government. NTP performs a valuable role in developing and evaluating new tests. It continues to elucidate structure-activity relationships in chemical carcinogenesis.

In most cases, NTP's process of evaluating nominated chemicals gathers available information to permit informed decisions on selection. It is not clear that the chemicals to which humans are significantly exposed are being selected adequately, in part because relatively few chemicals are nominated. NTP does not have direct control of nominations, other than by publishing information on the nominations process. It does consider human exposures in recommending chemicals for study.

Regardless of whether the chemicals that pose the greatest threat to humans are being nominated, more chemicals are nominated than can be tested given current budgets. It is possible that industry could perform more tests, as the BDZ case study suggests. There is little evidence to suggest that chemicals tested under NTP to "assist in defining groups of commercial chemicals that should be tested by industry" (see table 4-2, principle 3) are subsequently being tested by industry.

The time from nomination to selection is over 2 years for most chemicals. This time could be shorter. Whether the testing process itself can be shortened is problematic. The performance of chemical disposition and prechronic tests is necessary, and eliminating them would reduce the validity of the chronic bioassays. Testing had been completed by January 1987 on only one of the chemicals selected in fiscal year 1981 and 1982. Developing protocols, awarding contracts, and performing chemical disposition and prechronic and chronic tests takes at least 5 years; the evaluation of organs and microscopic sections adds at least an additional year; and preparation of the report, review, and publication add still more time. The time required is so intrinsically long that some chemicals presenting significant exposures may no longer do so by the time testing is completed. There should be a mechanism by which NTP is promptly informed of changes in the production status of chemicals, or of the substitution of analogs, so it can modify testing schedules and protocols accordingly. This aim may be accomplished to some extent by NTP's announcement of the completion of prechronic studies with a request for submission of relevant data. Chemical managers also attempt to obtain data on current production levels.

It is a grave oversimplification to maintain that animal testing takes 2 years. The current research by NTP and NCTR toward finding better biological markers of carcinogenicity may lead to better and more rapid means of detecting carcinogens in both humans and animals.

Agency Responses to the *Annual Report on Carcinogens* and NCI/NTP Test Results

CONTENTS

Figure

Tables

Agency Responses to the *Annual Report on Carcinogens* and NCI/NTP Test Results

INTRODUCTION

In this chapter, OTA examines Federal agency responses to the list of carcinogens in the *Annual Report on Carcinogens* and to positive results for chemicals tested in the carcinogenicity bioassays conducted by the National Cancer Institute (NCI) and National Toxicology Program (NTP). This analysis only provides summary information about the chemicals that may be considered carcinogenic based on the *Annual Report* or positive NCI/NTP bioassay results and describes the extent of agency actions on these chemicals. (As in the rest of this background paper the term "chemical" is used here broadly to include substances, mixtures, groups of substances, and exposures.)

Not all factors important in regulatory decisions are encompassed by the present analysis. In par-

ticular, estimates of the risk presented by these chemicals (including the qualitative weight of the evidence, quantitative potency estimates, and quantitative exposure estimates), agency judgments that these risks are reasonable or unreasonable, the availability of control technologies, and the costs of controls are not considered here. As discussed in chapter 3, another important issue is the time needed to develop, propose, issue, defend, and implement new regulations. This analysis, however, does not discuss the time agencies take to respond to identified carcinogenic chemicals. Finally, OTA did not attempt to evaluate the level of protection provided by the regulations that have been issued. The analysis focused just on whether or not regulations had been issued.

THE *ANNUAL REPORT ON CARCINOGENS*

History of the *Annual Report*

The *Annual Report on Carcinogens* was mandated by Congress in the 1978 amendments to the Community Mental Health Centers Act (Public Law 95-622). The idea of an annual report was first raised publicly in oversight hearings on the NCI held in March of that year. Witnesses testified that no agency kept a comprehensive list of carcinogenic chemicals, and that while some chemicals were regulated by one agency, the same chemicals were not regulated by other agencies. Representative Andrew Maguire introduced a bill to require that NCI publish a list of carcinogenic chemicals. He hoped that the report would educate the public, serve as a point of reference for scientists and regulators, and evaluate the activities of the regulatory agencies, who are not immune to pressure from the outside (113).

The bill first passed by the House of Representatives had specified that NCI would be responsi-

ble for the report and that the report should contain three elements:

- a list of all known or suspected carcinogens,
- information concerning the nature of exposure and number of individuals exposed, and
- an evaluation of the efficacy of existing regulatory standards designed to control suspected carcinogens (197).

In the final version, responsibility was given to the Department of Health and Human Services (DHHS), and "suspected carcinogens" was changed to "substances . . . reasonably anticipated to be carcinogens." As enacted, the law requires that the report include the following:

. . . a list of all substances (i) which either are known to be carcinogens or may reasonably be anticipated to be carcinogens and (ii) to which a significant number of persons residing in the United States are exposed (Public Law 95-622).

The provision mandating the evaluation of existing standards was modified to require that the *Annual Report* determine the following information:

(i) each substance . . . for which no effluent, ambient or exposure standard has been established by a Federal agency, and
(ii) for each [existing] standard . . . the extent to which, on the basis of available medical, scientific, or other data, such standard and the implementation of such standard by the agency decrease the risk to public health from exposure to the substance (Public Law 95-622).

Finally, the law requires that the *Annual Report* describe requests from Federal agencies for carcinogenicity testing and the responses of DHHS to those requests.

The original sponsor of this legislation, Andrew Maguire, and Paul Rogers, the chair of the subcommittee from which the legislation was originally reported, both argue that these changes did not alter the intent of the original legislation (113,177). Paul Rogers described the regulatory importance of the *Annual Report*:

The intention of the legislation was that listing in the annual report would be a first step in regulation, one triggering a review by the agencies responsible for enforcing various laws regulating carcinogens (175).

Development of the *Annual Report*

In the *Annual Report*, "known carcinogens" are defined to be human carcinogens, while animal carcinogens are deemed to be "substances . . . reasonably anticipated to be carcinogens." New chemicals are usually included in the *Annual Report* after testing positive in NCI/NTP bioassays in both sexes of one species and in at least one sex of a second species.[1]

The substances to be included in the *Annual Report* are selected by an interagency committee, with representatives from NCI, U.S. Consumer Product Safety Commission (CPSC), Environmental Protection Agency (EPA), Food and Drug Administration (FDA), National Institute of Environmental Health Sciences (NIEHS), National Institute for Occupational Safety and Health (NIOSH), National Library of Medicine (NLM), and Occupational Safety and Health Administration (OSHA). The committee compiles a list of chemicals based on the previous *Annual Report*, International Agency for Research on Cancer (IARC) reports, NTP animal testing results,[2] other peer-reviewed carcinogenesis studies, and data on chemical exposures from a variety of sources. The draft list is published for public comment, after which the committee makes its final selections.

While the legislation provided for a yearly report, in practice the reports have not been issued annually. Mandated by Congress in November 1978, the first report was dated July 1980; the second, December 1981; and the third, September 1983. The fourth report is dated "1985," although copies were not widely available until mid-1986. Much of the delay in issuing 1985 report was due to review within DHHS (26).

This year, for the first time, NTP held a public meeting to receive comments on chemicals to be listed in the fifth *Annual Report*. A number of interested trade associations, unions, public interest organizations, and individuals presented comments on the *Annual Report*. Many of the individuals and groups thought highly of the *Annual Report* and found it to be a useful reference. Some participants suggested that more chemicals be listed by using less stringent selection criteria. In addition, they wanted the *Annual Report* to focus attention on chemicals that should be subject to further regulatory activity (113,173,363).

A number of trade associations expressed concern, however, about the process used to develop the *Annual Report*. These groups suggested that NTP adopt written guidelines for determining listing in the *Annual Report* (specifying, in particular, the use of a "weight of the evidence" approach), that NTP develop more information on exposures (especially evaluating likely human exposures in relation to animal test exposures), and, in developing the report, that NTP give earlier public notice, more explanation of the rationale

---

[1]Using the terms of the next section, this means three or four positive experiments (clear evidence or some evidence) in an NTP bioassay.

[2]That these results are included leads to some overlap in this chapter between discussions of agency responses to *Annual Report* listings and positive results of NCI/NTP tests.

for chemical selection, and greater opportunity for public participation. Some participants also expressed concern that NTP was using the conclusions of an international organization, the IARC, to determine listing in the *Annual Report* (4,30,80).

Increased interest in the *Annual Report* has arisen in part because it is now being used to trigger regulatory requirements. OSHA is using the *Annual Report* as part of its hazard communication or "labeling" standard. That standard requires, first, that chemical manufacturers assess the hazards of the chemicals they produce and transmit this information to employers and employees and second, that employers provide hazard information to employees through a system of warning labels, employee training about hazardous chemicals, and employee access to material safety data sheets. The OSHA standard mandates that a chemical be considered a carcinogen (and hazardous) if it is included in the *Annual Report* or the IARC monograph series, or if it is regulated by OSHA as a carcinogen. The material safety data sheet for the chemical must also indicate this.

Several State worker and community "right to know" laws also use the *Annual Report* to trigger coverage. In addition, the recently enacted California proposition 65 ("the 1986 Safe Drinking Water and Toxic Enforcement Act") establishes rules and warning requirements for chemicals "known to cause cancer." To identify these substances, proposition 65 refers to the OSHA hazard communication standard, which in turn refers to the *Annual Report*.

Contents of the *Annual Report*

The *Annual Report* covers the reasons for listing substances; summaries of chemical properties; descriptions of production, uses, and exposures; and information on reported regulatory actions. Much of this information is provided by the participating agencies themselves. To a very limited extent, the *Annual Report* describes some of the exposure reductions associated with agency regulations.

The *Annual Report* has not attempted to identify regulatory "gaps"—areas where regulations appear to be needed—or to evaluate whether current standards are sufficiently protective. Instead, the *Annual Report* presents only descriptive information on the regulatory standards that have been issued.

Listed Chemicals

The first *Annual Report* listed only the 26 chemicals that IARC had determined to be human carcinogens. The second *Annual Report* listed 25 chemicals known to be carcinogenic based on human data[3] and 63 chemicals reasonably anticipated to be carcinogenic. The third *Annual Report* listed 22 chemicals known to be carcinogenic and 95 chemicals reasonably anticipated to be carcinogenic. In the fourth *Annual Report*, 29 chemicals are listed as human carcinogens and 119 as reasonably anticipated to be carcinogenic.

The fourth *Annual Report* lists 148 chemicals, chemical groups, mixtures, and exposures altogether. NTP has grouped these chemicals into 12 categories:

1. natural substances;
2. food or cosmetic additives;
3. pesticides;
4. drugs;
5. dyes, dye intermediates and pigments;
6. combustion products;
7. solvents;
8. metals and metal compounds occurring in mining, extraction, and refining processes;
9. analytical and research chemicals;
10. miscellaneous use chemicals;
11. various industrial chemicals and by-products; and
12. occupational exposures with unknown etiologic agents.

For this analysis, OTA has adjusted the total to eliminate double counting, yielding a total of 145.[4]

---

[3]IARC had reevaluated and reclassified Chloramphenicol, and for this reason, changes between the second and third reports, as well as between the third and fourth involve reevaluation of other chemicals by the committee compiling the *Annual Report*.

[4]Specifically, this adjustment affects "nickel" and "nickel refining," "Phenacetin" and "analgesic mixtures containing Phenacetin," and "certain combined chemotherapy" with some chemotherapeutic agents. In the discussion below, the number of actions does not always add correctly to the total indicated, usually because some chemical has been addressed in several different ways.

Some well-known carcinogens have not been included in the *Annual Report* lists, although the introduction briefly mentions several of them: tobacco smoke, alcohol, ionizing radiation, viruses, and ultraviolet radiation (including sunlight). Additionally, though underground hematite mining is listed, underground uranium mining, which exposes workers to radon daughters, is not.

NTP uses the IARC lists of human and animal carcinogens as a source, but several agents and processes on the IARC list have not been listed by NTP. With regard to the IARC listing of exposures under Boot and Shoe Manufacture and Repair and Furniture Manufacture, the report states that, while NTP "does not disagree with these judgments," it does not list these processes because the particular causes or process steps associated with cancer in these cases have not been isolated. NTP also notes that these processes vary significantly among countries and have also changed, thus changing the nature of the exposures.

NCI/NTP CHEMICAL TEST RESULTS

Classification of Test Results

Through June 1, 1987, 308 different substances and mixtures have been tested in long-term animal bioassays sponsored by NCI and NTP. The number of studies totals 327; 17 chemicals have been studied twice and one (trichloroethylene) has been studied three times (85).

Published reviews of these studies summarize the results for the nearly 200 substances tested when the program was under NCI (30,32,73) and the just over 100 conducted since then under NTP (84,85,91,92).

In this analysis, NCI/NTP test results are grouped by the level of evidence for carcinogenicity that they provide and then Federal agency responses to these results through risk assessments and regulations are examined. The "level of evidence" is determined for each particular species and sex. Separate results are given for each "experiment," that is, each combination of species and sex in a study: male rats, female rats, male mice, and female mice. Results of NCI-conducted bioassays and the early bioassays conducted by NTP were described in the technical reports as "positive," "negative," "equivocal," and "inadequate." Since June 1983, NTP has used five categories for levels of evidence, using either "clear evidence" or "some evidence" for positive results. Thus, in the current NTP scheme, the results of each experiment are classified as clear evidence, some evidence, equivocal evidence, inadequate evidence, or no evidence of carcinogenicity (see ch. 2).

NTP test results are examined by peer review, with the reviewing committee classifying the results. Results are then published as technical reports and in summary form. NTP has not developed any general classification or ranking that considers the results of all experiments (for all sexes and species) together.

OTA Grouping of Test Results

OTA has used the most recent summary of results to classify the chemicals tested in NCI/NTP bioassays (85).[5] In that summary, each study was classified by the scheme in use at the time of the study, relying on the conclusions published in the study's technical report. While the summary covers all test results up to June 1987, OTA has included only those chemicals for which the technical report had been printed, or which had already been subject to peer review and data audit, as of the September 1986 NTP Management Status Report.

Table 5-1 summarizes the number of substances in each evidence category. In total, 284 chemicals were tested in 298 separate studies.[6] In this analysis, 13 chemicals have been tested twice and 1 three times. In most cases, each study represents four "experiments": male rats, female rats, male mice, and female mice. In some cases, hamsters were the second species tested; in a few cases, only

[5]OTA will refer to the chemicals tested by NCI/NTP that tested positive in at least one experiment as "positive NCI/NTP chemicals."

[6]The number of studies here is fewer than that given above (306) because OTA's cutoff date was September 1986, while for the review above the date used was June 1, 1987 (85).

Table 5-1.— Summary of NCI/NTP Test Results

Level of evidence	Number of tests	Number of chemicals (Grouping duplicate tests in highest level of evidence)
4 positive	38	36
3 positive	25	25
2 positive	55	51
1 positive	33	32
Total positive	151	144
Equivocal evidence . . .	36	35
Inadequate test	11	9
All negative	100	96
Total	298	284

SOURCE: Office of Technology Assessment, 1987.

male and female animals of one species were tested.

For this analysis, OTA has grouped chemicals by the number of species and sex combinations that show a particular level of evidence. The first group consists of substances for which all four experiments showed positive results ("positive" for older tests, "clear evidence of carcinogenicity" or "some evidence of carcinogenicity" for later tests). The second group consists of chemicals testing positive in three of four experiments, that is, positive in both sexes of one species, but only one sex of the second species. The third group includes chemicals yielding two positive experiments: either positive results in both sexes of one species and equivocal evidence, inadequate evidence, or negative evidence in the second species, or positive results in a mixed fashion in two experiments (e.g., positive in male rats and female mice).[7] The fourth

---

[7]Because there is a high concordance of positive results within a species, a two-positive result in two species may be stronger evidence for carcinogenicity than a two-positive result in one species. For the chemicals analyzed by OTA, 8 of the 51 two-positive results are positive in two species. However, OTA has not analyzed these two kinds of two-positive results separately.

group covers chemicals with one positive experiment. Remaining test results are classified equivocal, inadequate, or negative ("no evidence").

When a chemical has been tested more than once, values on the table represent the highest level of evidence for the chemical. For example, test results for tetrachloroethylene (perchloroethylene) were published in 1977 and 1986. The 1977 results were inadequate in male and female rats and positive in male and female mice. The 1986 results consisted of clear evidence in male rats, some evidence in female rats, and clear evidence in male and female mice. The first test results are considered two positives; the second test results as four. Thus, tetrachloroethylene was grouped with the other chemicals yielding four positive results. When counting the number of chemicals for the second column of table 5-1, multiple tests of the same chemical were counted only once.

As shown in table 5-1, using this method of grouping, 36 chemicals (as opposed to studies) yielded four positive results, 25 three positives, 51 two positives, and 32 one positive. Chemicals that failed to yield at least one positive experiment were not included in analyzing agency responses to test results. The total number of chemicals analyzed by OTA was thus 144. Of the 144, 61 chemicals tested positive in three or four experiments.

Some factors have not been incorporated in the present analysis: affected tumor sites in the animals, whether both high and low doses (or all three doses in a three-dose experiment) produced a response, or the estimated potency of chemicals. Also, the grouping of chemicals here is based solely on results of NCI/NTP tests. OTA has not used results from other animal bioassays or from epidemiologic studies.

OTA ANALYSIS OF ACTIONS, EXPOSURES, AND AGENCY JURISDICTIONS

Agencies and Programs Analyzed

In this analysis, OTA covers the major agencies and programs authorized to regulate chemicals tested in NCI/NTP bioassays (EPA, FDA, OSHA, and CPSC), and two organizations with risk assessment responsibilities (NIOSH and EPA's Carcinogen Assessment Group (CAG)). EPA was analyzed by the following major program areas:

- hazardous air pollutants listed under section 112 of the Clean Air Act (CAA);
- chemicals covered by water quality criteria documents issued under the Clean Water Act (CWA);
- chemicals covered by interim drinking water standards issued and recommended and maximum contaminant levels (RMCLs and MCLs) proposed under the Safe Drinking Water Act (SDWA);
- pesticides canceled, regulated, or voluntarily removed from the market under the Federal Insecticide, Fungicide, and Rodenticide Act (FIFRA);
- chemicals evaluated, designated, or regulated under sections 4 and 6 of the Toxic Substances Control Act (TSCA);[8]
- chemicals listed as hazardous wastes and as hazardous constituents of wastes (Appendix VIII) by the Office of Solid Waste under the Resource Conservation and Recovery Act (RCRA);
- chemicals for which reportable quantities were established under the Comprehensive Environmental Response, Compensation, and Liability Act (CERCLA); and
- chemicals assessed by the Carcinogen Assessment Group (CAG).

Determining Agency Actions on *Annual Report* Chemicals

The source for the discussion below on carcinogen regulation is the *Annual Report* itself.[9] Regulations on carcinogens that are based on noncarcinogenic effects are also covered in the discussion.

Determining Agency Actions on NCI/NTP-Tested Chemicals

To determine which tested chemicals have been subject to regulatory action, OTA sent a list of all chemicals tested in NCI/NTP bioassays to

EPA, FDA, OSHA, and CPSC. OTA asked that they indicate which substances they had evaluated for carcinogenicity, which they had prepared a risk assessment for, which they had proposed to regulate, and which they had issued final regulations on. Federal agency responses were supplemented with information OTA gathered from other sources (see ch. 3).

Exposure Information

The present analysis faced one particularly difficult and important problem—determining exposures. Not every chemical tested by NCI/NTP is actually in commerce, and some have never been in commerce. Some are trace contaminants, chemical byproducts, or intermediates in closed industrial chemical processes, to which exposures may be limited. Others are found in consumer products and foods at relatively low concentrations, but because millions of people are exposed, the potential for harm may be great. Other substances analyzed here may be found in ambient air, surface water, or drinking water supplies.

Which agencies and programs (or statutes) should be concerned about a chemical depends on the nature and extent of exposure. Unfortunately, comprehensive data on toxic chemical exposures do not exist. For example, data on particular environmental media (e.g., drinking water) derive from studies measuring the concentrations of particular chemicals (e.g., EPA's 126 priority water pollutants), studies which of course do not determine the presence of all chemicals of interest.

Lacking information on exposures, agencies frequently use chemical production data to set priorities. To obtain such production data, OTA searched the Hazardous Substance Data Bank (HSDB) of TOXNET, a database maintained by NLM, which includes information on production levels estimated by the Stanford Research Institute (SRI).

While the SRI data on chemical production are frequently used, they also have limitations. First, SRI has not made production volume estimates for every chemical of interest in this analysis. Second, HSDB does not provide any information at all on some chemicals included in this analysis. (Even if SRI had prepared estimates in these cases,

---

[8]TSCA Interagency Testing Committee recommendations and EPA-issued test rules have been excluded.

[9]Throughout this discussion, references to the *Annual Report* are to the fourth report. OTA will refer to the chemicals as "*Annual Report* chemicals."

OTA was not able to use those estimates because the information was not contained in HSDB.) Third, production level is an imperfect proxy for what is really at interest—the actual levels of exposure. For example, a chemical can be produced and consumed in a closed system with relatively little exposure to workers or the environment. Production volume statistics by themselves do not reflect such situations. Nonetheless, production volume is frequently the only information available and is often used when describing and ranking the potential risks of different chemicals.

OTA also obtained information from the NIOSH-conducted National Occupational Hazard Survey (1972-74) and its update, the National Occupational Exposure Survey (1981-83). These surveys present the results of walk-through observations of chemicals found in workplaces representing the manufacturing and public utility sectors.

OTA also asked each agency to indicate which chemicals might be present in those media or exposure situations of interest to the agency. OTA used information on production volumes, potential worker exposures, uses of chemicals, and agency responses to narrow the list of chemicals for each agency or program to the chemicals of potential regulatory interest. This information was used to define a regulatory jurisdiction for each agency or program. In addition, OTA automatically included in an agency's jurisdiction those chemicals that the agency has already acted on.[10]

Regulatory Jurisdictions

In all cases, a chemical is included in the OTA-defined jurisdiction for an agency or program if the agency or program has already acted on that chemical. OTA supplemented this with other information to define the agency jurisdictions for chemicals they have not acted on.

FDA actions were analyzed in the following categories: 1) chemicals evaluated, regulated, or banned in foods, color additives, and cosmetics;

[10]A banned chemical will no longer be in production and thus would not be within a regulatory jurisdiction defined exclusively by production data. In such a case, the chemical would still be included in the regulatory jurisdiction by OTA analysis.

2) animal drugs; and 3) human drugs. Regulatory jurisdiction was based on information on chemical uses and responses from FDA staff concerning chemicals FDA had evaluated.

OSHA's and NIOSH's jurisdictions were defined based on whether a chemical was detected in the NIOSH occupational hazard and exposure surveys or is produced in quantities greater than 1 million pounds annually. OTA did not make distinctions based on the number of employees potentially exposed because that information is either fairly old (deriving from the 1972-74 survey) or still incomplete (data from the 1981-83 survey do not yet cover exposures to trade name products).

To determine a regulatory jurisdiction for CPSC, OTA obtained its staff's indications on the identities and levels of chemicals present in consumer products and on which chemicals present actual or possible consumer exposures.

For hazardous air pollutants, EPA has compiled a database on chemicals of interest and specified methods and developed a computer program for ranking pollutants. This system is called the Modified Hazardous Air Pollutant Prioritization System (MHAPPS) (167). OTA did not use the EPA priority-setting computer program. Rather, OTA searched the MHAPPS database of 609 chemicals for the positive NCI/NTP and *Annual Report* chemicals and for chemicals that were produced in quantities exceeding 1 million pounds per year with either of the following characteristics:

- a vapor pressure > 100 mm Hg or a boiling point $\leq 80°$ C, or
- a vapor pressure > 24 mm Hg and ≤ 100 mm Hg or a boiling point $> 80°$ C and $\leq 100°$ C.

These characteristics are those specified by MHAPPS methods (167). The vapor pressure and boiling point criteria were used to narrow attention to the most volatile chemicals. While dusts, such as those of arsenic or chromium, may also present problems as hazardous air pollutants, they were not included in this analysis.

To define the jurisdiction for EPA administration of the Clean Water Act, OTA used information from an EPA database—the Historical Frequency Database—prepared for EPA's Effluent Guidelines Division, with water samples collected from 1976 to 1979 and computer data entry from 1977 to 1981. That a chemical is in the database indicates it was detected in effluent streams associated with discharges into surface water. That a chemical is not included in the database does not necessarily indicate that it is not a potential water pollutant, only that it was not detected using a particular set of methods and water samples. The chemical's being present in the database, on the other hand, indicates that it is a potential pollutant in at least some locations (113).

To analyze actions on drinking water, OTA requested information from the staff of EPA's Office on Drinking Water on positive NCI/NTP chemicals and *Annual Report* chemicals known to be present or that might be present in drinking water. In addition, EPA staff indicated which chemicals on these two lists had been detected in drinking water, but at levels that they judged "not significant" (36).

For information on pesticides, OTA asked staff of the EPA pesticide program to indicate which chemicals on the two lists were used as active ingredients and inert ingredients in pesticides (21).

TSCA's jurisdiction is all toxic chemicals. Under TSCA, regulatory treatment differs if a chemical is in commerce or is not in commerce. Therefore, distinctions were made between chemicals that are produced in large quantity (more than 1 million pounds annually), produced in smaller quantity (less than 1 million pounds), not produced, or have unknown status because there were no entries on them in HSDB.

The regulatory jurisdictions for RCRA and CERCLA for this analysis are *Annual Report* and positive NCI/NTP chemicals.

Table 5-2 summarizes, by level of evidence, the number of positive NCI/NTP chemicals that have been acted on by the various agencies and programs.

Review of OTA Analysis

A first draft of this analysis was sent to the agencies in February 1987 in preparation for an OTA workshop in March 1987, to which agency staff and representatives of other groups were invited. Again, agencies were asked to indicate which of the *Annual Report* and NCI/NTP chemicals they had acted on and to provide information on regulatory jurisdictions. A second draft was sent to the agencies in May 1987, again requesting comments and providing an opportunity for errors to be corrected.

RESULTS OF OTA ANALYSIS

FDA Actions on Foods and Cosmetics

Annual Report Chemicals

Two *Annual Report* chemicals are in its category "food or cosmetic additives." But a number of chemicals in other *Annual Report* categories contaminate food or cosmetic additives or are considered indirect additives. These chemicals have also been acted on or evaluated by FDA. As described in chapter 3, the major types of materials FDA evaluates for potential hazards in foods and cosmetics are direct food additives, indirect food additives, color additives, other ingredients in cosmetics, and unavoidable environmental contaminants in food.

A few *Annual Report* chemicals are or were direct food additives: safrole and cadmium have been banned as additives to food (though the action taken on cadmium was not based on its carcinogenicity). FDA proposed to ban saccharin, but Congress acted to prevent the ban. Two more *Annual Report* chemicals, nitrilotriacetic acid and hydrazine, are added to boiler water and are considered secondary direct additives. FDA has not banned their use, but instead has set regulations specifying the safe use of these substances.

Fifteen *Annual Report* carcinogens are indirect food additives or were considered potential indirect additives. Indirect additives are usually substances that may migrate into food from packag-

Table 5-2.—Agency Actions on Positive NCI/NTP Chemicals

	Level of evidence				All positive NCI/NTP chemicals
	Four positives	Three positives	Two positives	One positive	
Number of chemicals	36	25	51	32	144
No activity	13	10	12	8	43
CAG assessment	6	3	11	2	22
RCRA listed or App. VIII	14	8	17	2	41
CERCLA					
Listed	14	8	19	6	47
Proposed RQ adjust.	11	4	15	2	32
CWA WQC or standards	6	1	5	2	14
SDWA					
Interim std. (1975 and 1979)	—	1	1	—	2
Proposed and final RMCL	5	—	5	1	11
CAA listed	1	—	—	1	2
TSCA					
Rule 8a or 8d or CHIP	14	8	20	11	53
4(f) review/SNUR/Sec. 6	2	—	2	1	5
FIFRA					
Susp. Canc.	2	—	6	1	9
Vol. Canc.	2	1	—	1	4
Food and Cosmetics (Ban, SUR, or action level)	4	3	7	3	17
FDA					
Animal drugs (revoked/withdrawn)	—	1	2	1	4
Human drugs (labeled/withdrawn)	3	2	—	1	6
NIOSH recommendation	11	2	15	3	31
OSHA					
Noncancer std.	11	4	10	2	27
Cancer std.	1	—	—	1	2
CPSC					
Ban/restricted	2	—	—	1	3
Voluntary reduction only	2	—	3	—	5

SOURCE: Office of Technology Assessment, 1987.

ing material. FDA prohibited the use of materials that contain three of these substances (4,4'-methylene bis (2-chloroaniline), ethylene thiourea, and 2,4-diaminotoluene). FDA proposed to ban the use of three other chemicals (hydrazine, 2-nitropropane, and chloroform [in food contact articles]), but final action was never taken.

For five *Annual Report* chemicals considered potential indirect additives, FDA chose not to ban the packaging materials, but issued rules for safe use of the material (1,2-dichloroethane, dimethyl sulfate, Epichlorohydrin, toluene diisocyanate, chromium [though the last was not for carcinogenicity]). Two *Annual Report* carcinogens, acrylonitrile and vinyl chloride, may migrate from certain plastic bottles. In the 1970s, FDA proposed to prohibit use of these bottles.[11] In the 1980s, FDA took action to permit their use. It is currently evaluating the risks presented by another two potential indirect additives (di(2-ethylhexyl)phthalate (DEHP) and 4,4'-methylene dianiline [MDA]).

Fifteen other *Annual Report* chemicals contaminate direct food or color additives. Urethane is a contaminant of diethylpyrocarbonate (DEPC), which was banned. Benzidine, 4-aminobiphenyl, 2-naphthylamine, o-toluidine, and polycyclic aromatic hydrocarbons contaminate various color

[11]Although at the time, FDA was not regulating acrylonitrile as a suspect carcinogen. Their action to prohibit this particular bottle was overturned in court.

additives.[12] As described in chapter 3, FDA has changed the approach it takes when a color additive is contaminated with carcinogenic impurities. Prior to 1982, FDA banned several such additives. After 1982, following its new policy on impurities, FDA has permitted these color additives.

In evaluating other cosmetic ingredients, FDA has banned the addition of chloroform to cosmetics and allows the use of lead acetate, 2,4-diaminoanisole sulfate, and 4-chloro-o-phenylenediamine in hair dyes. It has set a limit, however, on the amount of lead acetate permitted in these dyes. For the other two chemicals, FDA had attempted to require a product warning label on coal tar dyes. This requirement was stayed by court order. FDA has not taken action to reinstate the warning label.

Finally, with regard to environmental contaminants, FDA has set food tolerances for polychlorinated biphenyls (PCBs) and action levels for 15 *Annual Report* chemicals, including aflatoxins, several pesticides (DDT, ethylene dibromide, Kepone, Mirex, Toxaphene), polybrominated biphenyls, and cadmium (though this FDA action was not based on carcinogenicity). FDA is also currently considering a petition to reduce permissible levels of urethane (a product of the fermentation process) found in wine and other alcoholic beverages.

Considering all FDA activities together and eliminating double counting of chemicals yields a total of 52 different *Annual Report* carcinogens examined by FDA. Of these, 9 individual chemicals and one group of 10 chemicals, polycyclic aromatic hydrocarbons (PAHs), are associated with materials banned by FDA. Twelve chemicals are associated with materials FDA has issued safe use rules on or has permanently listed (as permissible color additives), and three *Annual Report* chemicals are permissible ingredients in hair dyes. For 16 chemicals, tolerances or action levels have been set. Some carcinogens in the *Annual Report* are subjects of proposed bans that were

never finalized. FDA is still evaluating at least three (DEHP, MDA, and urethane in alcoholic beverages).

Of the 52 chemicals, 46 have been subject to final FDA actions consisting of bans, safe use rules, permanent listing decisions, or the setting of tolerances or action levels. Ten chemicals have not been subject to final actions, although sometimes this lack of action concerns only some uses.[13]

Excluding environmental contaminants, 37 chemicals are associated with food or color additives, and potentially subject to bans under provisions of the Food, Drug, and Cosmetic Act. Of these, 19 (including 10 PAHs) are associated with materials actually banned by FDA. For most of the remaining chemicals, FDA has specified safe use rules.

Positive NCI/NTP Chemicals

Most of the bioassay information evaluated by FDA for food and cosmetic ingredients is obtained from testing FDA requires from the ingredients' sponsors. In a few cases, direct food or color additives or potential indirect additives have been tested by NCI/NTP. More frequently, the NCI/NTP bioassay program has tested chemicals that may be present as impurities in additives or cosmetics.

FDA actions on positive NCI/NTP chemicals may be broken down based on the use of the material. One direct food additive has been banned (cinnamyl anthranilate). Currently pending is a proposed safe use rule to allow use of methylene chloride to decaffeinate coffee. FDA has also proposed to ban use of trichloroethylene for coffee decaffeination and cosmetic uses. Although that proposal was never issued in final form, those uses of trichloroethylene have apparently stopped.

One color additive, D&C Red No.9, is in the group of 144 positive test result chemicals, having tested positive in male rats. As described in chapter 3, FDA has permanently listed this color additive.

---

[12]OTA has not determined which of the 10 polycyclic aromatic hydrocarbons listed in the *Annual Report* actually contaminated carbon black and graphite, the colors in question. For simplicity, all are included in this discussion.

[13]Hence these chemicals total 52 because several chemicals are in both groups.

FDA has taken some action on three positive NCI/NTP chemicals that are potential indirect additives. It has proposed to ban chloroform from food contact materials. Safe use rules have been issued for 1,4-dioxane and 1,1,2-trichloroethane.

Safe use rules have also been issued for three contaminants of color additives: aniline hydrochloride, azobenzene, and CI Vat Yellow No.4.

Ten other positive NCI/NTP chemicals have or had cosmetic uses. One has been banned from cosmetics (chloroform); FDA has proposed banning one other from cosmetics (methylene chloride). Seven positive NCI/NTP chemicals are used in hair dyes (2,4-diaminoanisole sulfate, 4-chloro-m-phenylenediamine, 4-chloro-o-phenylenediamine, 2-nitro-p-phenylenediamine, 4-amino-2-nitrophenol, HC Blue No.1, and 2,4-diaminotoluene). These dyes are all currently permitted. As mentioned above, FDA acted to require a warning label for coal tar hair dyes, but that requirement was overturned by court order. In addition, selenium sulfide is allowed for use in dandruff shampoo.

Action levels or food tolerances have been issued for eight positive NCI/NTP chemicals.

Considering these FDA actions together and eliminating double counting yields a total of 17 positive NCI/NTP chemicals on which FDA has taken some final action; for 2 other chemicals it has proposed action. A greater number of positive NCI/NTP chemicals, 31, have only been evaluated by FDA. These include 9 chemicals with four positive results, 3 with three positives, 10 with two positives, and 9 with one positive result. The evaluations include exposure assessments and risk assessments, which were conducted because FDA thought that the chemicals might be found in food additives or cosmetics.

The scope of FDA jurisdiction is thus 48 positive NCI/NTP chemicals. Of these, 19 had three or four positive experiments. FDA has issued final bans, safe use rules, or action levels or tolerances for seven of these.

FDA Actions on Human Drugs

Annual Report Chemicals

Thirty-four *Annual Report* chemicals have or had uses as human drugs. Thirty-one of these are listed in the *Annual Report* as "drugs." Several of these listings, however, may represent double counting: "phenacetin" and "analgesic mixtures containing phenacetin" are listed separately, and "certain combined chemotherapy for lymphomas" overlaps with the listing of specific drugs included in those therapies. For this discussion, therefore, OTA will count 29 chemicals as "drugs." In addition, five chemicals have or had drug uses, or might be found in drug products, but are listed in different categories: thorium dioxide (listed under "miscellaneous uses"), chloroform (listed under "solvents"), coal tar (listed under "occupational exposure with unknown etiologic agent"), urethane and vinyl chloride (both listed under "industrial chemicals and byproducts").

Of the 29 chemicals listed as "drugs," 25 are on the market with physicians' labeling information warning of carcinogenic effects, 2 have been removed from the market or were never approved (Phenacetin) and (Chlornaphazine), and 2 are approved drugs on the market (selenium sulfide and Propylthiouracil). Selenium sulfide is approved for use in dandruff shampoos and for topical treatment of fungal infections. The labeling in this case only indicates the negative results of skin-painting experiments in mice. There is no labeling information on propylthiouracil in the *Physician's Desk Reference*.

The remaining five chemicals in nondrug categories are the following:

1. Thorium dioxide, which FDA approved for x-ray imaging, although labeling restricts use to patients with limited life expectancy.
2. Chloroform, for which FDA banned drug uses in 1976.
3. Coal tar, which has medical use as a topical antifungal agent and in the treatment of psoriasis. It was declared to be unsafe for over-the-counter use by an FDA advisory panel in 1982. The final monograph on this decision is still being prepared for publication.

4. Urethane, which was removed from the market in 1970 because it was determined to be ineffective as a drug.

5. Vinyl chloride, for which FDA announced that a new drug application would be required for drug use to be permissible.

Thus there are 34 *Annual Report* chemicals with drug uses: 5 were removed from the market or never approved, 26 are approved for use but with the physicians' labeling information warning of potential carcinogenic effects, 2 were approved for topical use, and 1 was approved for use without any labeling information available.

Positive NCI/NTP Chemicals

Twelve positive NCI/NTP chemicals were indicated to be used in drugs, many as anticancer agents. All drugs are permitted to be on the market by virtue of some FDA regulatory action (e.g., through approval of a new drug application in the case of new drugs). But for this analysis, the actions of interest are regulatory and directed toward the carcinogenic risk that may be presented by these drugs. Thus only actions to remove a drug from the market, restrict its uses, or require warning labels are considered as regulatory actions. By this standard, two positive NCI/NTP chemicals have been regulated: chloroform was banned from drugs, and FDA required physician labeling for Reserpine (three positives) to warn of animal carcinogenicity. Physician labeling for 5 of the 11 remaining drugs warns of potential carcinogenicity. The final 6 drugs are not included in the latest edition of the *Physician's Desk Reference* and may never have gotten past the investigational stage. All were intended for use as anticancer drugs. Finally, although it is not a drug, DEHP, which is used to make blood bags, may migrate from those bags into the blood stored inside.

FDA Actions on Animal Drugs

Annual Report Chemicals

Six *Annual Report* chemicals are used as drugs for food-producing animals. FDA revoked approval for one of these—diethylstilbestrol (DES)—while for a second, Reserpine, the sponsor withdrew the application for approval. The remain-ing four substances and classes of substances—conjugated estrogens, nonconjugated estrogens, progesterone, and iron dextran complex—may be used in animals.

Positive NCI/NTP Chemicals

Five positive NCI/NTP chemicals have uses as animal drugs or are related to animal drugs. FDA has revoked approval for four of these. The fifth chemical, Zearalenone, is related to the animal drug Zeranol. Thus far, no action has been taken on Zeranol.

OSHA

Annual Report Chemicals

OSHA regulates exposures to 52 *Annual Report* chemicals, although for 35 of these chemicals, the standards were based on noncarcinogenic effects and were adopted as "startup" standards 15 years ago. These standards may not be sufficiently protective when potential carcinogenicity is considered. OSHA has issued "permanent" standards based on carcinogenicity for 17 *Annual Report* chemicals. Ten of these were part of the "14-carcinogen standard" issued in 1973, while the remaining 7 chemicals were regulated individually. One of these, asbestos, has been the subject of two different "permanent" standards. Two of OSHA's permanent standards were overturned by the courts. One of these (regulation of benzene) is currently the subject of a new proposal. In the meantime, however, the old startup standard continues to be used. The other (4,4'-methylene bis (2-chloroaniline)) never had a startup standard and is currently unregulated. New standards have been also proposed for two more *Annual Report* chemicals, formaldehyde and ethylene dibromide (EDB).

As mentioned above, OSHA is also using the *Annual Report* as part of its hazard communication standard. While this information will be valuable to employers and employees, it does not replace the need for standards that set exposure limits and require controls. OSHA itself is considering regulatory action on several *Annual Report* chemicals. The actions summarized here are only OSHA regulatory actions that set exposure limits or control requirements.

While OSHA has exposure standards and requirements, based on either carcinogenic effects or other toxicities, for 52 *Annual Report* chemicals, it has no exposure standards of either type for the other 93 *Annual Report* chemicals. These unregulated chemicals and the 35 chemicals with startup standards based on noncarcinogenic toxicities gives a total of 128 *Annual Report* chemicals that OSHA does not regulate for carcinogenicity.

Not all of these 128 chemicals are currently produced or used in the United States. Using the OTA-defined OSHA jurisdiction (potential worker exposure detected in NIOSH occupational surveys or production volume greater than 1 million pounds), 58 of the 93 *Annual Report* carcinogens lacking standards and 25 of the 35 with startup standards are of regulatory concern to OSHA. Thus, the OSHA jurisdiction includes 83 of the 128 *Annual Report* chemicals that lack standards based on carcinogenicity.

Considered another way, OSHA has issued standards based on carcinogenicity for 17 *Annual Report* chemicals and noncancer standards for another 35. Fifty-eight *Annual Report* chemicals lacking standards are in the OSHA jurisdiction. The total OTA-defined jurisdiction is 110 *Annual Report* chemicals.

Positive NCI/NTP Chemicals

As mentioned above, OSHA has issued permanent standards related to carcinogenicity for 2 of the 144 positive NCI/NTP chemicals—asbestos and 1,2-dibromo-3-chloropropane (DBCP). Neither of these actions, however, was based primarily on the NCI/NTP test results: asbestos is a human carcinogen, and DBCP was regulated by OSHA primarily because it caused sterility in male workers, although the NCI carcinogenicity data were available and were considered by OSHA when it set the standard.

Two other positive NCI/NTP chemicals have been proposed for new standards—benzene and EDB. Benzene had been the subject of a final standard in 1978 based on human evidence, but the standard was overturned by the courts. OSHA issued a proposal for EDB in 1983, but no standard has been issued in final form, although most pesticide uses of EDB were canceled by EPA in 1984.

EDB has actually been tested twice by NCI/NTP. The first results, positive in all four experiments using gavage exposure, were published in 1978, while the second results, positive in all four experiments in an inhalation study, were published in 1982. OSHA still has not issued a standard. In explaining its "cancer policy," OSHA described a bioassay result that it should not ignore, even if the test results were based on high doses:

> Those who would urge OSHA to reject data from tests conducted at "too toxic" doses would presumably wish OSHA to ignore data such as those on 1,2-dibromoethane [ethylene dibromide], which induced multiple-site tumors at both dose levels in both sexes of rats and mice within 60 weeks of exposure, at incidence rates up to 94 percent. . . . OSHA believes that it would be improper to ignore such overwhelming evidence of hazard (274).

This quotation refers to the 1978 NCI results.

Twenty-seven of 144 positive NCI/NTP chemicals (19 percent) are currently regulated under the OSHA startup standards based on noncarcinogenic toxicity. These include 15 among the 61 with three or four positive results (25 percent).

Some positive NCI/NTP chemicals are not present in the workplace in substantial quantities. As explained above, to develop a jurisdiction of chemicals of regulatory interest, OTA used information for chemical production and detection in the NIOSH surveys. Using these criteria, 17 of the 27 positive NCI/NTP chemicals currently regulated with startup standards are of potential regulatory interest to OSHA. Twelve of these have positive results in three or four experiments. As mentioned above, proposals are pending for two of these.

A total of 115 of the 144 positive NCI/NTP chemicals have no occupational exposure standard.[14] Of the 115 chemicals, 24 are in the OTA-defined OSHA regulatory jurisdiction. Forty-five

---

[14]Although positive test results may trigger coverage under standards for labeling and access to medical records.

of the 144 positive NCI/NTP chemicals were positive in three or four experiments. Of these, 14 are in OSHA's jurisdiction.

Considered another way, OSHA has noncancer standards for 27 positive NCI/NTP chemicals and carcinogenicity standards for 2. An additional 24 chemicals were positive in NCI/NTP bioassays and are in OSHA's jurisdiction, but have no occupational standard of either sort. Limiting attention to the chemicals that tested positive in three or four experiments, OSHA has noncancer standards for 15 chemicals and a carcinogenicity standard for 1 more. An additional 14 chemicals with three or four positive results are in OSHA's jurisdiction, but have no OSHA standards.

NIOSH

Annual Report Chemicals

NIOSH has prepared recommendations to OSHA for 59 Annual Report chemicals. For 18 of these, OSHA has issued standards based on carcinogenicity, although two of these standards were struck down by the courts (those for benzene and 4,4'-methylene bis (2-chloroaniline)). OSHA has proposed a new standard for benzene and for two additional Annual Report chemicals on which there are NIOSH recommendations (formaldehyde and EDB). In all, there are 20 chemicals with NIOSH recommendations for which OSHA has either proposed or issued standards for carcinogenicity. The remaining 39 chemicals include 24 covered by startup standards, which as already mentioned, were not based on carcinogenicity, and 15 currently lacking an OSHA standard.

NIOSH has not issued recommendations for 86 Annual Report chemicals. In this case, once again, not all of these chemicals are produced or present potential worker exposures. Based on the OTA-defined jurisdiction, 53 of the 86 chemicals lacking NIOSH recommendations are produced in quantities greater than 1 million pounds or were detected in the NIOSH occupational surveys and thus may present worker exposures.[15]

---

[15]OTA defined the NIOSH jurisdiction to include chemicals detected in NIOSH's occupational surveys as well as chemicals with annual production volume greater than 1 million pounds. The NIOSH and OSHA jurisdictions differ slightly because there are several chemicals that NIOSH has made recommendations on, but that do not satisfy any of the criteria for the OSHA jurisdiction.

Positive NCI/NTP Chemicals

NIOSH recommendations cover 31 of the 144 positive NCI/NTP chemicals (22 percent). NIOSH recommendations cover 13 of the 61 chemicals that tested positive in three or four experiments (21 percent). There are 113 positive NCI/NTP chemicals for which NIOSH has not issued recommendations, including 48 three- and four-positive results. Thirty-one of the 113 positive NCI/NTP chemicals and 22 of the 48 chemicals with positive results in three or four experiments are in the OTA-defined jurisdiction for NIOSH.

CPSC

Annual Report Chemicals

CPSC reported regulatory activity or voluntary control for 18 Annual Report chemicals. These include four chemicals banned from consumer products—carbon tetrachloride, tris(2,3-dibromopropyl)phosphate (tris), certain uses of asbestos, and vinyl chloride used as an aerosol propellant. In addition, for asbestos, a proposed ban of use in consumer hair dryers led to voluntary control. CPSC attempted to ban urea-formaldehyde foam insulation (UFFI) to prevent consumer exposures to formaldehyde from this source. This action was subsequently overturned by the courts, although use of UFFI has apparently ceased. In addition, products containing more than 1 percent formaldehyde must bear a label warning of irritation associated with formaldehyde.

Benzene products were already covered by a labeling requirement when the issue of their carcinogenicity was raised. After a ban of benzene in all consumer products (except gasoline) was proposed, the use of benzene in these products stopped. Exposures to lead acetate and lead phosphate (listed as Annual Report carcinogens, but grouped as one chemical) are indirectly regulated through CPSC limits on lead in paint. For five dyes related to benzidine and found in artist materials, the hazard was voluntarily reduced. Levels of six different N-nitroso compounds were restricted in children's pacifiers.

To define a regulatory jurisdiction, OTA asked CPSC staff to provide information on the occurrence of Annual Report chemicals in consumer products. Based on this information, 34 Annual

Report chemicals are present in consumer products for which no CPSC regulatory actions have been taken. Five of these are believed to present actual or possible consumer exposure—arsenic and arsenic compounds, chromium and chromium compounds, DEHP, 1,2-dichloroethane, and thiourea.

Limiting attention to chemicals believed to present actual or possible consumer exposures or that CPSC has already acted on yields a total of 24 chemicals in the OTA-defined CPSC jurisdiction. Eighteen of these have been subject to CPSC regulatory action or voluntary reductions or controls.

Positive NCI/NTP Chemicals

CPSC has addressed eight positive NCI/NTP chemicals. Three of these were the focus of some regulatory activity: tris (four positives) was banned from children's sleepwear, benzene (four positives) was voluntarily removed from consumer products after a proposed ban, and several consumer product uses of asbestos (one positive) were eliminated (through bans and voluntary actions). But again, all actions on benzene and asbestos were based on human evidence, not NTP bioassay results.

Five chemicals were subjects of some voluntary actions, for one of which a proposed labeling requirement is also pending. CPSC reports manufacturers voluntarily stopped using three different dyes (all yielding two positives) in consumer products. CPSC convened a chronic hazard advisory panel on DEHP (four positives) to consider regulatory action, but DEHP was voluntarily removed from children's pacifiers. CPSC has proposed that methylene chloride (four positives) be labeled a hazardous substance and has achieved some voluntary reductions in the use of this chemical in consumer products.

Among the 144 positive NCI/NTP chemicals, 65 are present in consumer products. Among these 65, 13 present what CPSC determines to be "actual or possible" consumer exposure.

Defining the CPSC jurisdiction to be those chemicals that present actual or possible exposures or that have been acted on by CPSC yields 14 positive NCI/NTP chemicals. Eight of these have been the subjects of CPSC regulations or volun-

tary reductions. Seven of the 61 NCI/NTP chemicals with three or four positive experiments fall in this CPSC jurisdiction. Four of these have been the subjects of CPSC regulations or voluntary reductions.

EPA Actions Under the Clean Air Act

Annual Report Chemicals

Six *Annual Report* chemicals have been listed as hazardous air pollutants, although the listing of one (beryllium) was based on noncarcinogenic effects. Emissions standards have been issued for five of the six listed (asbestos, beryllium, vinyl chloride, benzene, and arsenic), although proposed standards for benzene are still pending for other industries. Coke oven emissions standards have recently been proposed, but are not yet final. EPA has announced the "intent to list" five other *Annual Report* carcinogens: carbon tetrachloride, chloroform, chromium, cadmium, and ethylene oxide.

OTA defined a regulatory jurisdiction for EPA's regulation of hazardous air pollutants using information in the EPA-compiled MHAPPS database. Narrowing the jurisdiction to chemicals currently produced in quantities greater than 1 million pounds, OTA has selected the *Annual Report* carcinogens with relatively high vapor pressures, those greater than 100 mm Hg, and with vapor pressures between 24 and 100 mm Hg. The *Annual Report* carcinogens with relatively high volatility (vapor pressure greater than 100 mm Hg) are vinyl chloride, acrylonitrile, ethylene oxide, chloroform, and formaldehyde. One of these five (vinyl chloride) has been listed and regulated. Ethylene oxide and chloroform have been placed in the "intent to list" category. EPA has announced a plan to give local governments responsibility for addressing acrylonitrile exposures.

The *Annual Report* chemicals with lower volatility (vapor pressure between 24 and 100 mm Hg) and production greater than 1 million pounds are benzene, carbon tetrachloride, and 1,4-dioxane. One of these three chemicals (benzene) has been listed, and a second (carbon tetrachloride) is in the "intent to list" category.

The total jurisdiction in this case is 15 chemicals, consisting of the 8 chemicals above and 7 others that EPA has listed or has announced an intent to list. Of the 15, 6 have been listed.

Positive NCI/NTP Chemicals

Two NCI/NTP chemicals have been acted on under the Clean Air Act, although several more have been subject of pre-regulatory evaluations. Two NCI/NTP chemicals have been listed—benzene and asbestos—although both decisions to list were based on human evidence for carcinogenicity and were made before NCI/NTP test results were available. For six others, EPA has announced an "intent to list" (1,2-dichloroethane, tetrachloroethylene, chloroform, 1,3-butadiene, trichloroethylene, and methylene chloride).

Chloroform, 1,3-butadiene, methylene chloride, and 1,2-propylene oxide are chemicals within the OTA-defined jurisdiction because they occur in the MHAPPS database, have a production volume greater than 1 million pounds, and have a vapor pressure greater than 100 mm Hg. EPA has announced an intent-to-list decision for three of these four chemicals. NCI/NTP chemicals produced in volumes greater than 1 million pounds, but with lower vapor pressures (between 24 and 100 mm Hg) are benzene, 1,4-dioxane, ethyl acrylate, trichloroethylene, 1,2-dichloropropane, and 1,2-dichloroethane. Regarding these six chemicals, EPA has listed one, and announced its intent to list another.

Considering these chemicals along with those EPA has already listed or those it has announced its intent to list, and eliminating the double counting, yields a jurisdiction in this case of 12 positive NCI/NTP chemicals. Two of these have been listed. Limiting attention to positive NCI/NTP chemicals with three or four positive experiments, the jurisdiction is eight chemicals, of which one has been listed.

EPA Actions Under the Clean Water Act

Annual Report Chemicals

Under the Clean Water Act, water quality criteria documents have been issued for 47 *Annual Report* chemicals. During the 1970s, toxic effluent standards were issued for five of these chemicals.

But, as discussed in chapter 3, the "Flannery decree" replaced EPA development of toxic effluent standards with the use of technology-based effluent limitations, which are now used to regulate chemicals covered by the water quality criteria documents. In addition, discharges from the rubber industry (an *Annual Report* "chemical") are also regulated, although there is no specific water quality criteria document for this industry.

Beyond these 48 regulated chemicals, EPA's database shows that another 17 chemicals detected in effluent streams have not been regulated. Thus, the OTA-defined jurisdiction for the EPA clean water program is 65 chemicals.[16]

Positive NCI/NTP Chemicals

Water quality criteria documents under the Clean Water Act that consider carcinogenicity cover 14 chemicals from positive NCI/NTP bioassays, 7 with three- and four-positive results. Toxic effluent standards were issued for two of these chemicals—p,p'-DDE (related to DDT) and Toxaphene—although both actions took place prior to publication of NCI/NTP test results. All 14 chemicals are covered by technology-based standards because they are included in the list of 65 chemicals under the Clean Water Act.

In addition to the 14 positive chemicals covered by water quality criteria documents, EPA's database of chemicals detected in effluent streams shows another 13 positive NCI/NTP chemicals that have not been regulated. These include two chemicals with four positives, one with three positives, seven with two positives, and three with one positive. Thus, the OTA-defined jurisdiction of positive NCI/NTP chemicals for the clean water program includes a total of 27 positive test results. Fourteen of these 27 (52 percent) are covered in the water quality criteria documents that have been issued to date.

---

[16]By coincidence, this equals the number of classes of priority pollutants. But the actual overlap is limited to 47 chemicals plus the rubber industry.

EPA Actions Under the Safe Drinking Water Act

Annual Report Chemicals

Annual Report chemicals regulated in some way under the Safe Drinking Water Act include seven covered by interim standards: arsenic, cadmium, chromium, lead compounds (lead acetate and lead phosphate[17]), Lindane, Toxaphene, and chloroform. The first four of these, however, were not regulated for carcinogenic effects.

In the current process of setting standards, 11 *Annual Report* chemicals have been the subjects of proposed recommended maximum contamination levels (RMCLs): six standards were based on carcinogenicity for DBCP, EDB, Epichlorohydrin, Lindane, PCBs, and Toxaphene), while the remaining five were not (for arsenic, asbestos, cadmium, chromium, and lead compounds). For four more *Annual Report* carcinogens final RMCLs and final maximum contamination levels (MCLs) have been issued: benzene, carbon tetrachloride, 1,2-dichloroethane, and vinyl chloride.

For 19 *Annual Report* chemicals, EPA has issued health advisories. Of these chemicals, 14 are covered by interim standards or the current RMCL/MCL process. For the remaining five chemicals (1,4-dioxane, ethylene thiourea, hexachlorobenzene, nickel, and 2,3,7,8-tetrachlorodibenzo-p-dioxin), only the nonbinding health advisories have been issued.

To define a regulatory jurisdiction, OTA requested that the EPA Office of Drinking Water Standards indicate which *Annual Report* chemicals are found in drinking water and which present significant exposures. The office indicated that 120 of the chemicals in the *Annual Report* had been detected in drinking water. Of these, 31 were estimated to present significant human exposures, including 9 of the 10 polycyclic aromatic hydrocarbons. Finally, for eight *Annual Report* chemicals, information is inadequate to judge the magnitude of human exposure.[18]

EPA has interim standards in place for 7 of the 120 chemicals detected in drinking water. Fifteen

chemicals are included in the current RMCL/MCL process. For 18 chemicals, health advisories have been prepared, although the health advisory represents the only EPA action on 5 of those chemicals in drinking water. These actions cover 21 chemicals altogether; all but one of these were estimated to present significant human exposure.[19]

There has been no EPA action on drinking water exposures for the remaining 99 chemicals. Of these, 11 chemicals (beryllium, 2,4,6-trichlorophenol, and 9 of the 10 polycyclic aromatic hydrocarbons) were estimated to present significant known or potential human exposures.

OTA's jurisdiction for actions under the Safe Drinking Water Act consists of the 21 chemicals covered by interim standards or the current RMCL/MCL process and 11 additional chemicals with known or potential exposures in drinking water. Of these 32 *Annual Report* chemicals, 7 have interim standards, 15 are being considered in the RMCL/MCL process, and 11 have not been addressed by regulatory action.

Positive NCI/NTP Chemicals

Twelve positive NCI/NTP chemicals are addressed by either the interim standards under the act or the current RMCL/MCL standard-setting process. The two interim standards covered toxaphene (in 1975) and chloroform (in 1979). Eleven positive NCI/NTP chemicals are addressed by the current RMCL/MCL process: for three, final RMCLs and final MCLs have been issued; for one, a proposed MCL is pending; and for seven, proposed RMCLs are still pending. (Regulatory standards and proposals cover a total of 12 chemicals because chloroform is not now being addressed by the RMCL/MCL process.)

Fourteen of the positive NCI/NTP chemicals are found in drinking water and have been subjects of health advisories. Four of these chemicals have standards or proposed RMCLs. Thus 10 positive NCI/NTP chemicals found in drinking water are addressed by health advisories, but not by the standards-setting process.

---

[18]These chemicals were not included in the OTA-defined jurisdiction.

[19]The exception is 2,3,7,8-tetrachlorodibenzo-p-dioxin.

EPA staff indicated to OTA that another 24 positive NCI/NTP chemicals are found in drinking water, but in their judgment, 22 are not significant or data are not available on them. The two with a significant known or potential presence in drinking water that are also not addressed by the standards-setting process or health advisories are 2,4,6-trichlorophenol (three positives) and di-(2-ethylhexyl)-adipate (two positives).

The OTA-defined jurisdiction in this case includes only those chemicals EPA has acted on or that present significant potential drinking water hazards. Thus, the jurisdiction with regard to positive NCI/NTP chemicals consists of 14 chemicals. Of these, 2 are addressed by interim standards, 11 by current regulatory activity, and 2 by no action.[20]

EPA Actions Under the Federal Insecticide, Fungicide, and Rodenticide Act

Annual Report Chemicals

Twenty-four Annual Report chemicals are or were registered as active pesticide ingredients under FIFRA.[21] Thirteen of these are listed as "pesticides" in the Annual Report; the remaining 11 substances are listed in other categories but have or had pesticide uses. The 24 chemicals include 6 that were voluntarily canceled for some or all uses (acrylonitrile, Aramite, arsenic, benzene, Kepone, and Safrole).[22] Another eight were subject to complete or partial cancellation by EPA (Amitrole, carbon tetrachloride, DDT, DBCP, EDB, Mirex, Toxaphene, and vinyl chloride), and five were subject to special review but were not canceled or suspended (cadmium, chloroform, ethylene oxide, and Lindane). For Nitrofen and sulfallates, EPA has only set food tolerances. Food tolerances are also reported for six other Annual Report pesticides. A registration standard has been issued for formaldehyde.

No actions are reported for three chemicals used as active pesticide ingredients: 2,4,6-trichlorophenol, hexachlorobenzene, and 1,2-dichloroethane. The Annual Report, however, suggests that 2,4,6-trichlorophenol is no longer being produced because of the expense of removing dioxin contamination.

Positive NCI/NTP Chemicals

Twenty-seven positive NCI/NTP chemicals are or were used as active ingredients in pesticides: seven with four positive results, four with three positives, six with two positives, and five with one positive. NTP has also tested DDE, which is associated with DDT, and which yielded two positive results, for a total of 28 positive NCI/NTP chemicals.[23]

For about half of these chemicals some uses have been suspended, canceled, or voluntarily canceled, although nearly all of these chemicals remain on the market for at least some uses. EPA has canceled or suspended nine chemicals: DBCP (four positive results), EDB (four positive results), Chlordane (two positives), Heptachlor (two positives), chlorobenzilate (two positives), p,p'-DDE (a contaminant and metabolite of DDT) (two positives), propylene dichloride (two positives), Toxaphene (two positives), and Aldrin (one positive). Two of these (Chlordane and Heptachlor) are closely related chemically; EPA acted on them simultaneously. In addition, four chemicals were voluntarily canceled as active ingredients: benzene (four positive results), Chlordecone (Kepone) (four positives), Nitrofen (three positives), and Monuron (one positive). Of these 13 chemicals affected by regulatory and voluntary cancellations and suspensions, a number remain on the market for some uses. For example, Chlordane was canceled for food uses, but is still used for termite control.

The 15 remaining positive NCI/NTP chemicals are still on the market with their uses unchanged by suspension, cancellation, or voluntary cancellation. For one of these chemicals, Tetrachlorvinphos (three positives), EPA judged that the weight

---

[20]One chemical is addressed by both interim standards and current regulatory activity.

[21]Twelve Annual Report chemicals are used as inert ingredients in pesticide formulations. In 1987, EPA announced a policy covering some of these inert ingredients (see ch. 3).

[22]Non-Lindane isomers of hexachlorocyclohexane were also voluntarily canceled, although Lindane itself is still marketed.

[23]There are also 13 positive NCI/NTP chemicals that are used as inert ingredients, including 2 that are or were used as active ingredients.

of the evidence does not support regulation. Several of the other chemicals have been subjects of special reviews or registration standards and two have been proposed for cancellation (Captan and Dicofol). To some extent, EPA has issued requirements for labeling and use of protective equipment with these chemicals, but OTA did not evaluate these measures.

EPA Actions Under the Toxic Substances Control Act

Annual Report Chemicals

Under TSCA, 32 of the 145 *Annual Report* chemicals have received some attention. Most of this attention has consisted of developing information, including reporting requirements and Chemical Hazard Information Profiles (CHIPs). Under section 8, EPA can issue regulations requiring manufacturers to provide information on production, uses, exposures, environmental and health effects, and disposal of chemicals (sec. 8(a)) and requiring manufacturers to submit unpublished health and safety studies (sec. 8(d)).[24] CHIPs are medium-sized reviews (e.g., 20 to 70 pages) of physical properties, production and exposure information, health effects, environmental effects, and other existing standards and regulations for particular chemicals.

Eighteen *Annual Report* chemicals are subject to section 8(a) or 8(d) reporting rules. EPA has prepared CHIPs for 10 other chemicals: benzotrichloride, hydrazobenzene, Michler's base, 5-nitro-o-anisidine, o-toluidine, 1,4-dioxane, 2-nitropropane, thiourea, thorium dioxide, and toluene diisocyanate (TDI). The first five of these chemicals are used in manufacturing dyes, while 1,4-dioxane and 2-nitropropane are used as solvents. The remaining three chemicals are classed as miscellaneous chemicals and industrial chemicals.

EPA has issued final regulations banning production under section 6 for one *Annual Report* carcinogen—PCBs. As described in chapter 3, Congress specified this action under TSCA. A section 6 proposal on asbestos is pending. This proposed regulation would limit and eventually eliminate the use of asbestos.[25] For two other *Annual Report* chemicals—formaldehyde and MDA—EPA has initiated an expedited review under section 4(f) and then referred regulatory consideration to OSHA. EPA has issued significant new use rules for two *Annual Report* chemicals: hexamethylphosphoramide and urethane.

As defined by OTA, the TSCA jurisdiction consists of all chemicals in the *Annual Report*, although regulatory action under TSCA would differ depending on whether or not a chemical is produced. According to the information available to OTA, domestic production of 47 *Annual Report* chemicals exceeds 1 million pounds. Another 66 chemicals are produced, but in quantities less than 1 million pounds, 13 chemicals are currently not produced, and the production status of 19 chemicals is unknown.

Narrowing attention to just those *Annual Report* chemicals produced in quantities greater than 1 million pounds yields 6 chemicals with EPA-prepared CHIPs and 11 chemicals subject to section 8(a) or 8(d) reporting rules. The section 4(f) designations, significant new use rules, and the section 6 proposal on asbestos (mentioned above) all affect chemicals drawn from this high-production group. After eliminating multiple actions on the same chemical, there are 20 high-production chemicals addressed by some action and 27 high-production chemicals on which no action has been taken.

Under TSCA, for chemicals currently not produced EPA could require manufacturers to report significant new uses or provide production or exposure information prior to restarting production. Of the 145 carcinogens in the *Annual Report*, 13 chemicals are not currently produced. Because of

---

[24]Sec. 8(e) of TSCA requires manufacturers to report to EPA information on chemicals that present a "substantial risk of injury to health or the environment." While processing these reports has been an important TSCA activity, it is not directly regulatory and its purpose is to aid in identifying new hazards. This analysis discusses chemicals already identified as carcinogenic in the *Annual Report* or by NCI/NTP tests.

[25]In addition, EPA has issued a regulation concerning certain State and local government employees exposed to asbestos during removal operations, but whose working conditions are not regulated by OSHA or the States. This standard is similar to the asbestos standard issued by OSHA for other workers.

the congressional ban, PCBs are among these chemicals. The remaining 12 chemicals have no TSCA reporting requirements.

Positive NCI/NTP Chemicals

Three NCI/NTP chemicals have been reviewed under section 4(f) of TSCA: MDA, 1,3-butadiene, and methylene chloride. Methylene chloride and MDA had four-positive results in the NCI/NTP tests, while 1,3-butadiene showed two positives. However, 1,3-butadiene had already been shown to be carcinogenic in rats and thus NTP tested in mice only. Including the positive rat data would give this chemical a four-positive result as well.

The NTP test results for these chemicals led directly to TSCA activity. EPA has referred consideration of formaldehyde, 1,3-butadiene, and MDA to OSHA under section 9 of TSCA.

One positive NCI/NTP chemical, asbestos, has been proposed for regulation under section 6. Although the NTP tests found some evidence for carcinogenicity in male rats (one positive), the primary basis for all asbestos regulation is the human epidemiologic data.[26] EPA has also issued a significant new use rule for pentachloroethane (two positives).

Most of TSCA activity on existing chemicals tested by NCI/NTP has involved reporting requirements under section 8 of TSCA and the preparation of TSCA evaluation documents. The development of this information should be helpful to any future regulatory activity.

OTA considers all the positive NCI/NTP chemicals to be in the TSCA jurisdiction. Section 8(a) or 8(d) rules have been issued or CHIPs prepared for 53 of the 144 positive NCI/NTP chemicals.[27] Twenty-two of the 61 chemicals with four-positive and three-positive results have been the subjects of section 8(a) or 8(d) reporting rules or CHIPs.[28]

---

[26]Rules concerning manufacture and disposal of PCBs have also been issued under sec. 6. However, this action occurred because of congressional directive. The NCI/NTP test results for Arochlor 1254 (a PCB) were negative.

[27]These include 22 sec. 8(a) or 8(d) rules and 37 CHIPs. The total is 53 because not all of the chemicals are covered by both.

[28]These include 11 chemicals subjected to 8(a) or 8(d) rules and 17 chemicals with CHIPs.

EPA Actions Under the Resource Conservation and Recovery Act

Annual Report Chemicals

Compared to the regulation responses of other agencies to the *Annual Report* list, those of RCRA and CERCLA are the most comprehensive, although their corresponding programs do not address a number of *Annual Report* chemicals.

Two lists of chemicals are important for the RCRA program: a list of hazardous wastes (which lists commercial chemicals and waste streams) and a list of hazardous constituents of listed wastes (Appendix VIII of RCRA). The RCRA hazardous waste list currently includes 89 *Annual Report* carcinogens, while Appendix VIII includes 18. Because 10 chemicals are on both lists, the number of chemicals covered by the two lists together is 97. An additional 20 *Annual Report* chemicals, not included in either list, are proposed for inclusion in the list of hazardous wastes and 1 (iron dextran complex) is proposed for removal.

The RCRA lists should be prospective, allowing for the possibility that toxic chemicals currently not produced might be produced in the future and need to be disposed of safely. Therefore, as defined by OTA, the jurisdiction for RCRA consists of all chemicals in the *Annual Report*, whether or not they are currently produced.

The two RCRA lists currently include 97 of the 145 chemicals in the *Annual Report*; 48 *Annual Report* chemicals are not included. Some of the 48 *Annual Report* chemicals not listed under RCRA are not currently produced commercially —4 according to the SRI data reported in HSDB. Another six *Annual Report* chemicals were not found in HSDB or did not have production data reported and some of them may also not be produced. Thus, 38 *Annual Report* chemicals are produced but not included in the RCRA lists. Nine of these are produced in quantities greater than 1 million pounds.

Positive NCI/NTP Chemicals

Of the 144 positive NCI/NTP chemicals, 41 appear in either one of the RCRA lists (or both), while 103 positive NCI/NTP chemicals do not.

Limiting attention to chemicals testing positive in three or four experiments, or 61 chemicals, 39 chemicals (64 percent) have not been listed. Nineteen of the 103 positive NCI/NTP chemicals not included in the RCRA lists are not currently produced. Twenty-six chemicals were not found in HSDB or had no reported production data, and some of these may also not be produced. Fifty-eight of the 103 positive NCI/NTP chemicals are produced but not included in the RCRA lists. Sixteen of these are produced in quantities greater than 1 million pounds.

EPA Actions Under the Comprehensive Environmental Response, Compensation, and Liability Act

Annual Report Chemicals

The CERCLA list includes 95 *Annual Report* carcinogens, most of which RCRA also lists as hazardous wastes. Chemicals that were only included in RCRA Appendix VIII have not been incorporated in the CERCLA list, although EPA has issued an Advance Notice of Proposed Rule Making requesting comments about including them.

Activities under CERCLA may need to assess the hazards of chemicals no longer produced, but found in waste dumps from past production. OTA has not developed any information on which *Annual Report* chemicals have been found in dump sites or have been released into the environment.[29] The OTA-defined jurisdiction for CERCLA consists of all the chemicals in the *Annual Report*, whether or not they are currently produced.

Fifty of the 145 *Annual Report* chemicals are not included in the CERCLA list. Examining the chemicals currently produced in quantities greater than 1 million pounds yields nine not covered under CERCLA.

Positive NCI/NTP Chemicals

Of the 144 positive NCI/NTP chemicals, 47, or about one-third, are included in the CERCLA list. Two-thirds of the positive NCI/NTP chemicals (94) are thus not included. Of the 61 chemicals with three or four positive experiments, 22 are listed and 39 are not.

EPA recently proposed to adjust, based on evidence of carcinogenicity, the reportable quantities (RQs) for chemicals on its CERCLA list. (The EPA method for this is described in ch. 3.) The proposed adjustments do not add chemicals to the list, but change the RQ based on the classification of a chemical as a high hazard, medium hazard, or low hazard with regard to carcinogenicity. The 47 chemicals on the CERCLA list should thus be affected by the proposed adjustments. Only 32, however, are actually included in the list of chemicals evaluated for these adjustments. The 15 positive test result chemicals on the CERCLA list, but not evaluated for carcinogenicity, include several major industrial chemicals: methylene chloride, 1,2-propylene oxide, ethyl acrylate; 1,3-dichloropropene (Telone II), and TDI.

Assessments by EPA's Carcinogen Assessment Group

Annual Report Chemicals

CAG has prepared health assessments for 78 *Annual Report* chemicals. While selection of chemicals for CAG assessment depends on the needs of other programs within EPA, 67 of the chemicals and exposures listed in the *Annual Report* have not been covered by CAG's health assessments. These chemicals include 14 produced in quantities greater than 1 million pounds.

Positive NCI/NTP Chemicals

In all, CAG has conducted 22 assessments of the 144 positive NCI/NTP chemicals: 6 chemicals with four-positives, 3 with three-positives, 11 with two-positives, and 2 with one-positive. Grouping chemicals with four-positive and three-positive results together, the NCI/NTP test results cover

---

[29]EPA has recently published a list of 100 hazardous substances most commonly found at cleanup sites and which will be the subjects of toxicologic profiles required by sec. 110 of the 1986 Superfund amendments (302).

61 chemicals. CAG has prepared full assessments for nine of these, or about 15 percent.[30]

No Apparent Activity

Positive NCI/NTP Chemicals

Based on information available to OTA, 43 positive NCI/NTP chemicals appear not to have

been regulated or evaluated by any agency. These include 13 with four-positive results, 10 with three-positives, 12 with two-positives, and 8 with one-positive result. None of these chemicals is produced in a quantity greater than 1 million pounds. Ten are produced in quantities of less than 1 million pounds, 15 are not commercially produced, and the production status of 18 is unknown.

---

[30]As just discussed, CAG has also conducted analyses of available information for adjusting the CERCLA reportable quantities based on carcinogenicity.

AGENCY COMMENTS

In comments on a draft of this background paper, officials of Federal regulatory agencies emphasized their belief that they have acted appropriately in regulating the chemicals tested by NCI/NTP and the *Annual Report*. They pointed out that their statutes require that they assess risks and benefits of using chemicals, as well as the technical feasibility and costs of regulatory action. Agency responses to information on carcinogenicity sometimes involve requiring additional information enabling the agencies to make better decisions. In some cases, the agencies have decided that regulation is not necessary because a substance is no longer produced, does not present exposures, or the benefits of continued use exceed the risks. They stated that identification of a chemical as carcinogenic does not imply a need for regulation. EPA commented:

Our decision rules are just not so simple. Also, as the report basically tallies regulations and cannot readily assess decisions not to regulate, a biased picture emerges of the extent to which the Federal government has acted on carcinogens (104).

FDA commented:

We believe that FDA has acted responsibly and appropriately with regard to chemicals identified as carcinogens. Each purported carcinogen under the Agency's purview has been evaluated, and a determination of the appropriate course of action has been made. There is no backlog awaiting Agency review. Since many of the substances required no regulatory action, the Agency has made no formal public statements regarding those decisions (24).

SUMMARY

The *Annual Report on Carcinogens* is a useful compendium of information on carcinogenic chemicals, including its coverage of the uses of these chemicals and related regulatory actions. The NCI/NTP test results are useful for risk assessments of particular chemicals. Together, the NCI/NTP tests provide information useful for further development of risk assessment methods and exploration of topics in hazard identification. Such information has a research value in addition to its potential regulatory uses.

Table 5-3 summarizes the number of *Annual Report* chemicals and positive NCI/NTP chemicals acted on by each agency and program as well as the corresponding number of chemicals determined to be in the OTA-defined jurisdiction.[31] These tables separate the chemicals discussed in

---

[31]Two tables in app. B list the chemicals included in this analysis: table B-1 lists the chemicals that appear in the *Annual Report on Carcinogens*; table B-2 lists the chemicals tested by NCI/NTP and indicates the corresponding level of evidence.

Table 5-3.—Agency Actions on Positive NCI/NTP Chemicals and *Annual Report* Chemicals— Actions and Jurisdiction[a]

	Level of evidence					
	NCI/NTP chemicals				*Annual Report* chemicals	
	At least one positive experiment		Three and four positive experiments			
	Actions	Jurisdiction	Actions	Jurisdiction	Actions	Jurisdiction
No activity	—	43	—	23	—	—
CAG assessment	22	144	9	61	78	145
RCRA listed or App. VIII	41	144	22	61	97	145
CERCLA						
Listed	47	144	22	61	95	145
Proposed RQ adjust.	32	47	15	22	b	b
CWA WQC or standards	14	27	7	10	48	65
SDWA						
Interim std. (1975 and 1979)	2	14	1	7	7	32
Proposed and final RMCL ..	11	14	5	7	15	32
Total[c]	12		6		21	
CAA listed	2	12	1	8	6	15
TSCA						
Rule 8a or 8d or CHIP	53	144	22	61	28	145
4(f) review/SNUR/Sec. 6	5	144	2	61	6	145
Total[c]	56		24		33	
FIFRA						
Susp. Canc.	9	22	2	11	7	24
Vol. Canc.	4	22	3	11	5	24
Total[c]	13		5		12	
Food and Cosmetics (Ban, SUR, or action level)	17	48	7	19	46	52
FDA						
Animal drugs (revoked/withdrawn)	4	5	1	1	2	6
Human drugs (labeled/withdrawn)	6	12	5	6	26	31
NIOSH recommendation	31	62	13	39	59	112
OSHA						
Noncancer std.	27	53	15	30	35	110
Cancer std.	2	53	1	30	17	110
Total[c]	29		16		52	
CPSC						
Ban/restricted	2	14	1	7	11	23
Voluntary reduction only ...	57	14	3	7	8	23
Total[c]	8		4		18	

[a]Jurisdiction refers to the number of chemicals for which the agency is held responsible using the results of OTA's analysis. (See ch. 5).
[b]Not determined.
[c]Total after eliminating double counting.

SOURCE: Office of Technology Assessment, 1987.

this chapter into three groups: all NCI/NTP chemicals with at least one positive experiment, the NCI/NTP chemicals with three or four positive experiments, and the chemicals included in the *Annual Report*. Figure 5-1 summarizes these results by presenting graphically the number of these chemicals that have been acted on and the number not acted on for each agency and program included in the OTA analysis.

In general, while a number of regulatory actions appear to have been based directly on positive NCI/NTP results, there also appear to be substantial gaps in regulatory activity. Considering

194

Figure 5-1.—Agency Actions on *Annual Report* and Positive NCI/NTP Chemicals[a]

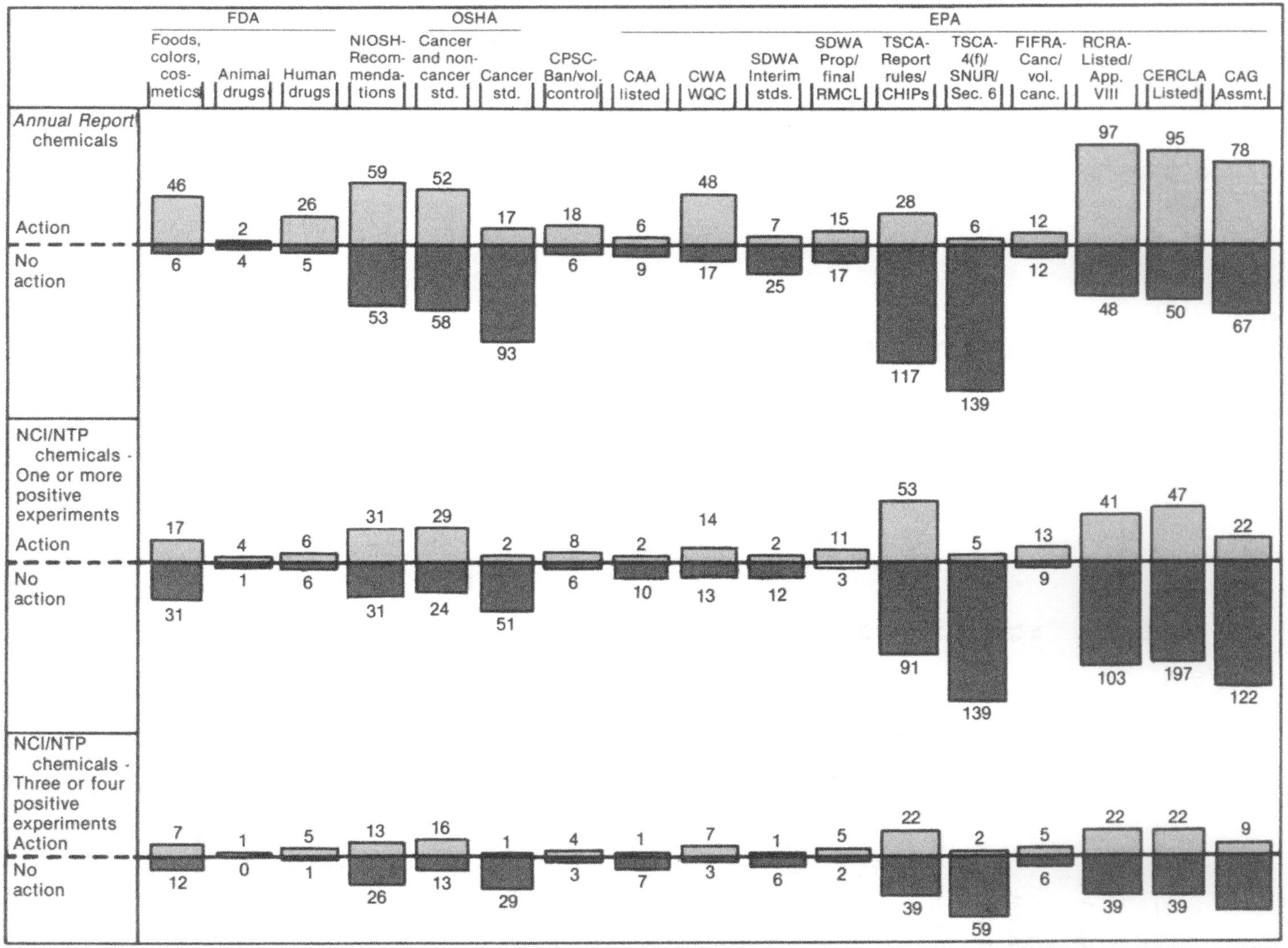

[a]For each agency or program, OTA included only chemicals in the OTA-defined jurisdiction for that agency or program. Agency decisions that regulation is not necessary or appropriate were included in the no action groups. Because of overlap between the three lists of chemicals, it is not appropriate to add them together. All actions through July 1987 are represented in this figure.

Key to acronyms: CAA—Clean Air Act; CAG—Carcinogen Assessment Group; CERCLA—Comprehensive Environmental Response, Compensation, and Liability Act; CHIPs—Chemical Hazard Information Profiles; CPSC—Consumer Product Safety Commission; EPA—Environmental Protection Agency; FDA—Food and Drug Administration; FIFRA—Federal Insecticide, Fungicide, and Rodenticide Act; NCI—National Cancer Institute; NIOSH—National Institute for Occupational Safety and Health; NTP—National Toxicology Program; OSHA—Occupational Safety and Health Administration; RCRA—Resource Conservation and Recovery Act; RMCL—recommended maximum contaminant level; SDWA—Safe Drinking Water Act; SNUR—Significant New Use Rule; TSCA—Toxic Substance Control Act; WQC—Water Quality Criteria.

SOURCE: Office of Technology Assessment, 1987.

each agency or program individually reveals that no agency has regulated more than a third of the positive test results. More typically, an agency will have acted out of concern for carcinogenicity on 5 to 30 of the 144 chemicals that tested positive in NCI/NTP bioassays.

As described in this chapter, OTA has attempted to focus on the chemicals of potential regulatory interest for each agency or program. However, as shown in table 5-3 and figure 5-1 and as discussed in this chapter, even when attention is limited to chemicals in the jurisdiction of the different agencies and programs, there appear to be omissions in regulatory coverage. The importance of these apparent regulatory gaps depends on factors not analyzed by OTA, including the extent and magnitude of exposures, the potency of the chemicals, as well as other exposures and risk factors. In some cases, voluntary industry actions may have reduced or eliminated risks in the absence of government regulation. OTA has not determined the extent of these voluntary actions.

Appendixes

Statutory Authority for Regulating Carcinogens

Introduction

By one accounting, 21 different laws may be used to regulate carcinogens (table A-1). However, this appendix describes only the major statutes providing for regulation of human exposure to carcinogens, and the significant judicial decisions affecting this regulation.

Most of the statutes do not single out carcinogens for specific consideration, but merely regulate them as a species of toxic substances. A few, however, have provisions aimed directly at carcinogens; one, the Food, Drug, and Cosmetic Act (FDCA), has special statutory provisions for regulating carcinogens as distinguished from other toxic substances, while several others—the Clean Water Act (CWA), the Toxic Substances Control Act (TSCA), and the Resource Conservation and Recovery Act (RCRA)—mention carcinogens specifically.

Some statutes require premarket review or approval of a substance before it can enter into commerce. This requirement is set in parts of FDCA and TSCA, and in the Federal Insecticide, Fungicide, and Rodenticide Act (FIFRA). Even for these three laws, however, the requirement for premarket review applies only to new pesticides or "new" chemicals being proposed for manufacture, although FDCA requires premarket approval of new uses for existing chemicals.

A much larger number of statutes, including parts of FIFRA and TSCA and the other statutes described in this chapter, provide for postmarket regulation of substances after they have been in commerce and people have been exposed to them. Such laws might require an agency to find that there is a health problem and then propose a regulation based on that finding,

Table A-1.—Statutes Authorizing Regulation of Carcinogens

Legislation	Agency	Area of concern
*Food, Drug, and Cosmetic Act (1906, 1938, amended 1958, 1960, 1962, 1968)	FDA	Foods, drugs, cosmetics, and medical devices
*Federal Insecticide, Fungicide and Rodenticide Act (1948, amended 1972, 1975, 1978)	EPA	Pesticides
Dangerous Cargo Act (1952)	DOT, USCG	Water shipment of toxic materials
Atomic Energy Act (1952)	NRC	Radioactive substances
*Federal Hazardous Substances Act (1960, amended 1961)	CPSC	Toxic household products
Federal Meat Inspection Act (1967)	USDA	Food, feed, color additives, pesticide residues
Poultry Products Inspection Act (1970)		
Egg Products Inspection Act		
*Occupational Safety and Health Act (1970)	OSHA	Workplace toxic chemicals
Poison Prevention Packaging Act (1970, amended 1977)	CPSC	Packaging of hazardous household products
*Clean Air Act (1970, amended 1974, 1977)	EPA	Air pollutants
Hazardous Materials, Transportation Act (1972)	DOT	Transport of hazardous materials
*Clean Water Act (formerly Federal Water Control Act) (1972, amended 1977, 1978)	EPA	Water pollutants
Marine Protection, Research and Sanctuaries Act (1972)	EPA	Ocean dumping
*Consumer Product Safety Act (1972, amended 1981)	CPSC	Hazardous consumer products
Lead-based Paint Poison Prevention Act (1973, amended 1976	CPSC, HHS, HUD	Use of lead paint in federal assisted housing
*Safe Drinking Water Act (1976)	EPA	Drinking water contaminants
*Resource Conservation and Recovery Act (1976)	EPA	Solid waste
*Toxic Substances Control Act (1976)	EPA	Hazardous chemicals not covered by other acts
*Federal Mine Safety and Health Act (1977)	DOL, NIOSH	Toxic substances in coal and other mines
*Comprehensive Environmental Response, Compensation, and Liability Act (1981)	EPA	Hazardous waste cleanup

*Discussed in this appendix.

SOURCE: Office of Sciency and Technology Policy, 1985.

as in the Clean Air Act (CAA), the Clean Water Act, the Safe Drinking Water Act (SDWA), and the Occupational Safety and Health Act. Still other laws might require an agency to find that there is a health problem, establish this fact in court, and seek some judicial remedy on that basis. Some sections of FDCA require this for foods contaminated by naturally occurring environmental carcinogens.

Except for the few parts of statutes that require court-ordered remedies, most agencies authorized to regulate toxic substances causing health problems must follow procedures mandated by the Administrative Procedure Act, or similar procedures. In regulating substances an agency must follow these procedures for agency "rulemaking" in order to "issue a rule." Such rules may be issued according to rulemaking procedures that range from the relatively informal to the formal, resembling proceedings in a court of law. In general, the agency must announce in the *Federal Register* that it is proposing to regulate a substance (or group of substances), and describe the nature of the proposed regulation (5 U.S.C. 553(b)). The agency must also give interested parties "an opportunity to participate in the rulemaking through submission of written data, views, or arguments . . ." (5 U.S.C. 553(c)). Following the comment period, the agency usually holds hearings during which interested parties may have their comments heard. After considering both written and oral comments, the agency issues a final rule.

Apart from these common features, informal and formal rulemaking are distinguished by the nature of evidence presented during the notice and comment period and at the hearings, the procedures followed at the hearing itself, and the standard of judicial review of agency action. Generally, under formal rulemaking an agency must conduct quasi-judicial proceedings with the opportunity to cross-examine witnesses, and agency decisions following such proceedings are in theory more closely scrutinized by the courts if the regulatory decisions are appealed (103). Under most of the statutes considered here, the agencies act under the requirements for informal rulemaking.

At least some of the statutes also provide for immediate emergency action, such as immediate suspension under FIFRA or the establishment of Emergency Temporary Standards under the Occupational Safety and Health Act.

Finally, in considering the statutes, the reader should be aware of some differences between the substantive requirements of the statutes. Different laws reveal different attitudes toward risk. Some statutes reflect attitudes quite averse to human health risks posed by chemical substances. The most extreme example is the Delaney clause of FDCA. According to that provision, if a food additive causes cancer in humans or in animal tests, it is declared "unsafe" and is not allowed as a food additive. The risk to human health is the only factor taken into account. This is a "no-risk" statute.

Other risk-based statutes use different statutory language. CAA makes risks to human health the primary factor by setting the goal of regulating with an "ample margin of safety" (42 U.S.C. 7412(a)(2)). Another approach is risk-risk balancing: weighing the risk to human health from exposure to a regulated substance against the risk to human health from not having the substance in commerce. The FDCA appears to permit this kind of risk-risk balancing for food additives approved by the Food and Drug Administration (FDA) prior to 1958 (127). For human drugs, FDA uses a "risk-benefit" approach, although again, the primary factor involves the benefits and risks to patient health.

Some statutes are "technology based" laws. These may require, for example, the agency to reduce emissions from a particular source to the extent this may be achieved by technological devices placed on the emitting source. Some such statutes require the "best practical technology" (BPT) or the "best available technology" (BAT). "Such regulations do not force new technology, but bring all control efforts up to standards established by existing control technologies" (217). Other technology-based statutes might be "technology forcing" because "new techniques may be required to achieve" some predetermined level of pollutant concentration (217).

Still other statutes permit agencies to balance the risks to human health from carcinogens against benefits to be obtained by consumers, manufacturers, and others by permitting the substance to be in commerce. This is a risk-benefit balancing statute. Congress used the term "unreasonable risk" in TSCA to refer to this kind of balancing.

Occupational Safety and Health Administration

The Occupational Safety and Health Act of 1970 established the Occupational Safety and Health Administration (OSHA) and the National Institute for Occupational Safety and Health (NIOSH). OSHA sets and enforces regulations to control occupational health and safety hazards, including exposure to carcinogens. NIOSH is a research agency that has contributed to the regulation of carcinogens by supporting epidemiologic research and recommending to OSHA changes in health standards.

The Occupational Safety and Health Act provides three statutory mechanisms for setting standards to protect employees from hazardous substances such as carcinogens. Section 6(a) authorized OSHA to adopt the health and safety standards already established by Federal agencies or adopted as national consensus standards. This authority was limited to the first 2 years after the act went into effect (April 1971 to April 1973). An unspecified number of carcinogens were regulated on the basis of their noncarcinogenic effects by these start-up standards.

Section 6(c) authorizes OSHA to issue emergency temporary standards (ETS) that require employers to take immediate steps to reduce workplace hazards. An ETS may be issued after OSHA determines that employees are exposed to a "grave danger" and that an emergency standard is "necessary to protect employees from such danger." An ETS, issued without opportunity for comments or for a public hearing, goes into effect immediately. The issuance of an ETS also initiates the process of setting a standard under section 6(b), with the published ETS ordinarily serving as the proposed standard. The act mandates that a final standard be issued within 6 months of publication of the emergency standard.

Finally, section 6(b) authorizes OSHA to issue new permanent exposure standards and modify or revoke existing ones by informal rulemaking. However, as a result of congressional compromise, OSHA's informal rulemaking is reviewed by the courts under the "substantial evidence test" normally reserved for formal rulemaking on the record (190).

OSHA's rulemaking can result in requirements for monitoring and medical surveillance, workplace procedures and practices, personal protective equipment, engineering controls, training, recordkeeping, and new or modified permissible exposure limits (PELs). Permissible exposure limits are the maximum concentrations of toxic substances permitted in the workplace air.

From 1971-1986 OSHA issued 23 separate health standards in 9 regulatory actions after rulemaking. Eight of OSHA's final actions on individual health standards established new PELs and other requirements on carcinogens (asbestos (1972), vinyl chloride, coke oven emissions, benzene, 1,2-dibromo-3-chloropropane (DBCP), arsenic, acrylonitrile, ethylene oxide, and asbestos (1986)). One OSHA action regulating a group of "14 carcinogens" did not establish a PEL, but created new requirements for work practices and medical surveillance for employees exposed to this group of carcinogens.

Significant Judicial Decisions

OSHA's regulation of carcinogens has been controversial. Of eight final actions on individual carcinogens, six have resulted in court challenges (asbestos (1972), vinyl chloride, coke oven emissions, benzene, arsenic, and ethylene oxide). Only the DBCP and acrylonitrile regulations were not challenged as final standards. In addition, there were 3 court challenges to the group regulation of 14 carcinogens: *Dry Color Manufacturing Association* v. *Department of Labor* (46) vacated temporary standards for ethyleneimine and DCB; *Synthetic Organic Chemical Manufacturers Association* v. *Brennan* (191) upheld the permanent standard for ethyleneimine; and *Synthetic Organic Chemical Manufacturers Association* v. *Brennan* II (192) vacated the standard for 4,4-methylenebil(2-chloraniline) MBOCA. OSHA's rules were upheld for asbestos (1972), for vinyl chloride, for coke oven emissions, and for arsenic. The courts vacated the temporary asbestos standard in 1984 and the permanent benzene standard in 1980.

The decisions resolving these disputes have focused on several major issues. Courts have had to provide interpretations of 1) the role of the courts in scrutinizing agency actions, and 2) the nature of OSHA's burden in demonstrating the merits of its standards. In setting standards the crucial section of the act states:

> [OSHA]. . . shall set the standard which most adequately assures, to the extent feasible, . . . that no employee will suffer material impairment of health or functional capacity even if such employee has regular exposure to the hazard dealt with by such standard for the period of his working life (U.S.C. 655 (b)(5)).

Early decisions by the Courts of Appeal clarified the courts' role in reviewing OSHA standards under the substantial evidence test and the standards of economic and technological feasibility. OSHA may impose regulations even if doing so means raising standards above those that exist in the status quo or above those achievable by present technology (186). In this sense, OSHA can force the development of new technology. Similarly, courts have held that although Congress did not intend to put whole industries out of business to protect their employees' health, it did foresee that health regulations would entail costs and that some businesses might close (5,98).

In a major decision, the Supreme Court invalidated OSHA's 1978 benzene exposure standard. The Court ruled that the 1 part per million (ppm) exposure limit was not supported by appropriate findings. A plurality of the Court said that the new standard did not rest on a finding that exposure to 10 ppm would cause leukemia while exposure to 1 ppm would not, and that OSHA acted on assumptions in claiming: a) that exposure to 10 ppm of benzene might cause leukemia and b) that the number of such cases might be reduced by lowering the permissible exposure level to 1 ppm (96).

According to the plurality opinion, section 3(8) defines occupational health and safety standards as requirements "reasonably necessary or appropriate to provide safe or healthful employment," and requires OSHA to make a threshold finding that a workplace is "unsafe" before issuing a standard (96). According to the plurality, "safe" does not mean risk free, and a workplace is not "unsafe" unless it it poses a "significant risk of harm" to the worker (96). In addition, the Court said that "the burden [is] on the Agency to show, on the basis of substantial evidence, that it is at least more likely than not that long-term exposure to [a toxic substance] . . . presents a significant risk of material health impairment" (96).

This review and interpretation of OSHA's benzene standard changed the way OSHA regulates. Prior to this decision, OSHA had refused to prepare quantitative risk assessments concerning substances it regulated. In proposing subsequent standards, the agency has had to demonstrate that exposure at the current permissible levels presents a "significant risk" to workers before it is justified in proposing lower exposure standards. OSHA now uses a four-step process for making decisions about health standards:

> First, the agency determines that the hazard in question poses a "significant risk . . ." [as required by the benzene decision]. Second, OSHA determines that regulatory action can reduce this risk. Third, it sets the regulatory goal (for health standards, this is the permissible exposure limit) based on reducing this risk "to the extent feasible." Finally, OSHA conducts a cost-effectiveness analysis of various options to determine which will achieve this chosen goal in the least costly manner (219).

In 1981, a second Supreme Court ruling provided a partial interpretation of "feasibility." In opposing new lower standards on exposure to cotton dust, industry had argued that the Occupational Safety and Health Act required cost-benefit analysis before OSHA could issue new standards. The Supreme Court, with a five-member majority comprised of the four dissenters from the benzene case and the author of the benzene plurality opinion, Justice Stevens, strongly rejected this claim, arguing that:

> Congress itself defined the basic relationship between costs and benefits, by placing the "benefit" of worker health above all other considerations save those making attainment of this "benefit" unachievable. Any standard based on a balancing of costs and benefits by the Secretary that strikes a different balance than that struck by Congress would be inconsistent with the command set forth in section 6(b)(5). Thus, cost-benefit analysis by OSHA is not required by the statute because feasibility analysis is (6).

The agency regards this decision as neither requiring nor permitting the use of quantified cost-benefit analysis for the purposes of setting standards (100,118). (See (219) for a more detailed discussion of these issues.)

Recently, a different kind of case has been litigated. Because of health concerns about employee exposure to ethylene oxide, the Public Citizen Health Research Group sued OSHA to issue an ETS. A district court ruled in favor of Public Citizen and ordered OSHA to issue such a regulation. On appeal, the District of Columbia Court of Appeals overruled the district court judge's decision on the grounds that he had impermissibly substituted his evaluation for that of OSHA. However, because OSHA had "unreasonably delayed" acting on ethylene oxide, the court ordered OSHA to issue a notice of proposed rulemaking within 30 days of the Court's decision and a final rule within 1 year's time (165).

The Mine Safety and Health Administration

The Mine Safety and Health Administration (MSHA) regulates the exposure of miners to carcinogens. The Federal Mine Safety and Health Act Amendments of 1977 consolidated the regulation of mine health and safety under one statute, and transferred responsibility from the Department of the Interior to the Department of Labor. Safety and health in coal mines had previously been regulated under the Federal Coal Mine Health and Safety Act of 1969 by the Department of the Interior's Mining Enforcement and Safety Administration (MESA). Safety and health in metal and nonmetalic mineral mines had been regulated by MESA under the Federal Metal and Nonmetallic Mine Safety Act (1966) (150).

As described in Chapter 3, much of MSHA's regulation of toxic exposures involves incorporating by reference the lists of standards of a private organization, the American Conference of Governmental Industrial Hygienists (ACGIH). The original intention was to update these standards automatically whenever ACGIH issued changes. However, this has not happened. The administration fears violating the Administrative Procedure Act because automatic updates would not provide an opportunity for public comment (190).

How much of a legal difficulty this situation presents is difficult to know. For example, OSHA's Hazard Communication (Labeling) Standard requires that employers provide information to employees on substances to which they are exposed and on the concentrations that are regarded as "not harmful." According to OSHA's regulations employees must be informed about the most recent ACGIH list of toxic substances (190).

Food and Drug Administration

The Food and Drug Administration regulates foods, drugs, and cosmetics under FDCA. FDCA is the result of several laws passed by Congress since the first Federal statute regulating food safety, the Food and Drug Act of 1906 (127). The Federal Food, Drug, and Cosmetic Act of 1938 established the general outlines of current FDA authority. Congress has amended it with the Pesticides Residue Amendment of 1954, the Food Additives Amendment of 1958, the Color Additive Amendments of 1960, the Drug Amendment of 1962, and the Animal Drug Amendments of 1968.

Main Statutory Provisions for Regulating Carcinogens in Foods and Cosmetics

Carcinogens in foods and cosmetics may be and have been regulated under many different provisions of FDCA, depending on whether they are considered food additives, food contaminants, naturally occurring parts of the food, or color additives in foods, drugs, and cosmetics. In addition, parts of the law have premarket approval provisions, while others have postmarket enforcement provisions.

Under FDCA a food is considered to be adulterated, and thus illegal to sell in interstate commerce, if it contains a food or color additive that is unsafe. An additive is regarded as unsafe if it "may be injurious to health" or is "ordinarily injurious to health" according to section 402(a)(1) (21 U.S.C. 342(a)(2)(c)); if it contains any added poisonous or deleterious substance that is unsafe according to section 406 (21 U.S.C. 342(e); 21 U.S.C. 346); if it contains a food additive which is carcinogenic according to the Delaney clause (sec. 409) (21 U.S.C. 348(a)); or if it contains a color additive that is carcinogenic according to the Delaney clause for color additives (21 U.S.C. 376(a)).

The Delaney Clause for Foods, Section 409.—In the Food Additives Amendment of 1958, Congress established a *premarket approval procedure* for food additives (127). This amendment contains the well-known "Delaney clause," named after Rep. James Delaney of New York, whose hearings led to the amendment (127).

According to FDCA, foods may not contain any intentional additive unless FDA has established conditions under which the additive may be safely used (the general safety provision of section 409) or has issued an exemption from this requirement. The Delaney clause applies to this process of approving the safe use of food additives and provides:

. . . that no additive shall be deemed to be safe if it is found to induce cancer when ingested by man or animal, or if it is found, after tests which are appro-

priate for the evaluation of the safety of food additives, to induce cancer in man or animal . . . (21 U.S.C. 348(c)(3)(a)).

Thus, if appropriate evidence indicates that a food additive is carcinogenic, FDA may not consider it safe and must prohibit its use in food.[1] The manufacturer has the burden of proving that food additives are safe before they receive approval to enter the market.

Food additives that are carcinogenic but that have previously been approved by FDA or the U.S. Department of Agriculture, or that are generally recognized as safe (GRAS) (because they were commonly used in food at the time the Delaney clause was passed or are generally recognized by experts as safe on the basis of toxicological tests), are not regulated under section 409. They are instead subject to section 402(a)(1) (127).

Provisions for color additives (21 U.S.C. 376) are similar to section 409 provisions for food additives, i.e., carcinogenic color additives are to be prohibited by FDA.

Although section 409 condemns food additives that are carcinogens, the agency, in an advance notice of proposed rulemaking (ANPR), has indicated an intention to rely on the general safety provision of section 409 rather than the Delaney clause for additives that contain carcinogens. In its 1982 ANPR, following the *Kennedy* v. *Monsanto* decision in 1979 (described below), the agency proposed that food and color additives that, taken as a whole, do not cause cancer in animals, but do contain small amounts of carcinogenic impurities should fall under the general safety clause, not the Delaney Clause. If the additive itself is not found to be harmful as shown by quantitative risk asessment procedures, then it is considered safe even though it contains carcinogenic impurities. The FDA impurities policy was upheld in the *Scott* v. *FDA* decision. Previously FDA had used the policy on a case-by-case basis for a number of food and color additives, including some color additives that are not ingested (245).

General Provision Concerning Food Adulteration (Section 402(a)(1)).—Section 402(a)(1) declares that a food is adulterated if it contains added substances that "may render it injurious to health," or if it contains any substance, added or not, that would "ordinarily render it injurious to health" (21 U.S.C. 342(a)(1)) (1982). The latter standard includes substances that are naturally occuring food constitutents (127). The difference between the two is that the first is more stringent, permitting FDA to establish its case with a lower prob-

---

[1]There are three different versions of the Delaney clause, the one described above for food additives, one enacted at the same time for drugs fed to animals, and one enacted in the Color Additives Amendments of 1960. While they differ in detail, they all prohibit animal and human carcinogens.

ability of risk. The first allows consideration of sensitive populations, while the second does not (128).

If FDA determines a food to be adulterated under either of these two clauses, it must go to court to remove the food from the market (127). In such a case FDA, not the manufacturer, has the burden of locating the contaminated food, analyzing its chemical makeup, and proving that the substance is harmful (127).

Unavoidable/Necessary Contaminants (Sections 406 and 402(a)(2)(A)).—Under sections 406 and 402(a)(2)(A), FDA is permitted to set tolerance limits for "unavoidable contaminants and other poisonous or deleterious substances that may be necessary aspects of food production" (127). These sections are used primarily for unavoidable contaminants of food such as aflatoxins in peanuts, mercury in fish, and PCBs in milk and fish (127). Together, sections 402(a)(2)(A) and 406 declare a food adulterated if it contains any added poisonous or deleterious substances except where they cannot be avoided or are required in the production of food. In this case, FDA may set tolerance levels for the protection of public health, and the food will be considered safe unless the tolerances for the added substances are exceeded (21 U.S.C. 348). Tolerances must be established by means of the most extreme version of formal rule making procedures (127). Foods that have levels of contaminants exceeding such tolerances are subject to postmarket judicial seizure under section 402(a)(2)(A).

Several considerations enter into the FDA's tolerance setting:
- the level of a contaminant not posing a risk to health,
- the ability of good manufacturing practices to reduce concentrations,
- analytical capabilities for detecting the contaminant, and
- the value of the food (127).

This section was not used until the mid 1970s, but since has been used to establish tolerances for some unavoidable environmental food contaminants (127). Even the tolerance setting provisions have not been widely used because they involve formal rulemaking. Thus, for most foods which might fall under sections 406 and 402(a)(2)(A), the agency has instead merely set "action levels" for contaminants: levels that, if exceeded, would lead the agency to bring court action to seize the food.

Section 408 provided FDA with authority to regulate pesticide residues on raw agricultural products, but this authority was transferred to EPA in 1970. The regulation of pesticide residues will be discussed under the regulation of pesticides.

The several categories of food constituents and color additives that are regulated under the authority of FDCA are summarized in table A-2.

The Regulation of Animal Drug Residues

Animal drug residues and animal feed additives that leave residues in human foodstuffs are subject to the Delaney clause, but with a qualification. Originally, animal drug residues were subject to the Food Additives Amendment of 1958, with its Delaney clause. In 1962, however, Congress amended FDCA to permit carcinogenic residues in animal tissues as long as the residues are in lower concentrations than those that it has set as safe and detectable by FDA-approved analytic techniques. If carcinogenic residues exceed FDA-specified levels, they are subject to the Delaney clause.

The Regulation of Drugs With Carcinogenic Potential

FDA evaluates both new and previously approved drugs. New drugs require premarketing approval, subject to risk-benefit considerations:

> A drug is approved only if the benefits are judged to exceed the risks (real and potential) under intended conditions of use. For drugs, vaccines, medical devices, and diagnostic aids, the term "safety" is never treated as an absolute but is thought of as inherently involving a weighing of benefit and risk (233).

In the approval process an applicant must submit two kinds of application: an Investigative New Drug (IND) application (an application to conduct an investigation into a new drug), and then a New Drug Application (NDA) (an application to conduct a more detailed investigation into a new drug). (For a more detailed discussion of this approval process see (218).)

An applicant first submits an IND application. FDA has 30 days to consider whether the preclinical investigations suggest an undue risk to research subjects that would preclude initiation of human studies. At any time during the research period FDA can terminate the research. In 1983, the Agency received approximately 2,000 IND applications (233). An initial IND application normally would include "chemical, manufacturing, and control information; pharmacologic and toxicologic information from animals and in vitro systems; a plan of clinical study . . ." (233). Animal carcinogenicity tests are required for marketing approval of drugs that would be administered chronically or intermittently in a large population (see discussion in ch. 2).

After the research period an applicant submits an NDA (data developed during the IND-NDA process). An NDA must include full reports of toxicological

Table A-2.—Foods and Drugs Regulated Under the Food, Drug, and Cosmetic Act

Category	Description	Applicable statutory scheme
A. Direct food additives		
1. Ordinary food additives	Substances intentionally added to processed foods	Section 409
2. Substances generally recognized as safe (GRAS)	Substances used in foods prior to 1958 or substances recognized by experts as safe based upon toxicological tests	Section 402(a)(1)
3. Substances previously sanctioned by FDA or USDA	Intentionally added food constituents previously sanctioned by either FDA or USDA	Section 402(a)(1)
B. Color additives	Substances used to color foods, drugs, and cosmetics	Section 706(b)(5) for foods, drugs and cosmetics
C. Indirect constituents of food		
1. Indirect food additives	Substances used in proximity with food in ways that may permit small amounts to migrate and to become part of the food, e.g., substances used in packaging or in equipment used to process or store food	Section 409, subject to qualifications of *Monsanto* v. *Kennedy* (described in text)
2. Animal drug residues	Compounds administered to food-producing animals as drugs or feed supplements	Section 409, but DES proviso allows this only if the amount of residue exceeds the detection limit set by FDA
3. Pesticide residues	Residues of pesticides on raw agricultural products or in processed agricultural products	Sections 408 and 409 (discussed under FIFRA)
D. Natural food constituents	Naturally occurring food constituents, e.g., oyster shell fragments, mushrooms, and mussels (not known as carcinogens)	Section 402(a)(1) (covering substances that would "ordinarily render [food] injurious to health")
E. Unavoidable "added" constituents of food	Substances not inherent in agricultural commodities which may unintentionally contaminate foods such as milk, grain, or fish during production or harvesting, e.g., aflatoxins in peanuts, polychlorinated biphenyls (PCBs) in milk and fish, and mercury in fish	Section 402(a)(1) ["May render injurious to health"] or sections 402(a)(2)(a) and 406 (which authorize the setting of tolerances); FDA has tended to set "action levels" which guide initiation of court action and seizure of such substances under section 402(a)(1) (see ref. 127)

SOURCE: Office of Technology Assessment, 1987.

studies and clinical investigations to show that the drug is safe and effective, a complete list of the drug composition, samples of the drug, information that may be required for subsequent FDA monitoring activity, and specimens of proposed labels (233). Any potential risks of inactive ingredients are also evaluated.

The agency shall not approve an application for a new drug if it is not shown to be safe, if the available information is not adequate to make that determination, or if the labeling is false or misleading. Otherwise, it "shall issue an order approving the application" (21 U.S.C. 355(d)).

Some drugs are not in chronic or widespread use, and have less potential for carcinogenicity. If a drug is chronically used, FDA requires long term carcinogenicity studies in rodents (233). For oral contraceptives, carcinogenicity tests in monkeys and dogs are required as well (68). In addition, since the agency uses a risk-benefit balancing test to evaluate the safety of drugs, "a drug . . . [that] has a significant effect on a fatal disease with no alternative therapy could be regarded as adequately safe despite major, even life threatening, side effects" (233). Thus, the agency would approve drugs taken chronically that have possible carcinogenic side effects only if "the benefits [were] judged to exceed the risks (real and potential) under intended conditions of use" (233).

Significant Judicial Decisions

One recent judicial decision is of note, for it may influence developments in the future. In 1977, FDA proposed an extreme procedure for regulating tiny amounts of carcinogenic substances that may migrate from packaging material into foods. The agency had found that under certain laboratory conditions acrylonitrile monomers migrated from beverage containers into the liquids contained inside. FDA argued that even if improved manufacturing methods would decrease the amount of acrylonitrile migrating into beverages, packaging material with acrylonitrile in it could be presumed "to become a component of food," even though present analytic methods could not detect it (237). The agency argued that the burden is on the manufacturer to prove that diffusion does not occur when packaging contains "lower residual levels of the material" (237). The Court of Appeals for the District of Columbia in *Monsanto* v. *Kennedy* rejected this argument (129). The Court was concerned

> . . . that the Commissioner may have reached his determination [concerning small amounts of acrylonitrile in beverages] in the belief that he was constrained to apply the strictly literal terms of the statute irrespective of the public health and safety considerations . . . [but] there is latitude inherent in the statutory scheme to avoid literal application of the statutory definition of "food additive" in those de minimis situations that, in the informed judgment of the Commissioner, clearly present no public health or safety concerns (129).

In particular, the Commissioner

> . . . has latitude . . . to find migration "insignificant" even giving full weight to the public health and welfare concerns that must inform his decision . . . [and] he would have latitude to consider whether acrylonitrile is generally recognized as safe at concentrations below a certain threshold, even though he has determined for higher concentrations that . . . acrylonitrile is not generally recognized as safe (129).

This case is important, not only because of its application to indirect food addtives, but because it introduces the idea of de minimis levels of risk into FDA regulation and into the regulatory community more generally. As discussed in chapter 3, it is generally believed that there are no "safe" threshold levels for carcinogens or that these levels have not been demonstrated. Under the Delaney clause, this belief would seem to require that any concentration of a carcinogen as a food additive would have to be banned, for there is no safe level. FDA, however, distinguishes legal arguments from scientific arguments and contends that the Delaney clause is a policy statement and that there are safe levels for carcinogens under the Delaney clause. Since *Monsanto* v. *Kennedy*, some in FDA support the idea that there may be de minimis levels of risk even with carcinogens.

A number of developments since *Monsanto* v. *Kennedy* indicate that FDA is adopting the idea of de minimis risks for carcinogens. The agency permitted lead acetate to remain in hair dye, even though it had been found to be carcinogenic in animal feeding studies, and even though some of it penetrated the scalp and was detectable in the blood stream when the dye was applied to hair (35). In addition, in 1982 the agency approved D&C Green No. 6, "a color additive for use in drugs and cosmetics that was not itself found to be carcinogenic, but contained a carcinogenic constituent" (35), on grounds that *Monsanto* allowed the agency some discretion in deciding how to deal

with color additives (39). A circuit court of appeals has upheld such an approval since that time (35). In 1982 and 1983 FDA considered six color additives, all of which had been found to be carcinogenic not merely to have carcinogenic components. Rather than banning them outright, as the Delaney clause would seem to require, the agency referred them to a panel of scientists at the U.S. Department of Health and Human Services. "The clear, but unstated, implication of this action is that the agency is now prepared to apply the de minimis principle to direct food and color additives that cause cancer in animals" (35). The agency may also be trying to build a broad base of scientific support for its approach. In a related development on December 18, 1985, "FDA proposed to ban the use of methylene chloride in cosmetics because it causes cancer in lab animals, but declined to lower the maximum residue of it permitted in decaffeinated coffee because that amount is considered safe" (35).

Monsanto v. *Kennedy* and subsequent developments contrast with an earlier case concerning the FDA's termination of provisional approval of Red No. 2 dye. In that instance, the agency had evidence that Red No. 2 caused cancer in rats at low doses and caused only a slight increase in cancer tumors (compared with those at low doses) at high doses. On this basis, FDA proposed terminating provisional approval of the dye and the industry sued. The court of appeals upheld the agency's action, (1) because the statistical relationships in the initial animal studies, while not providing "conclusive proof that Red No. 2 was a carcinogen, . . . [were] at least suggestive of it . . ." and (2) because these statistical relationships were later confirmed at low doses. More important from the Court's point of view the study *"could not* be used to *establish* safety . . ."* Thus the color's safety was sufficiently questionable to justify FDA terminating its provisional listing (27).

The Consumer Product Safety Commission

Created in 1970 by the Consumer Product Safety Act, the Consumer Product Safety Commission (CPSC) is an independent regulatory agency headed by five commissioners appointed by the President for staggered seven-year terms. Its authority to regulate carcinogens is established by both the Consumer Product Safety Act (CPSA) and the Federal Hazardous Substances Act (FHSA).

The Consumer Product Safety Act

CPSA gives the Commission power to regulate consumer products that pose "unreasonable risks" of injury or illness (15 U.S.C. 2051) (1984). CPSC regulates all consumer products except foods and drugs, pesticides, tobacco and tobacco products, motor vehicles, aircraft and aircraft equipment, and boats and boat accessories (15 U.S.C. 2052(a)) (1984). The statute also precludes CPSC from regulating risks of injuries associated with substances that are, or are contained by, a consumer product "if such risk could be eliminated or reduced to a sufficient extent by actions taken under the Occupational Safety and Health Act of 1970, the Atomic Energy Act of 1954, or the Clean Air Act" (15 U.S.C. Sec. 2052(a)) (1984), and it may not regulate "electronic product radiation emitted from an electronic product" (15 U.S.C. 2080) (1984).

Whenever a product poses an "unreasonable risk" of injury or illness, section 7(a) of CPSA authorizes CPSC to promulgate a consumer product safety standard (15 U.S.C. 2056(a)(1)) (1984). The safety standard may specify requirements for product performance or design, requirements for consumer instructions or warnings, or both (15 U.S.C. 2056(a)(1)) (1984). If "no feasible consumer product safety standard . . . would adequately protect the public from the unreasonable risk of injury" presented by a product, section 8 authorizes the Commission to ban the product from commerce (15 U.S.C. 2052(a)(3)) (1984).

Mere risk of injury, death, or serious illness does not by itself make a risk "unreasonable," and the Commission has considered both risks and offsetting benefits in regulating products. Over time, the courts have required that CPSC provide more extensive information to support its decisions concerning consumer product hazards. Prior to 1978, while the Commission indicated the benefits from a regulation in its regulatory rationales, it did not always provide a full description of the costs incurred by manufacturers and consumers because of regulation. In *Aqua Slide 'n' Dive* v. *the Consumer Product Safety Commission* the Court of Appeals for the Fifth Circuit invalidated a requirement that warning signs be attached to home swimming pools offered for sale. The court explained "[t]he Commission does not have to conduct an elaborate cost-benefit analysis . . . It does, however, have to shoulder the burden of examining the relevant factors and producing substantial evidence to support its conclusion that they weigh in favor of the standard" (10). Going on, the court said that "[t]he necessity for the standard depends upon the nature of the risk, and the reasonableness of the risk is a function of the burden a standard would impose on a user of the product." That burden can be measured by the "increases in price, decreased availability of a product, and also reductions in product usefulness . . . " Moreover, the Court suggested that CPSC had to show that con-

sumers were unaware of the risks before a product regulation would be warranted (10).

Two other sections of the act might be used to regulate carcinogens (126). Section 12 permits the agency to bring suit in Federal district court seeking the seizure of "an imminently hazardous consumer product" or injunctive relief against a distributor (15 U.S.C. 2061(a)) (1976). This section apparently has not been invoked against a product containing a carcinogen (126). Section 15 authorizes the Commission to order a variety of remedial actions with regard to any product that presents a "substantial product hazard," resulting from a product's defect. (15 U.S.C. 2064) (1984). The Commission intended to use this section to order the recall of hairdryers containing asbestos and had issued a preliminary conclusion that these hairdryers presented a "substantial product hazard," but the manufacturers voluntarily recalled their hairdryers (126). CPSA requires "informal rulemaking" to establish a product safety standard or to ban a product (126).

In 1981 Congress amended CPSA by requiring that CPSC convene a "Chronic Hazard Advisory Panel" (CHAP) before issuing any proposed rule designed to reduce exposures to a product presenting a risk of cancer, mutations, or adverse reproductive effects (15 U.S.C. 2081(b)(1)) (1984). Such a panel is appointed by the Commission, consists of seven members from a list of nominees submitted by the President of the National Academy of Sciences. The nominees cannot be Federal employees and must not have any "substantial financial interest in any manufacturer, distributor, or retailer of a consumer product" (15 U.S.C. 2077(b)(1)) (1984). The members must, in addition, be experts capable of critically assessing "chronic hazards and risks to human health" (15 U.S.C. 2077(b)(2)) (1984). CPSC cannot take action until it receives a report from such a panel. CPSC has convened CHAP's since 1981.

The Federal Hazardous Substances Act

FHSA was enacted in 1960 as a labeling statute intended to fill gaps in other statutes. The act was later amended to permit more drastic action to control hazards and expanded "to cover hazardous substances in general use in the home, and particularly to protect children from hazardous toys and products" (Poison Prevention Packaging Act of 1970, Public Law 91-601, 84 Stat. 1670 (1970); Child Protection and Toy Safety Act of 1969, Public Law 91-113, 83 Stat. 187 (1969); Child Protection Act of 1966, Public Law 89-756, 8 Stat. 1303 (1966)) (126). FHSA was administered by the Food and Drug Administration until the creation of CPSC, at which time the new agency took responsibility for the act.

Section 2(f)(1)(a) of FHSA defines "hazardous substance" as a substance or mixture which ". . . may cause substantial personal injury or substantial illness . . . as a proximate result of any customary or reasonably foreseeable handling or use, including reasonably foreseeable ingestion by children" (15 U.S.C. Sec. 1261(f)(1)(a)) (1984). The act excludes, among other things, pesticides, foods, drugs and cosmetics, certain radioactive materials, and tobacco and tobacco products. Under FHSA, CPSC may require a hazardous substance to bear a hazard label or, if CPSC determines that this step would be insufficient, to ban the product from commerce. CPSC has authority to seize banned substances and to require businesses to repurchase banned hazardous substances (15 U.S.C. 1264) (1984).

In contrast to CPSA, FHSA has slower and more complex rulemaking requirements. The agency issues a proposed rule, entertains comments, and publishes a "final order" (126). If parties adversely affected by the order file "legally sufficient" objections, the Commission is obligated to conduct an evidentiary hearing before an administrative law judge (126). In effect, this procedure is much like formal rulemaking on top of informal rulemaking. Any party whose products have been banned as hazardous under FHSA may petition a court of appeals within 60 days for review (126). A reviewing court must affirm the Commission's order if the order is "based on a fair evaluation of the hearing record" (126).

The Commission has preferred to rely on CPSA because of FHSA's more complex rulemaking and because it believes the informal procedures of the CPSA would better facilitate participation by diverse interests (229). This preference has recently been overruled for some kinds of cases, however. In the recent formaldehyde decision heard by the 5th Circuit Court of Appeals, the court noted that "[r]ulemaking under the Consumer Product Safety Act is to be the exception, not the rule . . ."(76).

Major Court Decisions

Four major legal decisions have affected CPSC regulation of carcinogens. Three cases concerned procedural matters: *Pactra Industries* v. *CPSC* (153), *Springs Mills* v. *CPSC* (190), and *Dow Chemical* v. *CPSC* (45). The major lesson from the cases is that the Commission must scrupulously follow due process requirements in issuing regulations (110).

In *Dow Chemical* v. *CPSC*, the Commission attempted to use its cancer policy, which had been issued as part of a proposed rulemaking, to classify substances according to evidence of their carcinogenicity. Using the policy, the Commission provisionally clas-

sified perchloroethylene as a suspect carcinogen. The Dow Chemical Company sued because it believed even such a provisional classification harmed Dow. The court held that CPSC could not rely on the cancer policy in this manner until it was adopted in rulemaking procedures (45). Subsequent to this decision CPSC formally withdrew its cancer policy from the rulemaking process and decided to use the guidelines adopted by the Interagency Regulatory Liaison Group (IRLG), and more recently the guidelines issued by the President's Office of Science and Technology Policy (OSTP). (Both are discussed more fully in ch. 2.) CPSC had intended to issue a legally binding cancer policy, hoping to foreclose some legal debates in issuing carcinogen regulations. The *Dow Chemical* court blocked the attempt. At present the Commission's position is that it may refer to guidelines such as the old IRLG or present OSTP cancer guidelines, but cannot use them to foreclose legal debates until they have been formally adopted as legal documents in rulemaking procedures. In addition, there are some legal issues that cannot be foreclosed by publishing a cancer policy, e.g., whether a particular animal bioassay is a valid scientific experiment or not.

A fourth case concerning CPSC's regulation of carcinogens could have more far-reaching impact. In 1982, after a 6-year investigation and rulemaking regarding urea-formaldehyde foam insulation (UFFI), the Commission issued a final rule banning UFFI in residences and schools (76). Four industry petitioners objected to the ban and convinced the 5th Circuit in *Gulf South Insulation* v. *Consumer Product Safety Commission* to overturn the ban on UFFI (76). In particular, the court objected to the Commission's "exclusive reliance" on a single animal study to support its risk assessment (76). About a large animal bioassay involving 300 animals, the court noted that

> . . . in a study as small as this one, the margin of error is inherently large. For example, had 20 fewer rats or 20 more developed carcinomas, the risk predicted by [the] Global 79 [risk assessment model] would be altered drastically (76).

This is very close to saying that if the victim of the gunshot wound had not died, the defendant wouldn't be guilty of murder.

The court went on to conclude that even if the study were valid for some purposes, it did not constitute substantial evidence to support CPSC's "precise" estimate of risk, without which CPSC could not validly conclude that UFFI posed an unreasonable risk of cancer.

The court believed that formaldehyde "should be presumed to pose a cancer risk to man," but regarded as "questionable" two assumptions the Commission relied upon: that at identical exposure levels the "effective dose for rats is the same as that for humans," and that the "risk of cancer from formaldehyde is linear at low dose—in other words that there is no threshold below which formaldehyde poses no risk of cancer" (76). Finally, as indicated above, the court held that CPSC had not properly justified its rulemaking under CPSA rather than FHSA, concluding that the Commission should have regulated UFFI under the Federal Hazardous Substances Act (76).

The Environmental Protection Agency

In 1970 by executive order, President Richard Nixon created the Environmental Protection Agency, merging 15 existing programs "managed by 5 different departments or councils . . . into . . . an organization headed by a single administrator . . . charged with regulating virtually all sources of pollution rather than a single industry" (114).

EPA is headed by an administrator, with assistant administrators in charge of its major divisions. (Table A-3 lists EPA units responsible for administrating the various environmental statutes.)

The Clean Air Act

One of the first major environmental statutes enacted in the early 1970s was the Clean Air Act Amendments of 1970. The statute provides an elaborate Federal-State scheme for controlling conventional pollutants, such as sulfur dioxide and carbon monoxide. Because of the emphasis on controlling conventional air pollutants, toxic pollutants were almost ignored. The House version of the bill did not contain a provision for hazardous air pollutants (although the Senate version did) and during the House-Senate Conference negotiations, the administration recommended deletion of the hazardous pollutants section (71). Despite this opposition, Congress approved a hazardous air pollution provision—section 112.

While sections 108-109 and 111 have been considered by EPA as possible statutory authority for regulating carcinogens, the agency has not relied on these and has used section 112 as its primary authority. However, in 1977 Congress amended CAA by adding, among other things, a section on the regulation of radioactive pollutants, cadmium, arsenic, and polycylic organic matter. The agency was ordered to review within one year all "available relevant information" on these substances to decide whether or not they "may reasonably be anticipated to endanger public health" (42 U.S.C. 7422(a)). If any did, it was to be considered for regulation under sections 108-109, 111 or 112, or a combination of them. The agency subsequently regulated two of these substances. (See ch. 4.)

Section 112 authorizes EPA to set emission stand-

Table A-3.—EPA Administration of Statutes

EPA office	Statute administered
Assistant Administrator for Air and Radiation	Clean Air Act
Assistant Administrator for Pesticides and Toxic Substances	
Office of Pesticide Programs	Federal Insecticide, Fungicide, and Rodenticide Act
Office of Toxic Substances	Toxic Substances Control Act
Assistant Administrator for Solid Waste and Emergency Response	Resource Conservation and Recovery Act, Comprehensive Environmental Response, Compensation, and Liability Act (Superfund)
Assistant Administrator for Water	Clean Water Act, Safe Drinking Water Act

SOURCE: Office of Technology Assessment, 1987.

ards for "hazardous" air pollutants, which include any air pollutant

. . . to which no ambient air quality standard is applicable and which in the judgment of the Administrator causes or contributes to, air pollution which may reasonably be anticipated to result in an increase in mortality or an increase in serious irreversible, or incapaciting reversible, illness (42 U.S.C. 7412(a)(2)).

The administrator must establish standards for each pollutant

. . . which in his judgment provides an ample margin of safety to protect the public health from such hazardous air pollutant. (42 U.S.C. 7412(b)(1)(b)).

Under this section EPA was required within 90 days of December 31, 1970, to publish, and from time to time revise, a list of hazardous air pollutants that EPA intends to regulate (42 U.S.C. 7412(b)(1)(a)). The idea is that a pollutant is first listed as hazardous based on pertinent scientific data, then national standards are established for each source category of such pollutants (71).

A substance can become a candidate for listing by agency nomination or by citizen petition (71). A substance is not listed until the EPA staff prepares a comprehensive health assessment document, the agency's Scientific Advisory Board (SAB) gives written comment, and the EPA administrator determines to list it (71).

Once a substance is listed, the administrator is to propose emission standards within 180 days, hold a public hearing within 30 more days, and publish final emission rules within 180 days of the proposal (42 U.S.C. 7412(b)(1)(b)). The pollutant must be regulated "unless [the agency] finds, on the basis of information presented at such hearings, that such pollutant clearly is not a hazardous air pollutant" (42 U.S.C. 7412(b) (l)(b))(1982). Once a substance is listed as a hazardous pollutant, the administrator has a duty within the deadlines to propose and issue national emission standards, and citizens may sue in Federal courts to force compliance with these procedures (42 U.S.C. 7604 (a)(2)).

States may issue their own standards for hazardous air pollutants "as long as they are at least as stringent as those required by EPA. If States submit adequate control programs to EPA, the administrator is authorized to delegate his implementation and enforcement authority to the States" (76).

However, Congress provided no explicit guidance for regulating carcinogens as compared with other hazardous substances under this section. This failure to address carcinogens explicitly has led to considerable controversy in the interpretation of the statute for application to carcinogens. For a substance with a toxic threshold, i.e., a level below which there are no harmful health effects to a group of people, setting a standard would involve determining a "no effects" threshold and providing for a margin of safety. However, for carcinogens there is no known safe threshold. Thus, providing an ample margin of safety as required by the statute would imply elimination of all exposures by setting an emissions standard of zero, or, possibly, a standard of no detectable concentrations. Faced with economically beneficial activities which produce such pollutants and with control equipment incapable of reducing emissions to zero, EPA's strategy has been to require

> . . . emission reduction to the lowest level achievable by use of the best available control technology in cases involving apparent nonthreshold pollutants, [where] complete emission prohibition would result in widespread industry closure and EPA has determined that the cost of such closure would be grossly disproportionate to the benefits of removing the risk that would remain after imposition of the best available control technology (325).

The definition of "best available technology" (BAT) differs for new and existing sources. For existing sources EPA considers "economic feasibility" and sets the requirements at "the most advanced level of technology that at least most members of an industry [can] afford without plant closures" (325). For new sources, BAT will be the "technology which in the judgment of the administrator, is the most advanced level of control adequately demonstrated, considering economic, energy, and environmental impacts" (324). In addition to requiring BAT, for new sources, if EPA finds that there is an "unreasonable residual risk, a more strin-

gent alternative would be required" (324). The administrator will base his judgment of "unreasonable residual risk" on:

- the range of additional cancer incidence;
- the range of health risks to the most exposed individuals;
- readily identifiable benefits of the substance or the activity;
- economic impacts of requiring additional control measures;
- the distributions of benefits of the activity versus the risks it causes; and
- other possible health and environmental risks (324).

Although this overall strategy was articulated in a proposed rule adopting a policy for airborne carcinogens (324) which has never been finalized by the agency, EPA staff say that the agency continues to follow the broad outlines of the policy (103). EPA has acknowledged that the BAT approach was not explicitly recognized by the statute and, at the time it regulated asbestos and vinyl chloride, the approach had not been tested in the courts.

The court test arose over the 1976 vinyl chloride standards (55). That case ended with a consent decree in which EPA agreed to issue proposed amendments to the standard for vinyl chloride with "the ultimate goal of zero vinyl chloride emissions." In addition, EPA agreed to lower the 10 ppm emission standard to 5 ppm as soon as "technology can achieve the lower standard" as a means of working toward the zero emission standard (55). EPA's interpretation of the "ample margin of safety" requirement of section 112 has been controversial. The General Accounting Office as well as others have disagreed with the agency's position (183,198). Nevertheless, it received tacit endorsement by the courts when the District of Columbia Court of Appeals approved the consent decree.

Subsequent to the consent decree in *EDF* v. *Train*, there has been further litigation concerning regulation of vinyl chloride and EPA's interpretation of "ample margin of safety." On January 9, 1985, EPA announced that it would not continue to pursue the goal of zero emissions for vinyl chloride and would not require a standard of 5 ppm (compared with the present 10 ppm). It supported its position on several grounds:

- EPA continues to hold that section 112 does not "express an intent to eliminate totally all risks from emissions of airborne carcinogens" (325),
- 10 ppm represents "the lowest level of control which has been consistently achieved" (325), and
- the proposed 5 ppm emission "was not based on data from a control technology different from that analyzed for the current standard" (325).

The Natural Resources Defense Council (NRDC), alleging that EPA had "reneged" on provisions of the 1977 consent decree, sued EPA on June 17, 1985, concerning its interpretation and procedure for setting standards under the "ample margin of safety" phrase (142). NRDC argued that the language of the statute does not specifically provide for cost-benefit analysis, and, additionally, section 112 does not permit EPA to impose cost-benefit or technological feasibility tests on proposed standards for toxic pollutants. NRDC argues that the Supreme Court has held that the agency cannot use cost-benefit analysis when setting health standards unless the statute provides for it (140). A panel of the D.C. Circuit ruled in 1986 that, since the statute provides EPA with some discretion in setting regulations under section 112 and since it does not specify precisely how this discretion is to be exercised, the court will permit the agency to exercise reasonable discretion in implementing the statute. However, the panel decision was subsequently vacated by a grant of rehearing for certain source categories by the entire court. Oral argument was held on April 29, 1987, and a decision is pending.

In addition, the Agency has been sued over its failure to issue benzene standards. NRDC argues that the statute requires that hazardous air pollutant regulations be based exclusively on public health considerations, not technology and cost tests which may compromise the protection of health. It also argues that EPA is prohibited by statute from dismissing as insignificant an increase in mortality that may be caused by benzene exposure, because the agency is required to issue regulations with an ample margin of safety, not simply to prevent "significant" health risks (145).

The agency responds that section 112 of the Clean Air Act permits "EPA not to regulate source categories of benzene that present an insignificant risk" (146). The contention is that Congress, in its 1977 amendments, "intended to codify an approach . . . [taken in an earlier legal case] . . . which plainly held that a finding of 'significant risk' is an appropriate test for regulating and stressed that the administrator may 'weigh risks and make reasonable projections of future trends' . . . " (146).

The outcome of this case will be important for the development of hazardous air pollutant regulations, and it may also indicate the extent to which the courts are willing to permit agencies not to regulate toxic substances because the agency has determined that the risk is "insignificant" or "de minimus."

The Clean Water Act

Since the Federal Water Pollution Control Act was first enacted in 1948, it has been amended nine times,

and is now generally referred to as the Clean Water Act. The most important of the amendments were made in 1972, 1977, and 1987. In 1972, Congress set the goal of achieving "fishable, swimmable" waters by 1983 and prohibiting the "discharge of toxic pollutants in toxic amounts . . ." by 1985 (33 U.S.C. 1251(a)) (1982). In the 1977 amendments, Congress endorsed a new method for regulating toxic pollutants that had been developed to settle a lawsuit between environmental organizations and EPA. In 1987, Congress continued its emphasis on control of toxic pollutants.

The CWA protections are less directly related to human health than are the protections of some other laws such as the Safe Drinking Water Act, which aims at securing the safety of drinking water supplies. Nonetheless, CWA has a number of sections aimed at regulating human exposure to carcinogens and other toxics.

Central to controlling all water pollutants under CWA are the National Pollution Discharge Elimination System (NPDES) permits for "direct" dischargers. It is lawful to discharge a pollutant only if the discharge is in compliance with an NPDES permit. Such permits can be issued by EPA or by States whose permit programs are approved by EPA. At present 37 of a possible 54 jurisdictions have been approved to administer their own NPDES programs; EPA administers the remainder (31).

An NPDES permit is written for a facility which may have a number of discharge pipes (typical facilities have from 1 to 3 such pipes, but large facilities like steel mills may have as many as 100) (74). For each discharge pipe a permit will contain the following:

- a list of pollutants that must be regulated according to Federal or State law, together with specified permissible amounts of each pollutant that may be discharged per unit of time;
- monitoring requirements and schedules for implementing the pollution concentration requirements; and
- special conditions regarding pollutants the permit writer thinks should be imposed on the permit holder, for example, additional testing and procedures for spills of pollutants into the water.

NPDES permits are written both for conventional pollutants (e.g. biological waste material) and for toxic substances.

In writing a permit for toxic substances such as carcinogens, the responsible State agency or EPA will consider the following:

- Federal toxic effluent standards and toxic effluent limitations,
- Federal water quality criteria for toxics,
- State water quality standards or effluent stand-

ards for toxics, and
* special conditions.

Sections 301, 304, and 307 of CWA are the center-piece, containing substantive conditions on NPDES permits for regulating toxic pollutants. Under it, EPA may issue binding regulations known as effluent limitations and effluent standards. These are legally binding Federal regulations limiting the concentrations of pollutants in point source discharges and they must be on NPDES permits.

An "effluent limitation" uses a technology-based approach to limit the amount of a toxic substance that can be discharged from a point source, such as a pipe. Toxic substances regulated in this manner have been regulated on an industry by industry basis.[2] An "effluent based standard," by contrast, is a control requirement based on the relationship between the discharge of a pollutant and the resulting water quality in a receiving body of water (62), but has not been used since 1977. Water quality-based effluent limitations which specify certain concentrations of a chemical in a point source of effluent, are typically more stringent than technology-based effluent limitations (62), and must be established "with an ample margin of safety" (33 U.S.C. 1317(a)(4)) (1982). Finally, CWA requires informal rulemaking (79).

Because only six pollutants had been regulated with effluent standards, EPA was sued by NRDC. The subsequent consent decree (discussed below) provided that EPA could instead set effluent limitations to regulate toxic pollutants. In 1977, Congress added a reference to a specific list of toxic substances that had been agreed to for the consent decree. (See table 1 of the House committee on Public Works and Transportation Committee Print 95-30.) An account of how this list of toxic pollutants was developed is provided in chapter 3. This list may be revised from time to time as the administrator deems appropriate (1311(b)(2)(C); 1317(a)) (1982).

For each listed toxic chemical the administrator must establish effluent limitations or standards. Substances controlled by either of these mechanisms have been regulated on a pollutant-by-pollutant basis. Effluent limitations are issued industry-by-industry, however, with specific requirements for each relevant pollutant. Effluent standards, when used, have been issued pollutant-by-pollutant, regardless of which industries might be affected.

Section 307(b) also requires pretreatment standards for toxic substances discharged from private pollution sources into publicly owned water treatment facilities. Pretreatment standards together with discharge limitations on publicly owned treatment facilities must produce as great a reduction of toxic pollutants as the use of effluent limitations would on a private polluting point source (33 U.S.C. 1317(b)).

Section 306 requires technology regulations on new facilities similar to effluent limitations on existing facilities (so-called new performance standards). New facilities must provide the best available demonstrated technology and, where it is practicable, there must be no discharge of pollutants (33 U.S.C. 1316).

Section 304 authorizes EPA to develop ambient water quality criteria for all pollutants, including toxics. These criteria are not legally binding on EPA or the States, but may be used as pollutant goals to be pursued in improving the quality of water courses. The production of the water quality criteria documents was a major risk assessment activity at EPA in the late 1970s (see ch. 4). NPDES permit writers, under the narrative criteria of the permits, may impose more stringent limitations on toxic pollutants than BAT effluent limitations would require. The extent to which permit writers have used this section of NPDES permits is unknown, but some in EPA believe that EPA regional offices do use the ambient water quality criteria in writing permits for facilities in States where there are not State-approved programs (15).

State water quality standards are to be developed for the amount of a pollutant permitted in a given course (sec. 303). The limitations established here must protect the public health or welfare, enhance the quality of water, and serve the purposes of the act. CWA requires that the States develop State water quality standards. The extent of this standards-setting activity, however, has been very limited (see ch. 3).

Water quality-based effluent limitations, adopted under section 302, might be used to impose "limitations more stringent than BAT for sources on a particular stream segment," such as the prohibiting of all discharges of toxic pollutants, "if the water quality in a stream will not attain the national goal of "fishable/swimmable' waters . . ." (62). To date this section has not been used, but EPA has announced its intention to begin to develop such standards (282). However, the agency will first develop limitations for fish and aquatic life and acknowledges that the human health limitations "lag behind" (15).

Special conditions imposed by an NPDES permit writer are designed to achieve a generalized goal, or narrative criteria, of the State water laws or CWA, e.g., preventing toxic pollutants in toxic amounts in the nation's waters (sec. 301).

---

[2]Best available technology provisions for regulating toxic substances were not included in the major revisions of CWA in 1972. Section 307 originally included only the nontechnology-based toxic effluent standards. Congress amended CWA in 1977 to provide for technology-based limitations as agreed to in the 1976 *NRDC* v. *Train* consent decree.

In addition to regulating toxic substances, CWA declares that there should be no discharges of "oil or hazardous substances" into or on the navigable waters, including shoreline coastal waters. Hazardous substances are those that present an "imminent and substantial danger to the public health or welfare . . ." (33 U.S.C. 1321(b)(2)(a)). EPA was required to list such substances, determine quantities which might be harmful, and issue regulations concerning these (33 U.S.C. 1321(b)(2)(a), 1321(b)(4)). The law also provides for liability in case of discharges of hazardous substances or oil. This provision of CWA anticipates some of the hazardous substances prevention and cleanup provisions of the later Resource Conservation and Recovery Act and the Comprehensive Environmental Response, Compensation, and Liability Act.

Significant Judicial Decisions.—A judicial decision led to the development of toxic effluent limitations under CWA. In 1975 a group of lawsuits were filed against EPA challenging its failure to regulate toxic pollutants (55). In 1976 those cases ended in a consent decree, with EPA agreeing to place specific "numerical limits on the quantities of 65 toxic pollutants in 21 industrial categories" (62). EPA was required to complete these regulations by June 30, 1983. The consent decree permitted EPA to regulate toxic substances through those sections of CWA designed to control ordinary nontoxic pollutants, in particular, the technology-based provisions of the statute. The agency has been in the process of issuing BAT effluent limitations for 28 industrial categories which may limit these 65 categories of toxics, including at least 29 carcinogens. Congress, in the 1977 amendments to the CWA, incorporated these changes, in effect creating the new category of toxic effluent limitations.

There have been several judicial developments related to the 1976 consent decree. The National Research and Demonstration Center sued EPA to show the agency in contempt of the agreement (this was settled out of court, extending EPA's original deadline from June 30, 1983, to June 30, 1984) (140). Eight times EPA requested Judge Flannery to modify the agreement to give it more time to complete the regulations. Except for the May 1982 request, which Judge Flannery denied, urging the agency to "work harder," (144) the requests were granted (144). These guidelines are in various stages of revision and litigation.

Another judicial development of note was Velsicol Corporation's attempt to overturn EPA's first regulations of two toxic substances, toxaphene and endrin, regulated under the Federal Water Pollution Control Act of 1972. Velsicol argued, among other things, that the technology-based effluent limitations added in the 1977 congressional amendments superceded the previous health-based authority of section 307 and that,

in addition, EPA was required to consider economic and technological factors in setting its regulations. The court ruled that the 1977 amendments were required to aid, not impede, EPA's health-based authority, and thus denied both pleas (and several others as well) (88).

One final judicial development of note is EPA's use of "fundamentally different factors" (FDF) variances. These have been used to modify, on a case-by-case basis, BAT or pretreatment limitations of pollutants, including toxics (62). However, such variances are not specifically authorized by the statute and have been the subject of some litigation. Although the Court of Appeals for the Third Circuit struck down the use of these variances, the Supreme Court in a 5-4 opinion upheld EPA'a authority to apply FDF variances to toxic pollutants (29).

The Safe Drinking Water Act

A second major statute for regulating carcinogens in water is the Safe Drinking Water Act (SDWA) of 1974. Although the Clean Water Act, discussed above, was designed to control water pollution, it provided no authority to regulate polluted water discharged into nonnavigable waters, such as groundwater, which often is a source of drinking water. Thus additional legislation was needed to "assure safe drinking water" (214). SDWA aims primarily to regulate water provided by public water systems, and it contains several provisions that may be used to regulate hazardous substances, including carcinogens in drinking water. In contrast to CWA, SDWA is more directly concerned with protecting human health.

Under SDWA, EPA is to regulate contaminants "which . . . may have an adverse effect on the health of persons . . ." (42 U.S.C. 300f(1)(b)). The act then prescribes the steps which the agency must go through over time to protect drinking water.

First, EPA was required to publish national interim primary drinking water regulations within 90 days of December 16, 1974. These regulations were to "protect health to the extent feasible, using technology, treatment techniques and other means, which the Administrator determines are generally available (taking costs into consideration)" on the date of enactment (42 U.S.C. 300g-1(a)(2)).

Second, Congress required that EPA request a National Academy of Sciences (NAS) study to determine the maximum contaminant levels that should be recommended as national standards, and to identify the "existence of any contaminants the levels of which in drinking water cannot be determined but which may have an adverse effect on the health of persons" (42 U.S.C. 300g-2(e)). In addition, revisions of the NAS study reflecting any new information "shall be reported to Congress each two years thereafter" (42 U.S.C. 300g-2(d)(2)). In considering whether contaminants have an adverse effect, the NAS study had to consider the impact of contaminants on groups or individuals in the population more susceptible to adverse effects than normal healthy individuals, as well as exposure to contaminants in other media, synergistic effects of contaminants, and body burdens of contaminants in exposed persons (42 U.S.C. 300g-1(3)). In 1977 NAS provided its first list of contaminants (chosen on the basis of its own criteria) that might have an adverse effect on health (36,134).

Third, within 90 days of the publication of the NAS study, EPA was required to establish by rulemaking "recommended maximum contaminant levels (RMCL) for each contaminant which . . . may have any adverse effect on the health of persons" (42 U.S.C. 300g-1(b)(1)(b). Each such RMCL was to be "set at a level at which . . . no known or anticipated adverse effects on the health of persons occur and which allows an adequate margin of safety" (42 U.S.C. Sec. 300g-1(b) (1)(b)).

The House report on the 1974 SDWA elaborated on the criteria for setting RMCLs:

> . . . the recommended maximum level must be set to prevent the occurrence of any known or anticipated adverse effect. It must include an adequate margin of safety, unless there is no safe threshold for a contaminant. In such a case, the recommended maximum contaminant level should be set at zero level (214).

RMCLs are nonenforceable health goals, which are used as guidelines for establishing enforceable drinking water standards. The agency also had to publish a list of contaminants whose levels cannot be measured accurately enough in drinking water to establish an RMCL, and which may have an adverse effect on the health of persons (42 U.S.C. 300g-1(b) (1)(b)).

Once the agency established RMCLs for each contaminant it was required to publish revised national primary drinking water regulations. These regulations establish the requisite enforceable health standards. The required regulations must specify a maximum contaminant level (MCL) or require the "use of treatment techniques for each contaminant" for which an RMCL is established. The established MCLs were to be as close to the RMCLs as is "feasible" (42 U.S.C. 300g-1(b)(3)). In determining feasibility, the administrator may consider "the use of the best technology, treatment techniques and other means, . . . [that] are generally available (taking cost into consideration)" (42 U.S.C. 300g-1(b)(3)).

In general, enforcement of MCLs rests with the States. EPA sets MCLs. The agency then has the responsibility of reviewing and approving State programs to achieve the mandated standards, and, once a State program is approved, it gives States the authority to enforce the MCLs. Until a State has an approved program, EPA has authority to regulate levels of contaminants in drinking water.

SDWA was modified by the 1986 Amendments which are discussed in chapter 3.

Significant Judicial Action.—To date only one case has been brought regarding EPA's regulation of carcinogens (concerning the regulation of trihalomethanes) under SDWA, and it was settled after briefs were filed. Since the agency issued a final rule for RMCLs and a proposed rule for MCLs for eight volatile organic compounds in November 1985, it has been sued by both industry and environmental organizations. Environmental organizations are challenging the classification of 1,1-dichloroethylene as category II in EPA's weight-of-the-evidence classification scheme and contend that that there should be zero concentration levels for RMCLs for carcinogens regardless of the weight of the evidence.

The Federal Insecticide, Fungicide, and Rodenticide Act

The Federal Insecticide, Fungicide, and Rodenticide Act was originally passed by Congress in 1947. This act was replaced by the Federal Environmental Pesticide Control Act (FEPCA) of 1972, although the name FIFRA continues to be used. Central to regulating the sale, shipment, and delivery of pesticides is a registration system: generally, it is unlawful to sell or distribute a pesticide which is not registered with EPA (7 U.S.C. 136j(a)(1)(a)).

Registration of New Pesticides.—An applicant for registration of a pesticide must file certain required information, including a statement of all claims made for the pesticide, directions for its use, a description of tests made upon it, and the test results used to support claims made for the substance with EPA (7 U.S.C. 136a(c)(1)(a-d)). Most important for this report is that an applicant must supply appropriate health and safety data for each pesticide.

In a typical registration procedure a prospective registrant, typically the pesticide manufacturer, submits an application for a registration. If a registration package contains all required material it goes on for an evaluation of toxicity studies, wildlife data, exposure information, etc. At the same time, if appropriate, the agency considers residue data for purposes of setting food-safety tolerances as required under sections 408 and 409 of FDCA (described below).

FIFRA requires that EPA "shall register" a pesticide if its composition warrants the proposed claims for it, and its labeling and other required material comply with the requirements of the Act (meaning "it will perform its intended function without unreasonable adverse effects on the environment," and "when used in accordance with widespread and commonly recognized practice it will not generally cause unreasonable adverse effects on the environment" (7 U.S.C. 136a(c)(5)). "Unreasonable adverse effects on the environment" means "any unreasonable risk to man or the environment, taking into account the economic, social and environmental costs and benefits of the use of any pesticide" (7 U.S.C.136(bb)). The agency may refuse to register a pesticide after giving the applicant notification of this intention and opportunity to correct the deficiencies in the application (7 U.S.C. 136a(c)(6)).

If EPA finds that a pesticide meets or exceeds any of several criteria for risk specified by EPA which includes carcinogenicity (40 CFR 154.7), it must initiate the special review process. During the special review, the risks and benefits of the pesticide are considered and public comments received. Unless the manufacturer can show EPA is wrong, or that exposures would not be significant, proceedings are initiated to deny, cancel, or modify the registration of the pesticide.

A pesticide may be registered for general use or restricted use. If a chemical will not "generally cause unreasonable adverse effects on the environment, [EPA] will classify" it for general use (7 U.S.C. 136a(d)(1)(b)). If it "may generally cause, without additional regulatory restrictions, unreasonable adverse effects on the environment, including injuries to the applicator, [EPA] shall classify" it for restricted use (7 U.S.C. 136a(d) (1)(c)). Nearly all pesticides are registered for particular uses, such as for particular crops (328). If a registrant desires to sell a product for a use not permitted by the registration, then he or she must submit it for agency approval and registration for that different use. If the agency classifies a pesticide for restricted use because "the acute dermal or inhalation toxicity of the pesticide presents a hazard to the applicator or other persons," then it shall be applied only by or under the direct supervision of a certified applicator (7 U.S.C. 136a(d)(1)(c)(i)). The States have the authority to certify pesticide applicators (7 U.S.C. 136b(a) (2)).

In registering a pesticide for use, EPA's approval requires the granting of a residue tolerance if pesticide residues are expected to remain on a raw foodstuff, or to issue an exemption from the tolerance requirement if appropriate. The setting of a tolerance is required by section 408 of the Food, Drug, and Cosmetic Act. Sections 408 and 402(a)(2)(b) of FDCA forbid the distribution of raw or processed foods bearing pesticide residues that have not been sanctioned by EPA (21 U.S.C. 342(a)(2), 346a(a)). Unlike the Delaney clause of section 409 of FDCA (concerning food additives), residues of pesticides used on foodstuffs are not precluded even when they induce cancer in laboratory animals (127). However, EPA must determine the quantity of a pesticide that may remain on a raw commodity when it enters commerce and must set an appropriate tolerance for the substance (21 U.S.C. 346a(a)). In setting a tolerance, the agency must take into account the necessity for an "adequate, wholesome, and economical food supply," other ways consumers may be affected by the same or other chemicals, and the opinion of the Secretary of Agriculture, "submitted with a certification of the usefulness" of the pesticide (21 U.S.C. 346a(a)).

EPA also must set tolerances for pesticide residues in processed agricultural products, such as grain flours or processed vegetable or fruit products, under section 409 of the Food, Drug, and Cosmetic Act. A pesticide residue becomes a "food additive" and thus may be subject to the Delaney clause, if the processing increases the residue concentration levels in the processed food above the tolerance established for the raw commodity (21 U.S.C. 342(a)(2)(c)). If the concentration after processing remains below that established for the raw commodity, the Delaney clause does not apply (127). EPA sets the pesticide residue limits, and FDA monitors and enforces the regulations.

When Congress amended FIFRA in 1978, it permitted "conditional registration" of a pesticide, even though some of the test data may not have been submitted to or evaluated by EPA. The Agency may conditionally register pesticides if "insufficient time has elapsed since the imposition of the data requirements for those data to have been developed, use of the pesticide product(s) containing the new active ingredient during the conditional period would not cause any unreasonable adverse effects, and conditional registration of the pesticide product and its uses are in the public interest" (203). Even though Congress intended conditional registrations to be "rarely exercised," EPA conditionally registered about half the pesticides submitted between 1978 and 1984 (203).

Pesticides contain both "active" and "inert" ingredients. Active ingredients are the components in pesticides that prevent, destroy, repel, or mitigate any pest, or disrupt the normal biological functioning of certain organisms such as insects, fungi, and plants (21 U.S.C. 136(a)(1-4)). An inert ingredient is one that is not active in this sense. Typically inert ingredients are used to dilute or deliver the active ingredients.

In the past most regulatory attention in registration has been focused on active ingredients, for they are typically the ones that damage pests or plants. However, there has been increasing concern about inert ingredients because they may have dangerous health effects (2). Problems with inert ingredients are described in chapter 3. EPA issued a policy statement for regulating inert ingredients on April 22, 1987 (295). EPA intends to encourage the use of the least toxic inert ingredients, require data to determine the conditions of safe use of particular inerts, require labeling, and hold hearings to determine if the use of certain inerts should continue.

Special Reviews, Cancellation, Suspension of Registered Pesticides.—In addition to the cancellation provisions of FIFRA, section 6, once a substance has been registered, if data indicate that it may present an acute toxicity or a chronic toxicity hazard, including oncogenicity, or if it lacks an emergency procedure in case of exposure, EPA may announce a "special review" (40 CFR 154.11 (a)(3)). Until 1983 these had been known as "rebuttable presumptions against registration (RPAR)" (2). Unless the data on which the special review is based are shown to be unreliable or invalid, or the estimated benefits of continued uses (possibly with additional restriction) outweigh the estimated risks, the pesticides are candidates for cancellation or suspension of their registration (360).

Reregistration of Pesticides Registered Prior to 1972.—When FIFRA was amended in 1972, a number of pesticides were in use that had neither been registered under the new statute nor subjected to more stringent data requirements. Congress required a review of all substances then registered, to reassess the safety of those pesticides. The "reregistration" was intially to be completed by 1976, but in 1975 Congress extended that deadline to 1977, and in 1978 dropped the deadline completely because of the large number of substances outstanding (203). EPA has identified some 600 active ingredients that are used in a large majority of pesticides, but only a small fraction of these have adequate health and safety data for reregistration purposes (2). Reregistration involves a number of steps described in chapter 3.

Significant Judicial Decisions.—The courts have heard a substantial number of cases concerning EPA's regulation of pesticides under FIFRA. Many of these may be summarized by saying that in general the courts have shown considerable deference to EPA's action, since it has considerable discretion to act under FIFRA (53,54). In particular, several court decisions have held that the burden of proof to establish the safety of a product remains at all times on the applicant and registrant of a product (53,54). Furthermore, a registrant must show that the benefits from a product outweigh risks, once the agency has identified risk to human health or the environment (54). Sometimes the courts have upheld the agency in refusing to suspend a substance (DDT) known to cause cancer in animals and "various injuries in man" (54), and sometimes they have upheld agency action (banning aldrin and

dieldrin) on the basis of animal test results showing the substances caused cancer in several strains of mice and rats (52). In addition, the courts upheld the agency when it ordered suspension of 2,4,5-T and Silvex based upon "inconclusive but suggestive evidence" that the pesticides caused a statistically significant increase in spontaneous abortions in women exposed to them (44). Finally, the court upheld EPA's reliance on "cancer principles" concerning use of animal tests, extrapolation from animals to humans, the presumption that there is no safe level for carcinogens, and use of both benign and malignant tumors in animal studies in determining cancer hazards to humans. The court noted that industry's scientific disagreement with these principles was not sufficient to rebut them (53).

The Toxic Substances Control Act

TSCA was enacted in 1976, and allows for the regulation of chemicals in commerce as well as before they even enter commerce. In TSCA, Congress set the policy that:

- chemical manufacturers and processors are responsible for developing data about the health and environmental effects of their chemicals;
- the government regulate chemical substances which pose an unreasonable risk of injury to health or the environment and act promptly on substances that pose imminent hazards; and
- regulatory efforts not unduly impede industrial innovation (15 U.S.C. 2601(b)).

Unlike some other statutes, which are aimed at regulating exposures to toxic substances through specific media, such as water or air, TSCA is directed toward hazardous substances wherever they occur. Section 4 singles out for special concern substances which present or will present significant risks of cancer, genetic mutations, or birth defects. Under TSCA, EPA must review data on "new" chemicals prior to their large-scale manufacture (sec. 5), may restrict or even ban uses of new or existing chemicals (secs. 5, 6, and 7), may require manufacturers to conduct toxicity tests (secs. 4 and 5), and may impose certain record keeping and reporting requirements (sec. 8).

TSCA permits EPA to do two important tasks in regulating new or existing toxic substances:

1. require testing of new or existing chemicals; and
2. restrict production and use of, or even ban, substances posing "unreasonable risks" to health or the environment.

The treatment of new and existing substances is somewhat different (as described below) and generates somewhat different pressures on the Agency.

TSCA is directed at chemical substances. Thus a number of substances are excluded from TSCA regulation, including pesticides when they are used as pesticides; tobacco and tobacco products; nuclear materials; foods, drugs, and cosmetics; and pistols, firearms, revolvers, shells, and cartridges (15 U.S.C. 2602).

Section 8 requires EPA to compile an "inventory of chemical substances" containing all chemicals subject to the provisions of TSCA manufactured or imported into the United States (15 U.S.C. 2607(b)). The initial inventory was published on June 30, 1979, and all chemicals that did not appear on that list and that are not exempted from TSCA are considered "new" chemicals. The treatment of "new" and "existing" chemicals on the inventory is somewhat different, although for both types EPA has broad authority to review available information, require testing, and regulate production and use.

The major criteria for EPA decisions under TSCA is whether use of a substance 1) presents an unreasonable risk to health or the environment or 2) may present such a risk. In deciding whether a substance poses an unreasonable risk to health or the environment, the agency must take into account:

- "the effects of such substance or mixture on health and the magnitude of the exposure of human beings" to it;
- "the effects of such substance or mixture on the environment and the magnitude of the exposure of the environment to such substance of mixture";
- "the benefits of such substance or mixture for various uses and the availability of substitutes for such uses"; and
- "the reasonably ascertainable economic consequences of the rule, after consideration of the effect on the national economy, small business, technological innovation, the environment, and public health" (15 U.S.C. 2606(c)).

It is not clear that the test of "unreasonable risk," however, is an explicit cost-benefit analysis, for at least one major committee, the House Committee on Interstate and Foreign Commerce, indicated that assessing the reasonableness of the risk does not require a quantitative cost-benefit analysis (212).

New Chemicals.—In general, anyone who intends to manufacture a "new" chemical must notify EPA of his or her intention 90 days before manufacture is to begin. The company must submit a "premanufacture notice" (PMN) which contains information about chemical identity, proposed uses of the chemical, the expected production volumes of the chemical for the various uses, expected byproducts, estimates of the numbers of people likely to be exposed in manufacture of the chemical, and methods for disposal (15 U.S.C. 2604(d) (1)(a), 2807(a)(2)). The PMN must also include information on any toxicity testing that the company has performed, although TSCA does not require that any testing be done prior to submission of a PMN.

EPA has 90 days to review the PMN, although this period may be extended for an additional 90 days. EPA's review can result in any of four actions:

1. the substance may be manufactured without restriction;
2. the substance may be manufactured for uses described on the PMN, but the agency can require that it be notified if any significant new use is considered (15 U.S.C. 2604(a)(2));
3. the manufacture, processing, use, distribution, or disposal of the new substance may be regulated pending the development of additional information about it (15 U.S.C. 2604(e)); or
4. the manufacture, etc. of it may be regulated because it presents or will present an unreasonable risk (15 U.S.C. 2604(f)) (222).

In the PMN process, often the mere threat that EPA might require additional testing or some kind of restrictive action is sufficient to cause a manufacturer to remove a substance from consideration or to agree to the proposed restrictions.

In addition, TSCA (sec. 6) gives EPA broad authority (for either new or existing substances) to regulate the manufacture, processing, distribution in commerce, use, or disposal of a toxic chemical. If the agency finds that "there is a reasonable basis to conclude" that any of these activities, alone or in combination, "presents or will present an unreasonable risk of injury to health or the environment" (15 U.S.C. 2605(a)), it may regulate the substance in a number of ways. For any substance the agency may:

- prohibit its manufacture, processing, or distribution in commerce;
- limit its amount;
- limit its uses or amounts;
- require certain labeling
- require maintenance of records and monitoring;
- prohibit or regulate any manner or method of commercial use;
- prohibit or regulate its disposal; or
- require manufacturers or processors to notify EPA of any unreasonable risks posed by it (15 U.S.C. 2605(a)).

Existing Chemicals.—For existing chemicals, EPA can require testing and can restrict production and use. For testing existing chemicals the statute establishes an Interagency Testing Committee to recommend substances to EPA for testing. Priority attention is to be given to substances that cause or contribute to cancer, gene mutations, or birth defects (15 U.S.C. 2603(e)1(a)). TSCA also lists a number of other factors to be considered in making testing recommendations (see ch. 4).

For substances on the list, EPA synthesizes existing exposure and hazard data to determine what data gaps must be filled. (It may also rely on section 8 record keeping and reporting provisions (described below) to get exposure information.) The aim is to try to obtain enough information to determine whether there is an "unreasonable risk" from the substance. If there is not sufficient data to make this decision, the agency can issue a regulation, using informal rulemaking, to require companies to test existing chemicals. EPA staff regard this as an inflexible and slow procedure (343).

Testing may be required when there is insufficient data and experience to determine whether a substance may present an unreasonable risk of harm to health or the environment, when testing is necessary to develop such data, and when one of two other circumstances occur: 1) the manufacture, distribution in commerce, processing, use, or disposal of a chemical substance or mixture "may present an unreasonable risk of injury to health or the environment" (15 U.S.C. 2603(a)(1)(a)(i)), or 2) the substance is or will be produced in large quantities and it will or may reasonably be anticipated to enter the environment in large quantities or there will or may reasonably be anticipated to be significant or substantial human exposure (15 U.S.C. 2603(a)(1)(b)(i)).

After receiving required test data or any other information available if EPA finds that a chemical "presents or will present an unreasonable risk of injury to health or the environment," it shall by rule require one of the actions permitted under section 6 (described above) using the least burdensome requirements consistent with preventing unreasonable risks (15 U.S.C. 2605(a)). In addition, if the agency receives information that indicates there "may be a reasonable basis to conclude" that a substance "presents or will present a significant risk of serious or widespread harm to human beings from cancer, gene mutations, or birth defects," the administrator may initiate appropriate action under sections 5, 6, or 7 of the act. Upon finding such risks from substances EPA has 180 days to initiate action to reduce the risks or explain why they are "not unreasonable" (15 U.S.C. 2603(f)).

For one group of substances, polychlorinated biphenyls (PCBs), the statute specifically requires EPA to issue rules to prescribe methods for their labeling and disposal (15 U.S.C. 2605(e)).

Other Provisions of TSCA.—EPA also has authority to regulate "imminent hazards," chemicals that present an imminent and unreasonable risk of serious or widespread injury to health or the environment, and are likely to occur "before a final rule under" section 6 could protect against it (15 U.S.C. 2606(f)). The statute permits EPA to initiate a civil action in Federal District Court for seizure of such substances or for judicial relief.

The agency is required under section 8 to issue rules requiring manufacturers and processors to keep and maintain certain records, which include information on the chemical, its byproducts, the quantity manufactured, exposures, and reports of adverse effects. In addition, manufacturers must report to EPA any information "which reasonably supports the conclusion that such substance or mixture presents a substantial risk of injury to health or the environment" (15 U.S.C. 2607(c)).

Finally, section 9 requires that if EPA finds that an unreasonable risk from a chemical may be better prevented or reduced by another Federal agency, it shall submit to that agency a report describing the risks and activities that lead to the risks, and it shall request the other agency to examine the risks and to prevent or reduce them, if this can be done by that agency's action. The other agency must report back to EPA within 90 days regarding its findings concerning the risks from the referred substance (15 U.S.C. 2608).

The Resource Conservation and Recovery Act

The Resource Conservation and Recovery Act of 1976 provides for regulating the treatment, transportation, and disposal of hazardous waste. The cornerstone of the hazardous waste management system is the identification and listing of hazardous wastes. Hazardous waste is defined as a solid waste that may cause death or serious disease, or may present a substantial hazard to human health or the environment if it is improperly treated, stored, transported, or disposed of (42 U.S.C. 6903(5)). The term "solid waste" includes solid, liquid, semisolid, or contained gaseous materials from various industrial and commercial processes. The definition excludes solid or dissolved materials in domestic sewage or in irrigation return flows, industrial discharges subject to the Clean Water Act or the Atomic Energy Act, and in situ mining waste (42 U.S.C. 6903(27)).

RCRA required EPA to develop and issue criteria for identifying the characteristics of hazardous waste and for listing hazardous waste within 18 months of the passage of the law.

Any waste which exhibits the characteristics of or which is listed as a hazardous waste is regulated under RCRA[3] (221).

Under RCRA, EPA sets standards concerning recordkeeping and reporting, as well as "proper" handling and management of hazardous wastes for generators,

---

[3]Any listed waste may be delisted by rulemaking procedure, upon petition from a particular generator or upon petition for a generic delisting. A waste from a particular generator may be delisted because under individual circumstances it does not meet criteria that caused it to be listed in the first place, or the generic waste itself may be removed from RCRA lists, if EPA erred in its original listing and the waste does not meet criteria for listing.

transporters, storers, and disposers of hazardous wastes. Generators must arrange for disposal of their wastes or shipment to a waste disposal site. Transporters must keep manifests. Transfer and disposal must be done under certain procedures. Because regulations take this form, EPA has not set specific ambient air or water standards or effluent concentrations for each hazardous substance, including carcinogens. There appears to be an assumption underlying RCRA and these regulations that once hazardous wastes are at a disposal site they will not escape into the environment—that "proper handling" will prevent escape.

EPA has decided that the defining characteristics of a hazardous waste are that it 1) pose a present or potential hazard to human health and the environment when it is improperly managed and 2) can be measured by a quick, available, standardized test method or reasonably detected by generators of solid waste through their knowledge of their waste (40 CFR 261.10). The idea is to provide a quick test to identify wastes that are capable of presenting a substantial present or potential hazard when improperly managed.

EPA has identified four characteristics of hazardous waste: ignitibility, corrosivity, reactivity, and extraction procedure (EP) toxicity. Only the fourth is important for this report. EPA designated "extraction procedure" as a method of chemical analysis to be used to detect the presence of certain toxic materials in wastes (listed at 40 CFR 261.24) at levels greater than those indicated in the regulation. This procedure is designed to identify wastes likely to leach hazardous concentrations of toxic substances into the groundwater under improper management (289). Constituents of waste materials are to be extracted in a manner designed to mimic the leaching action that occurs in landfills, and this test is used to determine whether the waste contained any toxic contaminants identified in the National Interim Primary Drinking Water Standards under the Safe Drinking Water Act. If an extract from a representative sample of waste contains any contaminants (listed in table I at 40 CFR 261.23) in concentrations equal to or greater than those indicated in the table, it exhibits EP toxicity. A person with such material must then follow RCRA regulations for handling, transport, disposal, and record keeping. In general, substances on the EP toxicity list are regulated under this procedure only if their EP concentrations are no greater than 100 times the maximum levels set by the National Interim Primary Drinking Water Standards. EPA believes that a variety of mechanisms in the soil and water, including dilution, adsorption, and absorption, will serve to attentuate the toxicity of hazardous wastes before they reach the intakes of underground water supplies, should they be improperly disposed of and escape from a facility (289).

Some properties of solid wastes that pose a threat to health or the environment, such as carcinogenicity, are not included in characteristics for identifying hazardous wastes because EPA does not know of generally available testing protocols for these effects (221).

Substances are also subject to regulation under RCRA if they are listed as hazardous wastes or are a mixture of solid wastes and listed wastes (42 U.S.C. 6921(b)). A substance is "listed" if it:

- exhibits ignitability, corrosivity, reactivity, or EP toxicity;
- has been found to be toxic to humans in human or animal studies;
- is otherwise capable of causing or significantly contributing to an increase in serious illness (in which case it is designated an "acute hazardous waste"); or
- contains any toxic constituents listed in Part 261, Appendix VIII which have been shown in scientific studies to have toxic, carcinogenic, mutagenic, or teratogenic effects on human or other life forms (such waste is designated "toxic waste") (221).

In May 1980 EPA published three generic lists of wastes, based on these criteria and available scientific and technical information, which were considered to be hazardous and subject to RCRA subtitle C regulation: "1) hazardous waste from nonspecific sources (40 CFR 261.31); 2) hazardous waste from specific sources (40 CFR 261.32); and 3) discarded commercial chemical products, off-specification species, containers, and spill residues thereof (40 CFR 261.33). The discarded commercial chemical products lists is further divided into wastes designated as *toxic wastes* (40 CFR 261.33 (f)) and as *acutely hazardous wastes* (40 CFR 261.33 (e))" (221). These lists contain 361 commercial chemicals and 85 industrial waste processes, with others proposed as additions. Of these substances 152 are suspected carcinogens.

Either the Federal Government, or a State entity if it has a federally approved hazardous waste program, may enforce regulations issued under RCRA. In addition, there is a provision to permit citizen suits against private parties and Federal or State agencies if they are in violation of RCRA permits or regulations.

There have been no significant judicial decisions concerning the regulation of carcinogens under RCRA.

The Comprehensive Environmental Response, Compensation, and Liability Act

While the Resource Conservation and Recovery Act was prospective—designed to prevent problems from hazardous wastes in the future—the Comprehensive Environmental Response, Compensation, and Liability Act (CERCLA), also known as "Superfund," was designed to address the problems of cleaning up existing hazardous waste sites.

These problems range from spills requiring immediate responses to hazardous waste dumps leaking into the environment and posing long-term health and environmental hazards. "Hazardous substances" includes substances specified by sections 307 and 311 of CWA, section 3001 of RCRA (42 U.S.C. 6921), section 112 of CAA, section 7 of TSCA (15 U.S.C. 2606) and any substance designated as hazardous under section 9602 of CERCLA (substances which "when released into the environment may present substantial danger to the public health or welfare or the environment . . .").

Through this definition of "hazardous substances," CERCLA establishes a list of substances which, when released in sufficient amounts, must be reported to EPA. CERCLA section 102 sets reportable quantities of hazardous substances at 1 pound, except when different reportable quantities have been set under section 311(b)(4) of the Clean Water Act, and authorizes EPA to adjust these amounts as appropriate (42 U.S.C. 9602).

Anyone in charge of an onshore or offshore facility is required to report immediately a release of more than the relevant "reportable quantity" of any hazardous substances to the National Response Center established under the Clean Water Act (33 U.S.C. 1251 et seq.) The National Response Center in turn must convey the information "expeditiously" to appropriate Federal and State agencies (42 U.S.C. 9603(a)). The site then becomes a candidate for cleanup action.

CERCLA provides EPA with "broad authority for achieving cleanup at hazardous waste sites" (173) and paying for the cleanup out of the Act's "Hazardous Substances Response Fund," financed jointly by industry and the government (Superfund), or forcing others "to do the cleanup by requesting an injunction in court or by itself issuing an administrative order" (14). This report considers only the the provisions affecting Federal regulation of carcinogens, not the many controversies surrounding the funding or administration of Superfund.

CERCLA section 105 requires EPA to establish procedures, standards, and criteria "for both EPA and private parties for responding to releases of hazardous waste and for cleaning up waste sites" (43). The basic design is contained in the EPA-issued "National Contingency Plan." This is based on a 5-step remedial response process:

1. site discovery or notification;
2. preliminary assessment and site inspection;
3. establishment of priorities for remedial action using a scoring process (the Hazard Ranking System (HRS)) for identifying sites to be included in the National Priorities List (NPL);

4. remedial investigation and feasibility study;
5. remedial design and construction (327).

HRS prescribes the method to be used to evaluate the relative potential of uncontrolled hazardous substance facilities to cause health or safety problems, or ecological or environmental damage. HRS is in part used to set cleanup priorities. HRS assigns scores for potential harms from migration of hazardous substances away from the site by means of groundwater, surface water or air, from explosion or fires, and from human contact. All assignments of scores must take into account "waste characteristics," which include waste quantities, toxicity, and persistence.

Carcinogens are assigned the highest toxicity scores, meaning they are among the most toxic substances according to the ranking system, and many of them tend to be judged quite persistent, thus less likely to biodegrade in the environment, according to the system (40 CFR 300, App. A, pp. 710-712). EPA has developed a quantitative system for deciding the reportable quantities of carcinogens based on the quality of evidence for the carcinogenicity of a substance and its potency. This procedure is described in chapter 4 under "Agency Actions."

Once a site has been identified for cleanup the appropriate agency (a Federal or State governmental entity or private party) must investigate the site and then remedy the problem. In studying the feasibility of cleanup, rather than proposing some target standards to answer the question, "How clean is clean?" EPA recommended that the lead agency consider at least five alternative cleanup strategies:

1. an alternative that considers treatment or disposal at an off-site facility, i.e., removal of the problem wastes;
2. an alternative that attains applicable or relevant Federal or State environmental and health standards;
3. an alternative that exceeds such standards;
4. an alternative that does not attain applicable or relevant standards, but will "reduce the likelihood of present or future threats to public health;" and
5. a no-action alternative (327).

The selection of remedy, however, is being reevaluated in light of the 1986 Superfund Amendments and Reauthorization Act (SARA), section 121, which requires that the degree of cleanup at superfund sites "assures the protection of human health and the environment" and achieves compliance with standards established under other Federal and State environmental laws. Other sections of SARA are discussed in chapter 3.

Chemicals Listed in *Annual Report on Carcinogens* and NCI/NTP Test Results

Table B-1.—Chemicals Listed in *Annual Report on Carcinogens*

CAS No.	Substance name	CAS No.	Substance name
53-96-3	2-Acetylaminofluorene	2602-46-2	Direct Blue 6
107-13-1	Acrylonitrile	1937-37-7	Direct Black 38
23214-92-8	Adriamycin	106-89-8	Epichlorohydrin
1402-68-2	Aflatoxins	50-50-0	Estrogens—conjugated:
82-28-0	1-Amino-2-methylanthraquinone		a) Estradiol benzoate
117-79-3	2-Aminoanthraquinone		b) Estradiol monopalmitate
92-67-1	4-Aminobiphenyl	50-28-2	Estrogens—not conjugated:
61-82-5	Amitrole		a) Estradiol 17 beta
90-04-0	o-Anisidine	53-16-7	Estrogens—not conjugated:
134-29-2	and o-Anisidine Hydrochloride		b) Estrone (metabolite of Estradiol 17
140-57-8	Aramite		beta)
7440-38-2	Arsenic	57-63-6	Estrogens—not conjugated:
1327-53-3	and certain arsenic compounds		c) Ethinylestradiol
1332-21-4	Asbestos	72-33-3	Estrogens—not conjugated:
446-86-6	Azathioprine		d) Mestranol
71-43-2	Benzene	96-45-7	Ethylene thiourea
92-87-5	Benzidine	75-21-8	Ethylene oxide
98-07-7	Benzotrichloride	50-00-0	Formaldehyde
7440-41-7	Beryllium and certain beryllium compounds	NA	Hematite underground mining
		118-74-1	Hexachlorobenzene
494-03-1	N,N-bis (2-chloroethyl)-2-naphthylamine (chlornaphazine)	680-31-9	Hexamethylphosphoramide
		302-01-2	Hydrazine &
542-88-1	Bis (chloromethyl) ether and technical grade chloromethyl methyl ether	10043-93-2	Hydrazine Sulfate
		122-66-7	Hydrazobenzene
154-93-8	Bischloroethyl nitro-sourea	9004-66-4	Iron dextran complex
55-98-1	1,4-Butanediol dimethy-sulfonate (myleran)	NA	Isopropyl alcohol manufacture (strong-acid process)
7440-43-9	Cadmium and certain cadmium compounds	143-50-0	Kepone (Chlordecone)
		301-04-2	Lead acetate
56-23-5	Carbon tetrachloride	7446-27-7	Lead phosphate
305-03-3	Chlorambucil	58-89-9	Lindane (y-hexachlorocyclohexane)
95-83-0	4-Chloro-o-phenylene-diamine	319-85-7	Lindane (b-Hexachlorocyclohexane)
13010-47-4	1-(2-Chloroethyl)-3-cyclohexyl-1-nitrosourea (CCNU)	608-73-1	Lindane (Hexachlorocyclohexane)
		NA	Manufacture of auramine
67-66-3	Chloroform	148-82-3	Melphalan
7440-47-3	Chromium and certain chromium compounds	NA	Methoxsalen with ultra-violet A therapy (PUVA)
NA	Coke oven emissions	74-88-4	Methyl iodide
120-71-8	p-Cresidine	75-55-8	2-Methylaziridine (propyleneimine)
135-20-6	Cupferron	101-61-1	4,4'-Methylenebis (N,N-dimethyl) benzenamine (Michler's base)
14901-08-7	Cycasin		
50-18-0	Cyclophosphamide	101-14-4	4,4'-Methylenebis (2-chloroaniline) (MBOCA)
4342-03-4	Dacarbazine		
50-29-3	DDT	101-77-9	4,4'-Methylenedianiline and its dihydrochloride
117-81-7	Di (2-ethylhexyl) phthalate		
391-41-7	2,4-Diaminoanisole sulfate	443-48-1	Metronidazole
95-80-7	2,4-Diaminotoluene	90-94-8	Michler's ketone
96-12-8	1,2-Dibromo-3-chloropropane	2385-85-5	Mirex
106-93-4	1,2-Dibromoethane (EDB)	505-60-2	Mustard gas
91-94-1	3,3'-Dichlorobenzidine	91-59-8	2-Naphthylamine
107-06-2	1,2-Dichloroethane	7440-02-0	Nickel and certain nickel compounds
1464-53-5	Diepoxybutane	139-13-9	Nitrilotriacetic acid
64-67-5	Diethyl sulfate	99-59-2	5-Nitro-o-anisidine

continued on next page

Table B-1.—Chemicals Listed in *Annual Report on Carcinogens*—Continued

CAS No.	Substance name	CAS No.	Substance name
56-53-1	Diethylstilbestrol (DES)	1836-75-5	Nitrofen
119-90-4	3,3'-Dimethoxybenzidine	55-86-7	Nitrogen mustard
77-78-1	Dimethyl sulfate	79-46-9	2-Nitropropane
60-11-7	4-Dimethylaminoazobenzene	759-73-9	N-Nitroso-N-ethylurea
119-93-7	3,3'-Dimethylbenzidine	684-93-5	N-Nitroso-N-methylurea
79-44-7	Dimethylcarbamoyl chloride	924-16-3	N-Nitrosodi-n-butylamine
57-14-7	1,1-Dimethylhydrazine	621-64-7	N-Nitrosodi-n-propylamine
123-91-1	1,4-Dioxane	1116-54-7	N-Nitrosodiethanolamine
55-18-5	N-Nitrosodiethylamine	671-16-9	Procarbazine
62-75-9	N-Nitrosodimethylamine	366-70-1	Procarbazine Hydrochloride
156-10-5	p-Nitrosodiphenylamine	57-83-0	Progesterone
4549-40-0	N-Nitrosomethylvinylamine	1120-71-4	1,3-Propane sultone
59-89-2	N-Nitrosomorpholine	57-57-8	beta-Propiolactone
16543-55-8	N-Nitrosonornicotine	51-52-5	Propylthiouracil
100-75-4	N-Nitrosopiperidine	50-55-5	Reserpine
930-55-2	N-Nitrosopyrrolidine	NA	Rubber industry (certain occupations)
13256-22-9	N-Nitrososarcosine	81-07-2	Saccharin
68-22-4	Norethisterone	394-59-7	Safrole
434-07-1	Oxymetholone	7446-34-6	Selenium sulfide
	PAHs:	8007-45-2	Soots, tars, and mineral oils
56-55-3	a) Benz (a) anthracene	1883-66-4	Streptozotocin
205-99-2	b) Benzo (b) fluoranthene	95-06-7	Sulfallate
50-32-2	c) Benzo (a) pyrene	1746-01-6	2,3,7,8-Tetrachlorodi-benzo-p-dioxin (TCDD)
226-36-8	d) Dibenz (a,h) acridine	62-55-5	Thioacetamide
226-36-8	e) Dibenz (a,j) acridine	62-55-5	Thiourea
53-70-3	f) Dibenz (a,h) anthracene	1314-20-1	Thorium dioxide
194-59-2	g) 7H-Dibenzo (c,g) carbazole	584-84-9	Toluene diisocyanate
189-64-0	h) Dibenzo (a,h) pyrene	95-53-4	o-Toluidine
189-55-9	i) Dibenzo (a,i) pyrene	636-21-5	o-Toluidine Hydrochloride
193-39-5	j) Indeno (1,2,3-cd) pyrene	8001-35-2	Toxaphene
62-44-2	Phenacetin and analgesic mixtures containing phenacetin	88-06-2	2,4,6-Trichlorophenol
		126-72-7	Tris (2,3-dibromopropyl) phospate
136-40-3	Phenazopyridine hydrochloride	52-24-4	Tris (1-aziridinyl) phosphine sulfide
57-41-0	Phenytoin and sodium salt of phenytoin	51-79-6	Urethane
		75-01-4	Vinyl chloride
36355-01-8	Polybrominated biphenyls		
1336-36-3	Polychlorinated biphenyls		

SOURCE: U.S. Department of Health and Human Services, Public Health Service, National Toxicology Program, *Fourth Annual Report on Carcinogens, 1985.*

Table B-2.—NCI/NTP Test Results

CAS No.	Substance	Test results Year	Test results Number positive	Results Rats M	Results Rats F	Results Mice M	Results Mice F	Technical report number
968-81-0	Acetohexamide	1978	neg.	N	N	N	N	[050]
7008-42-6	Acronycine	1978	two	P	P	I	I	[49]
9002-18-0	Agar	1982	neg.	N	N	N	N	[230]
116-06-3	Aldicarb	1979	neg.	N	N	N	N	[136]
309-00-2	Aldrin	1978	one	E	E	P	N	[21]
107-05-1	Allyl chloride	1978	equiv.	N	N	E	E	[73]
57-06-7	Allyl isothiocyanate	1982	one	P	E	N	N	[234]
2835-39-4	Allyl isovalerate	1983	two	P	N	N	P	[253]
82-28-0*	1-Amino-2-methylanthraquinone	1978	three	P	P	N	P	[111]
119-34-6	4-Amino-2-nitrophenol	1978	one	P	E	N	N	[94]
17026-81-2	3-Amino-4-ethoxyacetanilide	1978	one	N	N	P	N	[112]
121-66-4	2-Amino-5-nitrothiazole	1978	one	P	N	N	N	[53]
6109-97-3	3-Amino-9-ethylcarbazole hydrochloride	1978	four	P	P	P	P	[93]
117-79-3*	2-Aminoanthraquinone	1978	three	P	I	P	P	[144]
2432-99-7	11-Aminoundecanoic acid	1982	one	P	N	E	N	[216]
7177-48-2	Ampicillin trihydrate	NA	equiv.	E	N	N	N	[318]

KEY: P—phosphate; N—negative; I—inadequate; E—equivocal; NA—not available.
*Indicates listing in *Annual Report on Carcinogens.*

continued on next page

Table B-2.—NCI/NTP Test Results—Continued

CAS No.	Substance	Year	Number positive	Rats M	Rats F	Mice M	Mice F	Technical report number
101-05-3	Anilazine	1978	neg.	N	N	N	N	[104]
142-04-1	Aniline hydrochloride	1978	two	P	P	N	N	[130]
134-29-2*	o-Anisidine hydrochloride (2-Methoxyaniline)	1978	four	P	P	P	P	[89]
20265-97-8	p-Anisidine hydrochloride	1978	equiv.	E	N	N	N	[116]
118-92-3	o-Anthranilic acid	1978	neg.	N	N	N	N	[36]
11097-69-1	Aroclor 1254 (polychlorinated biphenyls)	1978	equiv.	E	E	I	I	[38]
12001-29-5	Asbestos —Chrysotile (IR)	1985	one	P	N	I	I	[295a]
12001-29-5	Asbestos —Chrysotile (SR)	1985	neg.	N	N	I	I	[295b]
12001-29-5	Asbestos —Chrysotile (IR) + Dimethyl Hydrazine	1985	inad.	I	I	I	I	[295c]
12182-73-5	Asbestos, Amosite	1983	neg.	Na	Na	I	I	[249]
50-81-7	L-Ascorbic acid	1983	neg.	N	N	N	N	[247]
8003-03-0	Aspirin, Phenacetin, and Caffeine (APC)	1978	equiv.	N	E	N	N	[67]
320-67-2	5-Azacytidine	1978	one	I	I	I	P	[42]
86-50-0	Azinphosmethyl	1978	equiv.	E	N	N	N	[69]
103-33-3	Azobenzene	1979	two	P	P	N	N	[154]
71-43-2*	Benzene	1986	four	P	P	P	P	[289]
119-53-9	Benzoin	1980	neg.	N	N	N	N	[204]
105-11-3	p-Benzoquinone dioxime	1979	one	N	P	N	N	[179]
95-14-7	1,2,3-Benzotriazole	1978	equiv.	E	E	N	E	[88]
140-11-4	Benzyl acetate	1986	two	E	N	P	P	[250]
2185-92-4	2-Biphenylamine hydrochloride	1982	one	N	N	E	P	[233]
108-60-1	bis(2-Chloro-1-methylethyl) ether	1982	two	I	I	P	P	[239]
		1979	neg.	N	N	I	I	[191]
80-05-7	Bisphenol A	1982	equiv.	E	N	N	N	[215]
106-99-0	1,3-butadiene	1984	two	I	I	P	P	[288]
85-68-7	Butyl benzyl phthalate	1982	one	I	P	N	N	[213]
128-37-0	Butylated hydroxytoluene (BHT)	1979	neg.	N	N	N	N	[150]
6459-94-5	C.I. Acid Red 14	1982	neg.	N	N	N	N	[220]
1936-15-8	C.I. Acid Orange 10	NA	neg.	N	N	N	N	[211]
569-61-9	C.I. Basic Red 9 Monohydrochloride	1986	four	P	P	P	P	[285]
2602-46-2*	C.I. Direct Blue 6	1978	two	P	P	I	I	[108]
1937-37-7*	C.I. Direct Black 38	1978	two	P	P	I	I	[108]
16071-86-6	C.I. Direct Brown 95	1978	two	I	P	I	I	[108]
2475-45-8	C.I. Disperse Blue 1	1986	two	P	P	E	N	[299]
2832-40-8	C.I. Disperse Yellow 3	1982	two	P	N	N	P	[222]
842-07-9	C.I. Solvent Yellow 14	1982	two	P	P	N	N	[226]
128-66-5	C.I. Vat Yellow 4	1979	one	N	N	P	N	[134]
156-62-7	Calcium cyanamide	1979	neg.	N	N	N	N	[163]
105-60-2	Caprolactam	1982	neg.	N	N	N	N	[214]
133-06-2	Captan	1977	two	N	N	P	P	[15]
77-65-6	Carbromal	1979	neg.	N	N	N	N	[173]
133-90-4	Chloramben	1977	one	N	N	E	P	[25]
57-74-9	Chlordane (technical grade)	1977	two	N	N	P	P	[8]
143-50-0*	Chlordecone (kepone)	NA	four	P	P	P	P	[NA]
115-28-6	Chlorendic acid	NA	three	P	P	P	N	[304]
63449-39-8	Chlorinated paraffins: C23, 43% chlorine	1986	one	N	E	P	E	[305]
63449-39-8	Chlorinated paraffins: C12, 60% chlorine	1986	four	P	P	P	P	[308]
56802-99-4	Chlorinated trisodium phosphate	NA	inad.	I	I	N	N	[294]
140-49-8	4-(Chloroacetyl) acetanilide	1979	neg.	N	N	N	N	[177]
563-47-3	3-Chloro-2-methylpropene	1986	four	P	P	P	P	[300]
5131-60-2	4-Chloro-m-phenylenediamine	1978	two	P	N	N	P	[85]
95-83-0*	4-Chloro-o-phenylenediamine	1978	four	P	P	P	P	[63]
95-79-4	5-Chloro-o-toluidine	1979	two	N	N	P	P	[187]
3165-93-3	4-Chloro-o-toluidine hydrochloride	1979	two	N	N	P	P	[165]
61702-44-1	2-Chloro-p-phenylenediamine sulfate	1978	neg.	N	N	N	N	[113]
95-74-9	3-Chloro-p-toluidine	1978	neg.	N	N	N	N	[145]
106-47-8	p-Chloroaniline	1979	equiv.	E	N	E	E	[189]
108-90-7	Chlorobenzene	1985	equiv.	E	N	N	N	[261]
510-15-6	Chlorobenzilate	1978	two	E	E	P	P	[75]

KEY: P—phosphate; N—negative; I—inadequate; E—equivocal; NA—not available.
*Indicates listing in *Annual Report on Carcinogens*.
aTested in hamsters.

continued on next page

Table B-2.—NCI/NTP Test Results—Continued

CAS No.	Substance	Year	Number positive	Rats M	Rats F	Mice M	Mice F	Technical report number
124-48-1	Chlorodibromomethane	1985	one	N	N	E	P	[282]
107-07-3	2-Chloroethanol (ethylene chlorohydrin)	1985	neg.	N	N	N	N	[275]
999-81-5	2-Chloroethyltrimethylammonium chloride	1979	neg.	N	N	N	N	[158]
67-66-3*	Chloroform	NA	three	P	I	P	P	NA
6959-47-3	2-Chloromethylpyridine hydrochloride	1979	neg.	N	N	N	N	[178]
6959-48-4	3-Chloromethylpyridine hydrochloride	1978	three	P	E	P	P	[95]
76-06-2	Chloropicrin	1978	inad.	I	I	N	N	[65]
1897-45-6	Chlorothalonil	1978	two	P	P	N	N	[41]
113-92-8	Chlorpheniramine maleate	1986	neg.	N	N	N	N	[317]
94-20-2	Chlorpropamide	1978	neg.	N	N	N	N	[45]
87-29-6	Cinnamyl anthranilate	1980	three	P	N	P	P	[196]
1420-04-8	Clonitralid	1978	equiv.	N	E	I	N	[91]
56-72-4	Coumaphos	1979	neg.	N	N	N	N	[96]
102-50-1	m-Cresidine	1978	two	P	P	I	N	[105]
120-71-8*	p-Cresidine	1979	four	P	P	P	P	[142]
135-20-6*	Cupferron	1978	four	P	P	P	P	[100]
21739-91-3	Cytembena	1981	two	P	P	N	N	[207]
69-65-8	D-Mannitol	1982	neg.	N	N	N	N	[236]
5160-02-1	D&C Red No. 9	1982	one	P	E	N	N	[225]
72-55-9	p,p'-DDE	1978	two	N	N	P	P	[131]
50-29-3	DDT	1978	neg.	N	N	N	N	[131]
1163-19-5	Decabromodiphenyl oxide	1986	two	P	P	E	N	[309]
103-23-1	Di(2-Ethylhexyl) adipate	1982	two	N	N	P	P	[212]
117-81-7*	Di(2-Ethylhexyl) phthalate	1982	four	P	P	P	P	[217]
72-56-0	Di(P-Ethylphenyl) dichloroethane	1979	equiv.	N	N	N	E	[156]
131-17-9	Diallyl phthalate	1983	equiv.	I	I	E	E	[242]
		1985	equiv.	N	E	I	I	[284]
39156-41-7*	2,4-Diaminoanisole sulfate	1979	four	P	P	P	P	[84]
95-80-7*	2,4-Diaminotoluene	1979	three	P	P	N	P	[162]
6358-85-6	Diarylanilide Yellow	1978	neg.	N	N	N	N	[30]
333-41-5	Diazinon	1979	neg.	N	N	N	N	[137]
262-12-4	Dibenzo-p-dioxin	1979	neg.	N	N	N	N	[122]
96-12-8*	1,2-Dibromo-3-chloropropane (DBCP)	1978	four	P	P	P	P	[28]
		1982	four	P	P	P	P	[206]
106-93-4*	1,2-Dibromoethane (ethylene dibromide)	1978	four	P	P	P	P	[86]
		1982	four	P	P	P	P	[210]
1067-33-0	Dibutyltin diacetate	1979	inad.	N	I	N	N	[183]
609-20-1	2,6-Dichloro-p-phenylenediamine	1982	two	N	N	P	P	[219]
106-46-7	1,4-Dichlorobenzene	NA	three	P	N	P	P	[319]
95-50-1	1,2-Dichlorobenzene (o-dichlorobenzene)	1985	neg.	N	N	N	N	[255]
33857-26-0	2,7-dichlorodibenzo-p-dioxin (DCDD)	1979	equiv.	N	N	E	N	[123]
75-43-3	1,1-Dichloroethane	1978	equiv.	N	E	N	E	[66]
107-06-2*	1,2-Dichloroethane	1978	four	P	P	P	P	[55]
75-09-2	Dichloromethane (methylene chloride)	1986	four	P	P	P	P	[306]
78-87-5	1,2-Dichloropropane (propylene dichloride)	1986	two	N	E	P	P	[263]
542-75-6	1,3-dichloropropene (Telone II)	1985	three	P	P	I	P	[269]
62-73-7	Dichlorvos	1977	neg.	N	N	N	N	[10]
115-32-2	Dicofol	1978	one	N	N	P	N	[90]
60-57-1	Dieldrin	1978	equiv.	N	N	E	N	[21]
		1978	neg.	N	N	I	I	[22]
DIESELFUEL	Diesel fuel marine	1986	equiv.	I	I	E	E	[310a]
DIESELFUEL	JP naval fuel	1986	neg.	I	I	N	N	[310b]
101-90-6	Diglycidyl resorcinol ether (DGRE)	1983	four	P	P	P	P	[257]
60-51-5	Dimethoate	1977	neg.	N	N	N	N	[4]
54150-69-5	2,4-Dimethoxyaniline hydrochloride	1979	neg.	N	N	N	N	[171]
91-93-0	3,3'-Dimethoxybenzidine-4,4'-diisocyanate	1979	two	P	P	N	N	[128]
868-85-9	Dimethyl hydrogen phosphite	1985	one	P	E	P	P	[287]
597-25-1	Dimethyl morpholinophos-phosphoramidate	1986	two	P	P	N	N	[298]
120-61-6	Dimethyl terephthalate	1979	equiv.	N	N	E	N	

KEY: P—phosphate; N—negative; I—inadequate; E—equivocal; NA—not available.
*Indicates listing in *Annual Report on Carcinogens*.

continued on next page

Table B-2.—NCI/NTP Test Results—Continued

CAS No.	Substance	Year	Number positive	Rats M	F	Mice M	F	Technical report number
513-37-1	Dimethylvinyl chloride	1986	four	P	P	P	P	[316]
121-14-2	2,4-Dinitrotoluene	1978	two	P	P	N	N	[54]
123-91-1*	1,4-Dioxane	1978	four	P	P	P	P	[80]
78-34-2	Dioxathion	1978	neg.	N	N	N	N	[125]
142-46-1	2,5-Dithiobiurea	1979	equiv.	N	N	N	E	[132]
15356-70-4	DL-Menthol	1979	neg.	N	N	N	N	[98]
316-42-7	Emetine hydrochloride	1978	inad.	I	I	I	I	[43]
115-29-7	Endosulfan	1978	inad.	I	N	I	N	[62]
72-20-8	Endrin	1979	neg.	N	N	N	N	[12]
134-72-5	Ephedrine sulfate	1986	neg.	N	N	N	N	[307]
229-66-79-6	Estradiol mustard	1978	two	N	N	P	P	[59]
536-33-4	Ethionamide	1978	neg.	N	N	N	N	[46]
140-88-5	Ethyl acrylate	NA	four	P	P	P	P	[259]
20941-65-5	Ethyl tellurac	1979	equiv.	E	N	E	E	[152]
150-38-9	Ethylenediamine tetraacetic acid (EDTA)	1977	neg.	N	N	N	N	[11]
97-53-0	Eugenol	1983	equiv.	N	N	E	E	[223]
2783-94-0	FD&C Yellow No. 6	1981	neg.	N	N	N	N	[208]
55-38-9	Fenthion	1979	equiv.	N	N	E	N	[103]
2164-17-2	Fluometuron	1980	equiv.	N	N	E	N	[195]
140-56-7	Formulated fenaminosulf	1978	neg.	N	N	N	N	[101]
105-87-3	Geranyl acetate	NA	neg.	N	N	N	N	[252]
9,000-30-0	Guar gum	1982	neg.	N	N	N	N	[229]
9,000-01-5	Gum arabic	1982	neg.	N	N	N	N	[227]
68916-39-2	Hamamelis Water (witch hazel)	1984	neg.	N	N	N	N	[286]
33229-34-4	HC Blue 2	1985	neg.	N	N	N	N	[293]
2784-94-3	HC Blue 1	1985	three	E	P	P	P	[271]
2871-01-4	HC Red 3	1986	equiv.	N	N	E	I	[281]
76-44-8	Heptachlor	1977	two	N	E	P	P	[9]
57653-85-7	1,2,3,6,7,8-Hexachlorodibenzo-p-dioxin	1980	three	E	P	P	P	[198]
		1980	neg.	I	I	N	N	[202]
67-72-1	Hexachloroethane	1978	two	N	N	P	P	[68]
70-30-4	Hexachlorophene	1978	neg.	N	N	I	I	[40]
122-66-7*	Hydrazobenzene	1978	three	P	P	N	P	[92]
148-24-3	8-Hydroxyquinoline	1985	neg.	N	N	N	N	[276]
21416-87-5	ICRF-159	1978	two	N	P	N	P	[78]
75-47-8	Iodoform (triiodomethane)	1978	neg.	N	N	N	N	[110]
3458-22-8	IPD (3,3'-Iminobis-1-propanol dimethanesulfonate (ester)hydrochloride)	1978	equiv.	E	E	E	E	[18]
78-59-1	Isophorone	1986	one	P	N	E	N	[291]
3778-73-2	Isophosphamide	1977	two	N	P	N	P	[32]
303-43-4	Lasiocarpine	1978	two	P	P	I	I	[39]
19010-66-3	Lead dimethyldithiocarbamate	1979	neg.	N	N	N	N	[151]
58-89-9	Lindane	1977	neg.	N	N	N	N	[14]
434-13-9	Lithocholic acid	1979	neg.	N	N	N	N	[175]
9,000-40-2	Locust bean gum	1982	neg.	N	N	N	N	[221]
1634-78-2	Malaoxon	1979	neg.	N	N	N	N	[135]
121-75-5	Malathion	1978	neg.	N	N	N	N	[24]
		1979	neg.	N	N	I	I	[192]
108-78-1	Melamine	1983	one	P	N	N	N	[245]
72-43-5	Methoxychlor	1978	neg.	N	N	N	N	[35]
80-62-6	Methyl methacrylate	1986	neg.	N	N	N	N	[314]
298-00-0	Methyl parathion	1979	neg.	N	N	N	N	[157]
129-15-7	2-Methyl-1-nitroanthraquinone	1978	four	P	P	P	P	[29]
101-61-1*	4,4'-Methylenebis (N,N-dimethyl) benzenamine	1979	three	P	P	E	P	[186]
13552-44-8*	4,4'-Methylenedianiline dihydrochloride	1983	four	P	P	P	P	[248]
315-18-4	Mexacarbate	1978	neg.	N	N	N	N	[147]
90-94-8*	Michler's ketone	1979	four	P	P	P	P	[181]
150-68-5	Monuron	1983	one	P	N	N	N	[266]
1465-25-4	N-(1-naphthyl) ethylenediamine dihydrochloride	1979	neg.	N	N	N	N	[168]

KEY: P—phosphate; N—negative; I—inadequate; E—equivocal; NA—not available.
*Indicates listing in *Annual Report on Carcinogens.*

continued on next page

Table B-2.—NCI/NTP Test Results—Continued

CAS No.	Substance	Year	Number positive	Rats M	F	Mice M	F	Technical report number
109-69-3	N-Butyl chloride	1986	neg.	N	N	N	N	[312]
86-30-6*	N-Nitrosodiphenylamine	1979	two	P	P	N	N	[164]
101-54-2	N-Phenyl-p-phenylenediamine	1978	neg.	N	N	N	N	[82]
2243-62-1	1,5-Naphthalenediamine	1978	three	N	P	P	P	[143]
139-94-6	Nithiazide.................................	1979	two	N	P	P	E	[146]
99-59-2*	5-Nitro-o-anisidine	1978	three	P	P	E	P	[127]
99-56-9	4-Nitro-o-phenylenediamine	1979	neg.	N	N	N	N	[180]
99-55-8	5-Nitro-o-toluidine........................	1978	two	N	N	P	P	[107]
1777-84-0	3-Nitro-p-acetophenetide	1979	one	N	N	P	N	[133]
5307-14-2	2-Nitro-p-phenylenediamine	1979	one	N	N	N	P	[169]
602-87-9	5-Nitroacenaphthene	1978	three	P	P	N	P	[118]
619-17-0	4-Nitroanthranilic acid	1978	neg.	N	N	N	N	[109]
94-52-0	5(6)-Nitrobenzimidazole	1979	two	N	N	P	P	[117]
1836-75-5*	Nitrofen	1978	three	E	P	P	P	[26]
		1979	two	N	N	P	P	[184]
86-57-7	1-Nitronaphthalene........................	1978	neg.	N	N	N	N	[64]
504-88-1	3-Nitropropionic acid	1978	equiv.	E	N	N	N	[52]
156-10-5	p-Nitrosodiphenylamine....................	1979	two	P	N	P	N	[190]
102-96-5	beta-Nitrostyrene	1979	neg.	N	N	N	N	[170]
1212-29-9	N,N'-Dicyclohexylthiourea..................	1978	neg.	N	N	N	N	[56]
105-55-5	N,N'-Diethylthiourea.......................	1979	two	P	P	N	N	[149]
139-13-9*	NTA (Nitrilotriacetic acid)	1977	four	P	P	P	P	[6a]
18662-53-8*	Trisodium nitrilotriacetate monohydrate	1977	equiv.	E	E	N	N	[6b]
18662-53-8*	Trisodium nitrilotriacetate monohydrate	1977	two	P	P	I	I	[6c]
101-80-4	4,4'-Oxydianiline	1980	four	P	P	P	P	[205]
2058-46-0	Oxytetracycline hydrochloride	NA	equiv.	E	E	N	N	[315]
56-38-2	Parathion	1979	equiv.	E	E	N	N	[70]
76-01-7	Pentachloroethane	1983	two	E	N	P	P	[232]
82-68-8	Pentachloronitrobenzene...................	1978	neg.	N	N	N	N	[61]
		NA	neg.	I	I	N	N	[325]
136-40-3*	Phenazopyridine hydrochloride	1978	three	P	P	N	P	[99]
3546-10-9	Phenesterin	1978	three	N	P	P	P	[60]
834-28-6	Phenformin hydrochloride..................	1977	neg.	N	N	N	N	[7]
108-95-2	Phenol	1980	neg.	N	N	N	N	[203]
63-92-3	Phenoxybenzamine hydrochloride	1978	four	P	P	P	P	[72]
103-85-5	1-Phenyl-2-thiourea.......................	1978	neg.	N	N	N	N	[148]
89-25-8	1-Phenyl-3-methyl-5-pyrazolone............	1978	neg.	N	N	N	N	[141]
624-18-0	p-Phenylenediamine dihydrochloride	1979	neg.	N	N	N	N	[174]
61-76-7	Phenylephrine hydrochloride	NA	neg.	N	N	N	N	[322]
90-43-7	o-Phenylphenol	1986	neg.	I	I	N	N	[301]
13171-21-6	Phosphamidon	1979	equiv.	E	E	N	N	[16]
13366-73-9	Photodieldrin	1977	neg.	N	N	N	N	[17]
85-44-9	Phthalic anhydride	1979	neg.	N	N	N	N	[159]
88-96-0	Phthalimide	1979	neg.	N	N	N	N	[161]
1918-02-1	Picloram.................................	1978	equiv.	N	E	N	N	[23]
51-03-6	Piperonyl butoxide	1979	neg.	N	N	N	N	[120]
120-62-7	Piperonyl sulfoxide	1979	one	N	N	P	N	[124]
1955-45-9	Pivalolactone	1978	two	P	P	N	N	[140]
67774-32-7	Polybrominated biphenyl mixture (Firemaster FF-1).	1983	four	P	P	P	P	[244]
366-70-1*	Procarbazine hydrochloride	1977	four	P	P	P	P	[19]
952-23-8	Proflavin hydrochloride	1977	equiv.	E	N	E	E	[5]
121-79-9	Propyl gallate	1982	equiv.	E	N	E	N	[240]
115-07-1	Propylene................................	1985	neg.	N	N	N	N	[272]
75-56-9	1,2-Propylene oxide	1985	four	P	P	P	P	[267]
98-96-4	Pyrazinamide	1978	inad.	N	N	N	I	[048]
58-14-9	Pyrimethamine	1978	inad.	N	N	I	N	[77]
50-55-5*	Reserpine................................	1982	three	P	N	P	P	[193]
7446-34-6*	Selenium sulfide (twice)	1980	three	P	P	N	P	[194]
		NA	neg.	I	I	N	N	[197]

KEY: P—phosphate; N—negative; I—inadequate; E—equivocal; NA—not available.
*Indicates listing in *Annual Report on Carcinogens*.

continued on next page

Table B-2.—NCI/NTP Test Results—Continued

CAS No.	Substance	Year	Test results Number positive	Rats M	F	Mice M	F	Technical report number
SELSUN	Selsun	1980	neg.	I	I	N	N	[199]
148-18-5	Sodium diethyldithiocarbamate	1979	neg.	N	N	N	N	[172]
7772-99-8	Stannous chloride	1982	equiv.	E	N	N	N	[231]
100-42-5	Styrene...	1979	equiv.	N	N	E	N	[185]
1596-84-5	Succinic acid 2,2-dimethyl-hydrazide (Daminozide) .	1978	one	N	P	E	N	[83]
95-06-7*	Sulfallate	1978	four	P	P	P	P	[115]
127-69-5	Sulfisoxazole...................................	1979	neg.	N	N	N	N	[138]
77-79-2	3-Sulfolene	1978	neg.	N	N	N	N	[102]
80-08-0	4,4'-Sulfonyldianiline (Dapsone)	1977	one	P	N	N	N	[20]
39300-88-4	Tara gum	1982	neg.	N	N	N	N	[224]
2438-88-2	2,3,5,6-Tetrachloro-4-nitro-anisole...............	1978	neg.	N	N	N	N	[114]
1746-01-6*	2,3,7,8-Tetrachlorodibenzo-p-dioxin	1982	four	P	P	P	P	[209]
		1982	one	I	I	E	P	[201a]
	2,3,7,8-Tetrachlorodibenzo-p-dioxin + (PDMBA)	1982	inad.	I	I	I	I	[201b]
72-54-8	Tetrachlorodiphenylethane (TDE).................	1978	equiv.	E	N	N	N	[131]
630-20-6	1,1,1,2-Tetrachloroethane	1983	two	E	N	P	P	
79-34-5	1,1,2,2-Tetrachloroethane	1978	two	E	N	P	P	
127-18-4	Tetrachloroethylene	1987	four	P	P	P	P	[311]
		1977	two	I	I	P	P	[13]
961-11-5	Tetrachlorvinphos	1978	three	N	P	P	P	[33]
97-77-8	Tetraethylthiuram disulfide	1979	neg.	N	N	N	N	[166]
55566-30-8	Tetrakis (hydroxymethyl) phosphonium sulfate.....	NA	neg.	N	N	N	N	[296]
139-65-1	4,4'-Thiodianiline..............................	1978	four	P	P	P	P	[47]
789-61-7	beta-Thioguanidine deoxyriboside...............	1978	one	E	P	I	I	[57]
13463-67-7	Titanium dioxide...............................	1979	neg.	N	N	N	N	[97]
1156-19-0	Tolazamide	1978	neg.	N	N	N	N	[51]
64-77-7	Tolbutamide	1977	neg.	N	N	N	N	[31]
26471-62-5	2,4-&2,6-Toluene diisocyanate	1986	three	P	P	N	P	[251]
15481-70-6	2,6-Toluenediamine dihydro-chloride.............	1980	neg.	N	N	N	N	[200]
6369-59-1	2,5-Toluenediamine sulfate	1978	neg.	N	N	N	N	[126]
636-21-5*	o-Toluidine hydrochloride	1979	four	P	P	P	P	[153]
8001-35-2*	Toxaphene.....................................	1979	two	E	E	P	P	[37]
79-00-5	1,1,2-Trichloroethane	1978	two	N	N	P	P	[74]
71-55-6	1,1,1-Trichloroethane (methyl chloroform)	1977	inad.	I	I	I	I	[3]
79-01-6	Trichloroethylene (without epichlorohydrin)	1976	two	N	N	P	P	[2]
		NA	two	I	N	P	P	[243]
		1986	inad.	I	I	I	I	[273]
75-69-4	Trichlorofluoromethane	1978	inad.	I	I	N	N	[106]
88-06-2*	2,4,6-Trichlorophenol	1979	three	P	N	P	P	[155]
1582-09-8	Trifluralin	1978	one	N	N	N	P	[34]
137-17-7	2,4,5-Trimethylaniline	1979	three	P	P	N	P	[160]
512-56-1	Trimethylphosphate	1978	two	P	N	N	P	[81]
2489-77-2	Trimethylthiourea	1979	one	N	P	N	N	[129]
76-87-9	Triphenyltin hydroxide.........................	1978	neg.	N	N	N	N	[139]
78-42-2	Tris(2-Ethylhexyl)phosphate	1984	one	E	N	N	P	[274]
126-72-7*	Tris(2,3-Dibromopropyl)phosphate	1978	four	P	P	P	P	[76]
52-24-4*	Tris(Aziridinyl)-phosphine sulfide................	1978	four	P	P	P	P	[58]
73-22-3	L-Tryptophan	1978	neg.	N	N	N	N	[71]
100-40-3	4-Vinylcyclohexene	1986	one	I	I	I	P	[303]
75-35-4	Vinylidene chloride (1,1-dichloroethylene)	1982	neg.	N	N	N	N	[228]
1330-20-7	Xylenes (mixed)...............................	NA	neg.	N	N	N	N	[327]
17924-92-4	Zearalenone...................................	1982	two	N	N	P	P	[235]
137-30-4	Ziram	1983	one	P	N	N	E	[238]

KEY: P—phosphate; N—negative; I—inadequate; E—equivocal; NA—not available.
*Indicates listing in Annual Report on Carcinogens.

SOURCE: Haseman, J.K., Huff, J.E., Zelger, E.E., et al., "Results of 327 Chemical Carcinogenicity Studies Conducted by the National Cancer Institute and the National Toxicology Program," Environ. Health Perspectives (in press).

Appendix C
Acknowledgments

OTA acknowledges the following individuals and groups for their assistance in reviewing drafts or furnishing information and materials:

Lois P. Adams
Food and Drug Administration
Department of Health and Human Services

Kulbi Bakshar*
Board on Environmental Studies and Toxicology
National Academy of Sciences

Donald G. Barnes
Office of Pesticides and Toxic Substances
Environmental Protection Agency

Diane D. Beal, Ph.D
Office of Pesticides and Toxic Substances
Environmental Protection Agency

John Blodgett*
Congressional Research Service

Norman Breslow, Ph.D.
School of Public Health and Community Medicine
University of Washington

Robert Brink
Toxic Substances Control Act Interagency Testing
  Committee

Gary J. Burin*
Office of Pesticides and Toxic Substances
Environmental Protection Agency

Daniel M. Byrd III, Ph.D., D.A.B.T.
Halogenated Solvents Industry Alliance

Hugh C. Cannon
Food and Drug Administration
Department of Health and Human Services

Dorothy A. Canter, Ph.D.*
National Toxicology Program
Department of Health and Human Services

Frank S. Chalmers
Occupational Safety and Health Administration
Department of Labor

Murray Cohn*
Consumer Product Safety Commission

Joseph A. Cotruvo, Ph.D.*
Office of Drinking Water
Environmental Protection Agency

Geraldine V. Cox, Ph.D.
Chemical Manufacturers Association

Jeff Davidson*
Office of Toxic Substances
Environmental Protection Agency

Philip Derfler
Food and Drug Administration
Department of Health and Human Services

Denny Dobbin
Environmental Protection Agency

David L. Dull
Office of Pesticides and Toxic Substances
Environmental Protection Agency

William Farland*
Carcinogen Assessment Group
Environmental Protection Agency

Tim Fields*
Office of Solid Waste and Emergency Response
Environmental Protection Agency

Gary Flamm
Food and Drug Administration
Department of Health and Human Services

David W. Gaylor, Ph.D.
National Center for Toxicological Research
Food and Drug Administration
Department of Health and Human Services

Clarice Gaylord
Environmental Protection Agency

Richard E. Geyer
Food and Drug Administration
Department of Health and Human Services

Vera C. Glocklin, Ph.D.*
Food and Drug Administration
Department of Health and Human Services

Michael Gough*
Environ Corp.

Murray Grant
General Accounting Office

---

*Participated in a workshop held on Mar. 9, 1987 to discuss the first draft of this background paper.

Martin P. Halper
Office of Toxic Substances
Environmental Protection Agency

Bryan D. Hardin, Ph.D.
National Institute for Occupational
Safety and Health

Edward D. Harrill
Consumer Product Safety Commission

Joseph K. Haseman
National Institute of Environmental Health
Sciences
Department of Health and Human Services

Michael D. Hogan, Ph.D.
National Institute of Environmental Health
Sciences
Department of Health and Human Services

John Jennings, M.D.
Pharmaceutical Manufacturers Association

James E. Huff, Ph.D.*
National Toxicology Program
Department of Health and Human Services

Boris Kachura
General Accounting Office

Robert Kellam*
Environmental Protection Agency

Thomas E. Kelly
Office of Standards and Regulations
Environmental Protection Agency

Joanne Kla
Office of Toxic Substances
Environmental Protection Agency

Edward Klein
Environmental Protection Agency

Jack Kooyoomjian*
Office of Solid Waste and Emergency Response
Environmental Protection Agency

Walter Kovalick, Jr.
Office of Emergency and Remedial Response
Environmental Protection Agency

Philip J. Landrigan, M.D., M.Sc.
Department of Community Medicine
The Mount Sinai Medical Center

Dr. Richard Lemen*
National Institute for Occupational
Safety and Health

John Martonik*
Health Standards Program
Occupational Safety and Health Administration

Carl Mazza*
Environmental Protection Agency

Ernest E. McConnell, D.V.M.
Division of Toxicology Research and Testing
Department of Health and Human Services

Tom McGarity
University of Texas Law School

J. Michael McGinnis, M.D.
Office of Disease Prevention and Health
Promotion
Department of Health and Human Services

Mortimer L. Mendelsohn
Lawrence Livermore National Laboratory
University of California

Bill Menza
Consumer Product Safety Commission

Joseph Meranda
Environmental Protection Agency

Robert J. Moolenaar
Health and Environmental Sciences
DOW Chemical Company

Barbara Ostrow
Office of Toxic Substances .
Environmental Protection Agency

George H. Pauli, Ph.D.
Food and Drug Administration
Department of Health and Human Services

Frederica P. Perera, Dr., P.H.
Natural Resources Defense Council

James R. Petrie
Mine Safety and Health Administration
Department of Labor

Henry C. Pitot, M.D., Ph.D.
McArdle Laboratory for Cancer Research
Medical School
University of Wisconsin

J. Winston Porter
Office of Solid Waste and Emergency Response
Environmental Protection Agency

Peter Preuss*
Office of Health and Environment Assessment
Environmental Protection Agency

*Participated in a workshop held on Mar. 9, 1987 to discuss the first draft of this background paper.

Tom Purcell
Office of Water Regulations and Standards
Environmental Protection Agency

Patricia A. Roberts
Office of the General Counsel
Environmental Protection Agency

Laurence S. Rosenstein*
Office of Pesticides and Toxic Substances
Environmental Protection Agency

Mark Rothstein
A. Rubin
Food and Drug Administration
Department of Health and Human Services

Mark Rushefsky
Department of Political Science & Philosophy
Southwest Missouri State University

Robert A. Scala, Ph.D.
Exxon Biomedical Sciences, Inc.

Robert J. Scheuplein
Food and Drug Administration
Department of Health and Human Services

Scott Schneider
Workers Institute for Safety and Health

Randal P. Schumacher*
Chemical Manufacturers Association

Dave Scudder
PEMD
General Accounting Office

Dan Sigelman
Human Resources and Intergovernmental Relations
  Subcommittee
House of Representatives

Ellen Silbergeld*
Environmental Defense Fund

Michael Simpson
Congressional Research Service

Donald E. Stevenson*
American Industrial Health Council

William H. Sutherland
Mine Safety and Health Administration
Department of Labor

Gary Timm
Office of Pesticides and Toxic Substances
Environmental Protection Agency

Andrew Ulsamer
Consumer Product Safety Commission

Georgia Valaoras
Office of Policy, Planning, and Evaluation
Environmental Protection Agency

Jamie Walters
Environmental Protection Agency

Nicholas E. Weber, Ph.D.
Office of New Animal Drug Evaluation
Food and Drug Administration

James L. Weeks, Sc.D.
United Mineworkers of America

Elizabeth K. Weisburger, Ph.D.
National Cancer Institute
National Institute of Health
Department of Health and Human Services

Dr. Sidney Wolfe*
Health Research Group

Mike Wright
United Steelworkers of America

---

*Participated in a workshop held on Mar. 9, 1987 to discuss the first draft of this background paper.

List of Acronyms

List of Acronyms

ACGIH —American Conference of Governmental Industrial Hygienists
AADI —adjusted acceptable daily intake
AEC —Atomic Energy Commission
AFSCME —American Federation of State, County, and Municipal Employees
ALARA —as low as reasonably achievable
ANPRM —Advanced Notice of Proposed Rulemaking
ASCP —American Society of Clinical Pathologists
ATSDR —Agency for Toxic Substances and Disease Registry
BAT —best available technology
BDZ —benzodiazepine
BPT —best practicable technology
CA —chromosome aberration
CAA —Clean Air Act
CAG —Carcinogen Assessment Group (EPA)
CCERP —Committee to Coordinate Environmental and Related Programs
CDC —Centers for Disease Control
CEC —Chemical Evaluation Committee (NTP)
CERCLA —Comprehensive Environmental Response, Compensation, and Liability Act
CFC —chlorofluorocarbon
CHAP —Chronic Hazard Advisory Panel
CHIP —Chemical Hazard Information Profile
CHO —Chinese hamster ovary
CMA —Chemical Manufacturers Association
CMP —3-Chloro-2-methylpropene
CPSA —Consumer Product Safety Act
CPSC —Consumer Product Safety Commission
CSPI —Center for Science in the Public Interest
CSWG —Chemical Selection Working Group
CVM —Center for Veterinary Medicine (FDA)
CWA —Clean Water Act
DBCP —1,2-dibromo-3-chloropropane
DCB —1,4-Dichloro-2-butene
DDT —1,1,1-trichloro-2,2-bis(chlorophenyl)ethane
DEHP —di(2-ethylhexyl)phthalate
DEPC —diethylpyrocarbonate
DES —diethylstilbestrol
DHEW —Department of Health, Education, and Welfare
DHHS —Department of Health and Human Services
DOE —Department of Energy
DOL —Department of Labor
DOT —Department of Transportation
EDB —ethylene dibromide

EDF —Environmental Defense Fund
EMTD —estimated maximum tolerated dose
EP —extraction procedure
EPA —Environmental Protection Agency
ERC —Environmental Reporter Cases
ETS —emergency temporary standards
FDA —Food and Drug Administration
FDCA —Food, Drug, and Cosmetic Act
FDF —fundamentally different factors
FEPCA —Federal Environmental Pesticide Control Act
FHSA —Federal Hazardous Substances Act
FIFRA —Federal Insecticide, Fungicide, and Rodenticide Act
FRC —Federal Radiation Council
FYI —for your information
GAO —General Accounting Office
GRAS —generally recognized as safe
HA —Health Advisory
HEEP —Health and Environmental Effects Profile
HRG —Health Research Group
HRS —Hazardous Ranking System
HSDB —Hazardous Substance Data Bank
HT —Histology Technician
HUD —Department of Housing and Urban Development
IARC —International Agency for Research on Cancer
ICRP —International Commission on Radiological Protection
IND —Investigative New Drug
IOC —inorganic compound
IRLG —Interagency Regulatory Liaison Group
ITC —Interagency Testing Committee
IUMRRG —Interagency Uranium Mining Radiation Review Group
LOAEL —lowest observed adverse effect level
MBOCA —4,4-methylenebis(2-chloroaniline)
MCL —maximum contaminant level
MCLG —maximum contaminant level goal
MDA —4,4'-methylenedianiline
MESA —Mining Enforcement and Safety Administration
MHAPPS—Modified Hazardous Air Pollutant Prioritization System
ML —mouse lymphoma
MMWR —Morbidity and Mortality Weekly Report

MOPP — mechlorethamine [nitrogen mustard] oncovine [vincristine], procarbazine, prednisone
MSHA — Mine Safety and Health Administration
MTD — maximum tolerated dose
NADA — New Animal Drug Application
NAS — National Academy of Sciences
NCAB — National Cancer Advisory Board (NCI)
NCAMP — National Coalition Against the Misuse of Pesticides
NCI — National Cancer Institute
NCTR — National Center for Toxicological Research
NDA — New Drug Application
NIEHS — National Institute of Environmental Health Sciences
NIH — National Institutes of Health
NIOSH — National Institute for Occupational Safety and Health
NLM — National Library of Medicine
NOAEL — no observed adverse effects level
NOEL — no observed effects level
NPDES — National Pollution Discharge Elimination System
NPDWR — National Primary Drinking Water Regulations
NPRM — Notice of Proposed Rulemaking
NRC — National Research Council
NRDC — Natural Resources Defense Council
NTP — National Toxicology Program
OCAW — Oil Chemical and Atomic Workers International Union
OMB — Office of Management and Budget
OSHA — Occupational Safety and Health Administration
OSTP — White House Office of Science and Technology Policy
OTA — Congressional Office of Technology Assessment
P,P'-DDE — 1,1'-(2,2-dichloroethenylidine)
PAH — polycyclic aromatic hydrocarbon

PBBs — polybrominated biphenyls
PCBs — polychlorinated biphenyls
PCE — pentachloroethane
PEL — Permissible Exposure Limit
PMA — Pharmaceutical Manufacturers Association
PMN — premanufacture notice
POTWs — publicly owned treatment works
PVN — predicted value negative
PVP — predicted value positive
PWG — Pathology Working Group
RCRA — Resource Conservation and Recovery Act
RMCL — recommended maximum contaminant level
RPAR — rebuttable presumptions against registration
RQ — reportable quantity
SAB — Scientific Advisory Board (EPA)
SAR — structure-activity relationship
SARA — Superfund Amendments and Reauthorization Act
SCE — sister chromatid exchange
SDWA — Safe Drinking Water Act
SNARL — suggested no adverse response level
SNUR — Significant New Use Rule
SOC — synthetic organic chemical
SOM — sensitivity of method
SR — special review
SRI — Stanford Research Institute
SUR — Safe Use Rule
TCDD — tretrachlorodibenzo-p-dioxin
TCE — trichloroethylene
TDI — toluene diisocyanate
TDRC — Toxic Design Research Committee
TLV — threshold limit value
TSCA — Toxic Substances Control Act
UAW — United Auto Workers
UFFI — urea-formaldehyde foam insulation
USCG — United States Coast Guard
USDA — United States Department of Agriculture
USPHS — United States Public Health Service
VOC — volatile synthetic organic chemical
WLM — working level month

References

References

1. Administrative Conference of the United States, Office of the Chairman, *OSHA Rulemaking Procedures*, report to the Assistant Secretary for Occupational Safety and Health, Occupational Safety and Health Administration (Washington, DC: Feb. 19, 1987).

2. Aidala, J.V., *Federal Insecticide, Fungicide, and Rodenticide Act: Reauthorization Issues Before the 99th Congress*, Congressional Research Service, Library of Congress, Feb. 27, 1985.

3. Albert, R.E., Train, R.R., and Anderson, E., "Rationale Developed by the Environmental Protection Agency for the Assessment of Carcinogenic Risks," *J. Nat. Cancer Inst.* 58:1537-1541, May 1977.

4. American Industrial Health Council, "Statement of Donald E. Stevenson," presented at public meeting on National Toxicology Program, *Annual Report*, Washington, DC, Apr. 21, 1987.

5. *American Iron and Steel Institute* v. *Occupational Safety and Health Administration*, Federal Reporter, Second Series 577 (1978), p. 825.

6. *American Textile Manufacturers Institute* v. *Donovan*, U.S. Reports, Second Series 577 (1981), p. 490.

7. Ames, B.N., "Dietary Carcinogens and Anticarcinogens," *Science* 221:1256-1264, 1983.

8. Ames, B.N., Magaw, R., and Gold, L.S., "Ranking Possible Carcinogenic Hazards," *Science* 236:271-280, 1987.

9. Anderson, E.L., and U.S. Environmental Protection Agency, Carcinogen Assessment Group, "Quantitative Approaches in Use To Assess Cancer Risk," *Risk Analysis* 3(4):77-295, 1983.

10. *Aqua Slide 'n' Dive* v. *Consumer Product Safety Commission*, Federal Reporter, Second Series 569 (1978), p. 83.

11. Auer, C.M., "Approaches to Chemical Hazard Assessment in the Presence of Limited Data," typescript, July 12, 1985.

12. Auer, C.M. and Francis, E.Z., *Executive Summary—The PMN Process: A Discussion and Analysis of the New Chemicals Program in the Office of Toxic Substances*, U.S. Environmental Protection Agency, Office of Pesticides and Toxic Substances, 1987 (forthcoming).

13. Bailar, J., Office on Disease Prevention and Health Promotion, U.S. Department of Health and Human Services, personal communication, 1987.

14. Blaymore, A., "Retroactive Application of Superfund: Can Old Dogs Be Taught New Tricks?" *Boston College Environmental Affairs Law Review* 12:1-50, 1978.

15. Brandes, R., U.S. Environmental Protection Agency, Office of Water Enforcement and Permits, Division of Municipal Pollution Control, personal communication, July 1986.

16. Brink, R.H., Executive Secretary, Toxic Substance Control Act Interagency Testing Committee, letter to OTA, Dec. 2, 1986.

17. Brink, R.H., Executive Secretary, Toxic Substance Control Act Interagency Testing Committee, letter to OTA, Mar. 24, 1987.

18. Brink, R.H., Executive Secretary, Toxic Substance Control Act Interagency Testing Committee, personal communication, 1987.

19. Bristol, D.W., "National Toxicology Program Approach to Monitoring Study Performance," *Managing Conduct and Data Quality of Toxicology Studies*, B.K. Hoover, et al. (eds.) (Princeton, NJ: Princeton Scientific Publishing Co., 1986).

20. Brown, A., "Summary of the Revised Preamble," presentation at the Winter Toxicology Forum, Washington, DC, Feb. 16-18, 1987.

21. Burin, G., U.S. Environmental Protection Agency, Office of Pesticide Programs, personal communications, 1986 and 1987.

22. Byrd, D., Halogenated Solvents Industry Alliance, personal communication, 1987.

23. Calkins, D.R., Dixon, R.L., Gerber, C.R., et al., "Identification, Characterization, and Control of Potential Human Carcinogens: A Framework for Federal Decision-Making," staff paper, U.S. Office of Science and Technology Policy, Executive Office of the President, Feb. 1, 1979; printed in *Journal of the National Cancer Institute* 64: 169-176, January 1980.

24. Cannon, H.C., Associate Commissioner for Legislative Affairs, U.S. Department of Health and Human Services, Food and Drug Administration, comments on unpublished OTA draft report, "Identifying and Regulating Carcinogens" (1987).

25. Canter D., Assistant to the Director, National Toxicology Program, U.S. Department of Health and Human Services, letter to OTA, Jan. 27, 1987.

26. Canter D., Assistant to the Director, National

Toxicology Program, U.S. Department of Health and Human Services, personal communication, 1987.

27. *Certified Color Manufacturers Association* v. *Mathews*, Federal Reporter, Second Series 543 (1976), p. 284.

28. Chemical & Engineering News, "Court Rules Standards To Be Based on Health," *Chemical & Engineering News* 65(31):6, 1987.

29. *Chemical Manufacturers Association* v. *NRDC*, U.S. Reports 470 (1985), p. 116.

30. Chemical Manufacturers Association, National Agricultural Chemicals Association, and Pharmaceutical Manufacturers Association, "Comments on the National Toxicology Program's Annual Reports on Carcinogens," presented at public meeting on National Toxicology Program *Annual Report*, Washington, DC, Apr. 21, 1987.

31. Cherney, C.T., and Wadzinski, K.M., "State and Federal Rules Under the Clean Water Act," *Nat. Resources Environ.* 1:1922, 1968.

32. Chu, K.C., Cueto, C., and Ward, J.M., "Factors in the Evaluation of 200 National Cancer Institute Carcinogen Bioassays," *J. Toxicol. Environ. Health* 8:251-280, 1981.

33. Cohn, M., Consumer Product Safety Commission, personal communication, 1987.

34. Coodley, C.F., "Risk in the 1980's: New Perspectives on Managing Chemical Hazards," *San Diego Law Review* 21(5):1015-1044, September-October 1984.

35. Cooper, R.M., "Stretching Delaney Till It Breaks," *Regulation* 9:11-17, November-December 1985.

36. Cotruvo, J., Director, Criteria and Standards Division, U.S. Environmental Protection Agency, Office of Drinking Water, personal communication, 1986 and 1987.

37. Crump, K.S., Hoel, D.G., Langely, C.H., et al., "Fundamental Carcinogenic Processes and Their Implications for Low Dose Risk Assessment," *Cancer Res.* 36:2973-2979, 1976.

38. DeMuth, C.C., and Ginsburg, D.H., "White House Review of Agency Rulemaking," *Harvard Law Review* 99:1075-1088, 1986.

39. Derfler, P., U.S. Department of Health and Human Services, Food and Drug Administration, Office of General Counsel, comments on unpublished OTA draft report, "Identifying and Regulating Carcinogens" (1987).

40. Di Carlo, F.J., and Fung, V.A., "Summary of Carcinogenicity Data Generated by the National Cancer Institute/National Toxicology Program," *Drug Metabol. Rev.* 15(5&6):1251-1273, 1984.

41. Doll, R., and Peto, R., "The Causes of Cancer: Quantitative Estimates of Avoidable Risks of Cancer in the United States Today," *J. Nat. Cancer Inst.* 66(6):1193-1308, June 1981, reprinted with the same title by Oxford University Press, 1981.

42. Doniger, E., Senior Staff Attorney, Natural Resources Defense Council, personal communication, 1987.

43. Dorge, Carol L., "After 'Voluntary' Liability: The EPA's Implementation of Superfund," *Boston College Environmental Affairs Law Review* 11:443-479, 1984.

44. *Dow Chemical* v. *Blum*, Federal Supplement 467 (1979), p. 892.

45. *Dow Chemical* v. *Consumer Product Safety Commission*, Federal Supplement 459 (1978), p. 378.

46. *Dry Color Manufacturing Association* v. *Occupational Safety and Health Administration*, Federal Reporter, Second Series, 486 (1979), p. 98.

47. Dull, D., U.S. Environmental Protection Agency, Office of Toxic Substances, comments on unpublished OTA draft report "Identifying and Regulating Carcinogens" (1987).

48. Environmental and Energy Study Institute and the Environmental Law Institute, *Statutory Deadlines in Environmental Legislation: Necessary But Need Improvement* (Washington, DC: September, 1985).

49. Environmental Defense Fund and Boyle, R.H., *Malignant Neglect* (New York: Vintage Books, 1980).

50. *Environmental Defense Fund* v. *Environmental Protection Agency*, Federal Reporter Second Series 465 (1972), p. 528.

51. *Environmental Defense Fund* v. *Environmental Protection Agency*, Federal Reporter, Second Series 489 (1973), p. 1247.

52. *Environmental Defense Fund* v. *Environmental Protection Agency*, Federal Reporter, Second Series 510 (1975), p. 1292.

53. *Environmental Defense Fund,* v. *Environmental Protection Agency*, Federal Reporter, Second Series 548 (1978), p. 998.

54. *Environmental Defense Fund* v. *Ruckelshaus*, Federal Reporter, Second Series 439 (1971), p. 584.

55. *Environmental Defense Fund* v. *Train*, 8 Environmental Reporter Cases (BNA) 2120 (D.D.C. Cir. 1976), as modified sub. nom. *NRDC* v. *Costle*, 12 Environmental Reporter Cases (BNA) 1833 (D.D.C. Cir. 1979).

56. Environmental Monitoring Services, Inc., *Tech-*

nical Background Document To Support Rule-making Pursuant to CERCLA Section 102: Volume 3 (Proposed Rulemaking), report to the U.S. Environmental Protection Agency, Contract No. 68-03-3182, December 1986.

57. Epstein, S.S., and Swartz, J.B., letter, *Science* 224:660-666, 1984.

58. Farland, W., Director, Carcinogen Assessment Group, U.S. Environmental Protection, personal communication, 1987.

59. Feron, V.J., Grice, H.C., Griesemer, R., et al., "Report 1: Basic Requirements for Long-Term Assays for Carcinogenicity," *Long-Term and Short-Term Screening Assays for Carcinogens: A Critical Appraisal*, R. Montesano, H. Bartsch, and L. Tomatis (eds.), IARC Monographs, Supplement 2 (Lyons, France: International Agency for Research on Cancer, 1980).

60. Fredrickson, D.S., "Chairman's Overview," *Health Research Activities of the Department of Health, Education and Welfare: Current Efforts and Proposed Initiatives* (Washington, DC: U.S. Department of Health, Education, and Welfare, 1979).

61. Funk, W., "The Exception That Proves the Rule: FDF Variances Under the Clean Water Act," *Boston University Journal of Environmental Affairs* 13:1-60, 1985.

62. Gaba, J.M., "Regulation of Toxic Pollutants Under the Clean Water Act: NPDES Toxics Control Strategies," *Journal of Air Law and Commerce* 50:761-791, 1985.

63. Gaylor, D.W., and Kodell, R.L., "Linear Interpolation Algorithm for Low Dose Risk Assessment of Toxic Substances," *J. of Environ. Path. and Toxicol.* 4:305-312, 1980.

64. Gaylor, D.W., National Center for Toxicology Research, U.S. Department of Health and Human Services, Food and Drug Administration, comments on unpublished OTA draft report, "Identifying and Regulating Carcinogens" (1987).

65. Glocklin, V.G., "Current FDA Perspectives on Animal Selection and Extrapolation," paper presented at the Monsanto Environmental Health Laboratory Toxicology Symposium, St. Louis, MO, Oct. 27, 1985.

66. Glocklin, V.G., "Drug Regulation," paper presented at Toxicology Forum, Annual Meeting, Aspen, CO, July 28, 1980.

67. Glocklin, V.G., "Justification for Requiring 12-month Chronic Toxicity Studies for New Drug Substances," paper presented at Centre for Medicines Research Workshop, London, U.K., Oct. 2, 1984.

68. Glocklin, V.G., U.S. Food and Drug Administration, comments on unpublished OTA draft report, "Identifying and Regulating Carcinogens" (1987).

69. Glocklin, V.G., and Temple R., U.S. Food and Drug Administration, personal communication, 1987.

70. Gold, L.S., Sawyer, C.B., Magaw, R., et al., "A Carcinogenic Potency Database of the Standardized Result of Bioassays," *Environ. Health Perspectives* 58:9-319, 1984.

71. Graham, J.D., "The Failure of Agency-Forcing: The Regulation of Airborne Carcinogens Under Section 112 of the Clean Air Act," *Duke Law Journal* 19:100-150, 1985.

72. Gray, K., Office of General Counsel, U.S. Environmental Protection Agency, personal communication, 1986.

73. Griesemer, R.A., and Cueto, C., "Toward a Classification Scheme for Degrees of Experimental Evidence for the Carcinogenicity of Chemicals for Animals," *Molecular and Cellular Aspects of Carcinogen Screening Tests*, R. Montesano, H. Bartsch, and L. Tomatis (eds.) (Lyons, France: International Agency for Research on Cancer, 1980).

74. Grubbs, G., Bureau Chief, Technical Support, Permits Division, Office of Municipal Pollution Control, U.S. Enviromental Protection Agency, personal communication, June 1986.

75. Grundfest, C., Pharmaceutical Manufacturers Association, personal communication, 1986.

76. *Gulf South Insul.* v. *U.S. Consumer Product Safety Commission*, Federal Reporter, Second Series 701 (1983), p. 1137.

77. Hall, K., "The Control of Toxic Pollutants Under the Federal Water Pollution Control Act Amendments of 1972," *Iowa Law Review* 63: 609-648, 1978.

78. Hall, R., Jr., "Regulation of Problem Pollutants Under the Federal Water Pollution Control Act: The 1976 Consent Decree," *Proceedings of the Second Open Forum on Management of Petroleum Refinery Waste Water*, EPA Doc. No. 600/2-78-078, March 1978.

79. Hall, R., Jr., "The Evolution and Implementation of EPA's Regulatory Program To Control the Discharge of Toxic Pollutants to the Nation's Waters," *Natural Resources Lawyer* 10:507-529, 1977.

80. Halogenated Solvents Industry Alliance, "Comments," statement presented at public meeting on National Toxicology Program *Annual Report*, Washington, DC, Apr. 21, 1987.

81. Harrill, E.D., Director of Congressional Relations, Consumer Product Safety Commission, comments on unpublished OTA draft report, "Identifying and Regulating Carcinogens" (1987).

82. Haseman, J.K., "Comments," paper presented at the Winter Toxicology Forum, Washington, DC, Feb. 16-18, 1987.

83. Haseman, J.K., "Statistical Issues in the Design, Analysis and Interpretation of Animal Carcinogenicity Studies," *Environ. Health Perspectives* 58:385-392, 1984.

84. Haseman, J.K., Crawford, D.D., Huff, J.E., et al., "Results From 86 Two-Year Carcinogenicity Studies Conducted by the National Toxicology Program," *J. Toxicol. Environ. Health* 14: 621-639, 1984.

85. Haseman, J.K., Huff, J.E., Zeiger, E.E., et al., "Results of 327 Chemical Carcinogenicity Studies Conducted by the National Cancer Institute and the National Toxicology Program," *Environ. Health Perspectives* (in press).

86. Hattis, D., and Smith, J.A., "What's Wrong With Quantitative Risk Assessment?" paper presented at the Conference on Moral Issues and Public Policy Issues in the Use of the Method of Quantitative Risk Assessment, Georgia State University, Sept. 26-27, 1985.

87. Haworth, S., Lawlor, T., Mortelman, K., et al., "Salmonella Mutagenicity Test Results for 250 Chemicals," *Environ. Mutagen.* 5 (Suppl. 1):3-142, 1983.

88. *Hercules, Inc.* v. *Environmental Protection Agency*, Federal Reporter, Second Series 598 (1978) pp. 91-131.

89. Hoel, D.B., Kaplan, N.L., and Anderson, M.W., "Implication of Nonlinear Kinetics on Risk Assessment in Carcinogenesis," *Science* 219:1032-1037, 1983.

90. Hudson, E.J., "National Toxicology Program," Office of Program Planning and Evaluation, National Institute of Environment Health Sciences (unpublished paper), July, 1981.

91. Huff, J.E., "Carcinogenesis Bioassay Results From the National Toxicology Program," *Environ. Health Perspectives* 45:185-198, 1982.

92. Huff J.E., Eustis, S.L., and Haseman, J.K., "Chemicals Causing Malignant, Malignant and Benign, or Benign Neoplasia in 113 Recent NTP Long-Term Carcinogenesis Studies in Rodents," unpublished manuscript, May 1986.

93. Huff J.E., McConnell, E.E., Haseman, J.K., et al., "Carcinogenesis Studies Results of 398 Experiments on 104 Chemicals From the U.S. National Toxicology Program," proceedings of the Collegium Ramazzini Conference, Bologna, Italy, October 1985, *Annals of the New York Academy of Sciences* (in press).

94. Huff, J.E., Pharmacologist, Division of Toxicology Research and Testing, National Toxicology Program, Department of Health and Human Services, National Institute of Environmental Health Sciences, memorandum to Dr. John Moore, May 9, 1980.

95. Huff, J., Pharmacologist, Division of Toxicology Research and Testing, National Toxicology Program, Department of Health and Human Services, National Institute of Environmental Health Sciences, personal communication, 1986 and 1987.

96. *Industrial Union Department AFL-CIO* v. *American Petroleum Institute*, U.S. Reports 448 (1980), p. 607.

97. *Industrial Union Department AFL-CIO* v. *Hodgson*, Federal Reporter, Second Series 467 (1974), p. 467.

98. International Agency for Research on Cancer, *Approaches To Classifying Chemical Carcinogens According to Mechanism of Action*, Technical Report No. 53/001 (Lyons, France: International Agency for Research on Cancer, 1983).

99. International Agency for Research on Cancer, *IARC Monographs on the Evaluation of Carcinogenic Risks to Humans: Preamble*, IARC Technical Report No. 87/001 (Lyons, France: IARC, 1987).

100. Jacoby, D., Counsel for Standards, U.S. Department of Labor, Office of the Solicitor, personal communication, 1986.

101. Jaffe, L.L., and Nathanson, N.L., *Administrative Law* (Boston, MA: Little, Brown and Co., 1968).

102. Juskevich J., Department of Health and Human Services, Food and Drug Administration, Center for Veterinary Medicine, personal communication, July 27, 1987.

103. Kellam, R., Chief, Program Analysis and Technology Division, Office of Air and Radiation, U.S. Enviromental Protection Agency, personal communication, July 1986.

104. Kelly T.E., Acting Director, Office of Standards and Regulations, U.S. Environmental Protection Agency, comments on unpublished OTA draft report, "Identifying and Regulating Carcinogens" (1987).

105. Keyworth, G.A., Science Advisor to the President, statement before the Subcommittee on Commerce, Transportation, and Tourism, Committee on Energy and Commerce, U.S. House of Representatives, Mar. 17, 1983.

106. Kooyoomjian, J., Office of Solid Waste and

Emergency Response, U.S. Environmental Protection Agency, personal communication, July 1986.

107. Kovalick W., Deputy Director, Office of Emergency and Remedial Response, U.S. Environmental Protection Agency, letter to OTA, Mar. 16, 1987.

108. Kover F., Chief, Chemical Screening Branch, Office of Toxic Substances, U.S. Environmental Protection Agency, personal communication, 1987.

109. Lehman, A.J., Patterson, W.I., Davidow, B., et al., "Procedures for the Appraisal of the Toxicity of Chemicals in Foods, Drugs and Cosmetics," *Food Drug Cosmetic Law J.* 10:679-748, 1955.

110. Lemberg, S., Assistant General Counsel, U.S. Consumer Product Safety Commission, personal communication, 1986.

111. Lowrance, W., *Of Acceptable Risk: Science and the Determination of Safety* (Los Altos, CA: William Kaufman, 1976).

112. Lynn, F.M., "The Interplay of Science and Values in Assessing and Regulating Environmental Risks," *Sci. Technol. Human Values* 11(2):40-50, Spring 1986.

113. Maguire, A., statement at public meeting on National Toxicology Program Annual Report, Washington, DC, Apr. 21, 1987.

114. Marcus, A., "The Environmental Protection Agency," in *The Politics of Regulation*, James Wilson (ed.) (New York: Basic Books, 1980).

115. Maronpot, R.R., Haseman, J.K., Boorman, G.A., et al., "Liver Lesions in B6C3F1 Mice: The National Toxicology Program, Experience and Position," *Arch. Toxicol. Suppl.* 10:10-26, 1987.

116. Marshall, E., "Carcinogenesis Without Controversy," *Science* 224:1078, June 8, 1984.

117. Marshall, E., "Revisions in Cancer Policy," *Science* 220:36-37, Apr. 1, 1983.

118. Martonick, J., Deputy Director, Health Standards, Occupational Safety and Health Administration, personal communication, 1986 and 1987.

119. Mazuzan, G.T., and Walker, J.S., *Control of the Atom* (Berkeley, CA: University of California Press, 1984).

120. McConnell E., Director Toxicology Research and Testing Program, U.S. Department of Health and Human Services, National Institute of Environmental Health Sciences, personal communication, 1986.

121. McConnell, E., Director Toxicology Research and Testing Program, U.S. Department of Health and Human Services, National Institute of Environmental Health Sciences, comments on unpub-lished OTA draft report, "Identifying and Regulating Carcinogens" (1987).

122. McGarity, T.O., "OSHA's Generic Carcinogen Policy: Rule Making Under Scientific and Legal Uncertainty," *Law and Science in Collaboration*, J.D. Nyhart and M.M. Carrow (eds.) (Lexington, MA: Lexington Books, 1983).

123. McGaughy, R., Carcinogen Assessment Group, U.S. Environmental Protection Agency, personal communication, 1987.

124. McGinnis, J.M., Director, Office of Disease Prevention and Health Promotion, U.S. Department of Health and Human Services, comments on unpublished OTA draft report, "Identifying and Regulating Carcinogens" (1987).

125. McGuire, J., Supervisory Research Chemist, Measurements Branch, Environmental Research Laboratory, U.S. Environmental Protection Agency, personal communication, 1987.

126. Merrill, R.A., "CPSC Regulation of Cancer Risks in Consumer Products: 1972-1981," *Virginia Law Review* 67:1261-1375, 1981.

127. Merrill, R.A., "Regulating Carcinogens in Food: A Legislator's Guide to the Food Safety Provisions of the Federal Food, Drug, and Cosmetic Act," *Michigan Law Review* 72:172-250, 1979.

128. Merrill R.A. and Schewel, "FDA Regulation of Environmental Contaminants of Food," *Virginia Law Review* 8:1357, 1980.

129. *Monsanto* v. *Kennedy*, Federal Reporters, Second Series 613 (1979), p. 947.

130. Moore, J., Assistant Administrator, Office of Pesticides and Toxic Substances, U.S. Environmental Protection Agency, letter to Representative Henry Waxman, Oct. 2, 1985.

131. Morbidity and Mortality Weekly Report, "NIOSH Recommendations for Occupational Safety and Health Standards," *Morbidity and Mortality Weekly Report* 34(1S):22S, July 19, 1985.

132. Morrison, A.B., "OMB Interference With Agency Rulemaking: The Wrong Way To Write a Regulation," *Harvard Law Review* 99:1059-1074, 1986.

133. Mortelmans, K., Haworth, S., Lawlor, T., et al., "Salmonella Mutagenicity Tests, II: Results From the Testing of 270 Chemicals," *Environ. Mutagen.* 8(Suppl. 7):1-119, 1986.

134. National Academy of Sciences/National Research Council, Board on Toxicology and Environmental Health Hazard, *Drinking Water and Health*, vols. 1-8 (Washington, DC: National Academy Press, 1977-86).

135. National Academy of Sciences/National Research Council, *Meat and Poultry Inspection:*

The Scientific Basis of the Nation's Program (Washington, DC: National Academy Press, 1985).

136. National Academy of Sciences/National Research Council, *Regulating Pesticides in Food—The Delaney Paradox* (Washington, DC: National Academy Press, 1987).

137. National Academy of Sciences/National Research Council, *Risk Assessment in the Federal Government: Managing the Process* (Washington, DC: National Academy Press, 1983).

138. National Academy of Sciences/National Research Council, *Toxicity Testing: Strategies To Determine Needs and Priorities* (Washington, DC: National Academy Press, 1984).

139. *National Association of Metal Finishers v. Environmental Protection Agency*, Federal Reporter, Second Series 719 (1983), p. 624, revised sub nom. *Chemical Manufacturers Association v. Natural Resources Defense Council*, Supreme Court Reports 105 (1985), p. 1102.

140. *Natural Resources Defense Council v. Costle*, Environmental Reporter Cases 12 (D.D.C.Cir., Mar. 9, 1979).

141. *Natural Resources Defense Council v. Environmental Protection Agency*, Environmental Reporter Cases 17 (D.D.C.Cir., July 8, 1982), p. 1721.

142. *Natural Resources Defense Council v. Environmental Protection Agency, et al.*, No. 85-1150 (D.D.C.Cir., July 28, 1987).

143. *Natural Resources Defense Council v. Gorsuch*, Environmental Reporter Cases 17 (1982), p. 2013.

144. *Natural Resources Defense Council v. Gorsuch*, October 1982; *Natural Resources Defense Council v. Ruckelshaus*, August 1983; *Natural Resources Defense Council v. Ruckelshaus*, January 1984; *Natural Resources Defense Council v. Ruckelshaus*, July 1984; *Natural Resources Defense Council v. Thomas*, January 1985; and *Natural Resources Defense Council v. Thomas*, April 1986, Civil Actions Nos. 73-2153, 75-0172, 75-1698, 75-1267.

145. *Natural Resources Defense Council v. Thomas*, U.S. Court of Appeals, District of Columbia Circuit, Civil Actions 84-1387, 85-1567, brief for petitioners (1985).

146. *Natural Resources Defense Council v. Thomas et al.*, U.S. Court of Appeals, District of Columbia Circuit, Civil Actions 84-1387, 85-1567, brief for respondents, 1985.

147. *Natural Resources Defense Council v. Thomas*, No. 86 CIV 0603 (CSH), pending in S. Dist. of N.Y.

148. *Natural Resources Defense Council v. Train*, Environmental Reporter Cases 8(D.D.C.Cir., 1976), p. 2120.

149. Nichols, A.L., and Zeckhauser, R.J., "The Dangers of Caution: Conservatism in Assessment and the Mismanagement of Risk," Harvard University Energy and Environmental Policy Center, John F. Kennedy School of Government, discussion paper series (Cambridge, MA, November 1985).

150. Nothstein, G.Z., *The Law of Occupational Safety and Health* (New York: The Free Press, 1981).

151. Ohanian, E.V., and Cotruvo, J.A., "Summation From a Regulatory Perspective," *Environmental Health Perspectives* 69:281-284, 1986.

152. Olson, E.D., "The Quiet Shift of Power: Office of Management and Budget Supervision of Environmental Protection Agency Rulemaking Under Executive Order 12,291," *Virginia Journal of Natural Resources Law* 4:1-80, 1984.

153. *Pactra Industries v. Consumer Product Safety Commission*, Federal Reporter, Second Series 55 (1977), p. 677.

154. Page, T., "A Framework for Unreasonable Risk in the Toxic Substances Control Act (TSCA)," *Management of Assessed Risk for Carcinogens*, W.J. Nicholson (ed.), *Annals of the New York Academy of Sciences*, vol. 363 (New York: New York Academy of Sciences, 1981).

155. Perera, F.P., and Petito, C., "Formaldehyde: A Question of Cancer Policy?" *Science* 216:1285-1291, 1982.

156. Perera, F.P., "Quantitative Risk Assessment and Cost-Benefit Analysis for Carcinogens at EPA: A Critique," *J. Environ. Health Pol.* (in press).

157. Perera, F.P., "The Genotoxic/Epigenetic Distinction: Relevance to Cancer Policy," *Environ. Res.* 34:175-191, 1984.

158. Perera F.P., Senior Science Advisor, Natural Resources Defense Council, comments on unpublished OTA draft report, "Identifying and Regulating Carcinogens" (1987).

159. Peto, R., and Schneiderman, M., *Quantification of Occupational Cancer*, Banbury House Report 9 (Cold Spring Harbor, NY: Cold Spring Harbor Laboratory, 1981).

160. Petrie, J., U.S. Department of Labor, Mine Safety and Health Administration, personal communication, 1986 and 1987.

161. Pharmaceutical Manufacturers Association,

"Guidelines for the Assessment of Drug and Medical Device Safety in Animals," typescript, February 1977.

162. *Physician's Desk Reference* (Oradell, NJ: Medical Economics Co., 1987).

163. Pitot, H.C., *Fundamentals of Oncology*, 3d ed. (New York: Marcel Dekker, 1986).

164. Preuss P., Director, Office of Health and Environmental Assessment, U.S. Environmental Protection Agency, personal communication, 1986.

165. *Public Citizen Health Research Group v. Auchter*, Federal Reporter, Second Series 702 (1983), p. 1150.

166. Purcell T., Office of Water Regulations and Standards, Criteria and Standards Division, U.S. Environmental Protection Agency, comments on unpublished OTA draft report, "Identifying and Regulating Carcinogens" (1987).

167. Radian Corp., *Draft User's Manual for Modified Hazardous Air Pollutant Prioritization System (MHAPPS)*, prepared for U.S. Environmental Protection Agency, Office of Air Quality Planning and Standards, Pollutant Assessment Branch, Contract No. 68-02-3889, December 1985.

168. Reagan, R., "Federal Regulation," Executive Order No. 12291, *Fed. Reg.* 46(33):13193, Feb. 17, 1981.

169. Reagan, R., "Regulatory Planning Process" Executive Order No. 12498, *Fed. Reg.* 50:1036, Jan. 4, 1985.

170. Renwick, A.G., "Pharmacokinetics in Toxicology," *Principles and Methods of Toxicology*, A.W. Hayes (ed.) (New York: Raven Press, 1982).

171. Rikleen, L.S., "Negotiating Superfund Settlement Agreements," *Boston College Environmental Affairs Review* 10:639-714, 1982-83.

172. Rinsky, R.A., Smith, A.B., Hornung, R., et al., "Benzene and Leukemia," *New England J. Med.* 316:1044-1050, Apr. 23, 1987.

173. Robbins, A., Professor, Boston University School of Public Health, statement presented at public meeting on National Toxicology Program *Annual Report*, Washington, DC, Apr. 21, 1987.

174. Rodgers, W., *Handbook of Environmental Law* (St. Paul, MN: West Publishing Co., 1977).

175. Rogers, P., quoted in *Occupational Health & Safety Letter*, May 22, 1987.

176. Rogers, P., statement presented at public meeting on National Toxicology Program *Annual Report*, Washington, DC, Apr. 21, 1987.

177. Rubin, M., Chief, Chemicals Industry Branch, Industrial Technology Division, Office of Water, U.S. Environmental Protection Agency, personal communication, 1986.

178. Ruckelshaus, W.D., "Risk in a Free Society," *Risk Anal.* 4(3):157-162, 1983.

179. Rushefsky, M.E., "Assuming the Conclusions: Risk Assessment in the Development of Cancer Policy," *Pol. Life Sci.* 4(1):31-66, August 1985.

180. Rushefsky, M.E., *Making Cancer Policy* (Albany, NY: State University of New York Press, 1986).

181. Ruttenberg, R., and Bingham, E., "A Comprehensive Occupational Carcinogen Policy as a Framework for Regulatory Activity," *Management of Assessed Risk for Carcinogens*, W.J. Nicholson (ed.), *Annals of the New York Academy of Sciences*, vol. 363 (New York: New York Academy of Sciences, 1981).

182. Scheuplein, R., Food and Drug Administration, personal communication, 1987.

183. Schroeder, C., "A Decade of Change in Regulating the Chemical Industry," *Law and Contemporary Problems* 46(4):1, 1983.

184. *Scott v. Food and Drug Administration*, Federal Reporter, Second Series, 728 (1984), p. 322.

185. Shackelford, W.M., and Cline, D.M., "Organic Compounds in Water," *Environ. Sci. Technol.* 20(7):652-657, 1986.

186. Silbergeld E., Environmental Defense Fund, letter to OTA, Mar. 11, 1987.

187. *Society for Plastics Industries v. Occupational Safety and Health Administration*, Federal Reporter, Second Series 509 (1975), p. 1301.

188. Springer, J., Food and Drug Administration, personal communication, 1987.

189. *Springs Mills v. Consumer Product Safety Commission*, Federal Supplement 434 (1977), p. 416.

190. Sutherland, W., Mine Safety and Health Administration, U.S. Department of Labor, personal communication, 1986 and 1987.

191. *Synthetic Organic Chemical Manufacturers Association v. Brennan* I, Federal Reporter, Second Series 503 (1974), p. 1155.

192. *Synthetic Organic Chemical Manufacturers Association v. Brennan* II, Federal Reporter, Second Series 506 (1974), p. 385.

193. Temple R., Director, Office of Drug Research and Review, Center for Drugs and Biologics, U.S. Department of Health and Human Services, Food and Drug Administration, personal communication, Feb. 2, 1987.

194. Tennant, R.W., Margolin, B.H., Shelby, M.D., et al., "Prediction of Chemical Carcinogenicity in Rodents From In Vitro Genetic Toxicity Assays," *Science* 236:933-941, 1987.

195. Timm, G., Office of Toxic Substances, U.S. Environmental Protection Agency, personal communication, Apr. 2, 1987.

196. Todhunter, J.A., "Review of Data Available to the Administrator Concerning Formaldehyde and Di(2-ethylhexyl) Phthalate (DEHP)," memo to A.M. Gorsuch, Feb. 10, 1982.

197. U.S. Congress, "Joint House-Senate Summary and Explanation," legislative history for Public Law 95-622, *Congressional Record* 124:H13566, Oct. 14, 1978.

198. U.S. Congress, General Accounting Office, *Delays in EPA's Regulation of Hazardous Air Pollutants*, RCED-83-199 (Washington, DC: U.S. Government Printing Office, Aug. 26, 1983).

199. U.S. Congress, General Accounting Office, *Federal Efforts To Protect the Public From Cancer-Causing Chemicals Are Not Very Effective* (Washington, DC: U.S. Government Printing Office, 1976).

200. U.S. Congress, General Accounting Office, *Hazardous Waste: EPA Has Made Limited Progress in Determining the Wastes To Be Regulated*, GAO/RCED-87-27 (Washington, DC: U.S. Government Printing Office, Dec. 1986).

201. U.S. Congress, General Accounting Office, *National Toxicology Program: Efforts To Improve Oversight of Contractors Testing Chemicals*, HRD-85-86 (Washington, DC: U.S. Government Printing Office, 1985).

202. U.S. Congress, General Accounting Office, *Nonagricultural Pesticides: Risks and Regulation* (Washington, DC: U.S. Government Printing Office, Apr. 18, 1986).

203. U.S. Congress, General Accounting Office, *Pesticides: EPA's Formidable Task To Assess and Regulate Their Risks* (Washington, DC: U.S. Government Printing Office, 1986).

204. U.S. Congress, House of Representatives, Committee on Agriculture, Subcommittee on Department Operations, Research, and Foreign Agriculture, *Hearings on the Federal Food Irradiation Development and Control Act of 1985* (H.R. 696), Hearings, Nov. 11, 1985 (Washington, DC: U.S. Government Printing Office, 1986).

205. U.S. Congress, House of Representatives, Committee on Appropriations, Subcommittee on Labor and Health Education and Welfare, *Department of Labor, Health, Education and Welfare, and Related Agencies Appropriations for 1981, Part 4, National Institutes of Health*, Hearings, Feb. 25, 28, 29, Mar. 3-6, 1980 (Washington, DC: U.S. Government Printing Office, 1980).

206. U.S. Congress, House of Representatives, Committee on Energy and Commerce, Subcommittee on Oversight and Investigations, *EPA's Asbestos Regulations: Report on a Case Study on OMB Interference in Agency Rulemaking* (Wash-

ington, DC: U.S. Government Printing Office, 1985).

207. U.S. Congress, House of Representatives, Committee on Energy and Commerce, Subcommittee on Oversight and Investigations, *OMB Review of CDC Research: Impact of the Paperwork Reduction Act* (Washington, DC: U.S. Government Printing Office, 1986).

208. U.S. Congress, House of Representatives, Committee on Energy and Commerce, Subcommittee on Oversight and Investigations, Investigation of the Environmental Protection Agency, *Report on the President's Claim of Executive Privilege Over EPA Documents, Abuses in the Superfund Program, and Other Matters*, Committee Print 98AA (Washington, DC: U.S. Government Printing Office, 1984).

209. U.S. Congress, House of Representatives, Committee on Government Operations, *FDA's Regulation of Zomax*, H. Rpt. 98-584 (Washington, DC: U.S. Government Printing Office, 1983).

210. U.S. Congress, House of Representatives, Committee on Government Operations, *HHS' Failure To Enforce the Food, Drug, and Cosmetic Act: The Case of Cancer-Causing Color Additives*, H. Rpt. 99-151 (Washington, DC: U.S. Government Printing Office, 1985).

211. U.S. Congress, House of Representatives, Committee on Government Operations, *Human Food Safety and the Regulation of Animal Drugs*, H. Rpt. 99-461 (Washington, DC: U.S. Government Printing Office, 1985).

212. U.S. Congress, House of Representatives, Committee on Interstate and Foreign Commerce, *Committee Report To Accompany H.R. 14032*, H. Rpt. 94-1341 (Washington, DC: U.S. Government Printing Office, 1976).

213. U.S. Congress, House of Representatives, Committee on Science and Technology, Subcommittee on Investigations and Oversight, *National Toxicology Program*, Hearings, July 15, 1981 (Washington, DC: U.S. Government Printing Office, 1981).

214. U.S. Congress, House of Representatives, *Report To Accompany H.R. 13002 (Safe Drinking Water Act)*, H. Rpt. 93-1185 (Washington, DC: U.S. Government Printing Office, July 10, 1974).

215. U.S. Congress, House of Representatives, *Report on the Safe Drinking Water Act Amendments of 1985*, Rpt. No. 99-168.

216. U.S. Congress, Office of Technology Assessment, "Smoking-Related Deaths and Financial Costs," staff memorandum, September 1985.

217. U.S. Congress, Office of Technology Assessment, *Assessment of Technologies for Determin-*

ing Cancer Risks From the Environment (Washington, DC: U.S. Government Printing Office, 1981).

218. U.S. Congress, Office of Technology Assessment, *Postmarketing Surveillance of Prescription Drugs* (Washington DC: U.S. Government Printing Office, 1982).

219. U.S. Congress, Office of Technology Assessment, *Preventing Illness and Injury in the Workplace* (Washington, DC: U.S. Government Printing Office, 1985).

220. U.S. Congress, Office of Technology Assessment, *Reproductive Health Hazards in the Workplace* (Washington, DC: U.S. Government Printing Office, 1985).

221. U.S. Congress, Office of Technology Assessment, *Technologies and Management Strategies for Hazardous Waste Control* (Washington, DC: U.S. Government Printing Office, 1983).

222. U.S. Congress, Office of Technology Assessment, *The Information Content of Premanufacture Notices* (Washington, DC: U.S. Government Printing Office, 1983).

223. U.S. Congress, Office of Technology Assessment, *Transportation of Hazardous Materials* (Washington, DC: U.S. Government Printing Office, 1986).

224. U.S. Congress, Office of Technology Assessment, *Wastes in Marine Environments* (Washington, DC: U.S. Government Printing Office, 1987).

225. U.S. Congress, Senate, Committee on Environment and Public Works, *Office of Management and Budget Influence on Agency Actions*, S. Rpt. 99-156 (Washington, DC: U.S. Government Printing Office, 1986).

226. U.S. Consumer Product Safety Commission, Commission decision not to convene a Chronic Hazard Advisory Panel Regarding Formaldehyde Emissions, October 1986.

227. U.S. Consumer Product Safety Commission, "Children's Pacifiers Containing Nitrosamines; Enforcement Policy," *Fed. Reg.* 48:56989, 1983.

228. U.S. Consumer Product Safety Commission, "Classifying, Evaluating, and Regulating Carcinogens in Consumer Products: Interim Statement of Policy and Procedure," *Fed. Reg.* 43:58-25665, 1978.

229. U.S. Consumer Product Safety Commission, "Consumer Products Containing Benzene as an Intentional Ingredient or as a Contaminant Under the Consumer Product Safety Act," Proposed Rule, *Fed. Reg.* 43:21838, 1978.

230. U.S. Consumer Product Safety Commission, "Federal Hazardous Substance Act Regulations; Part 1500—Hazardous Substances and Articles; Administration and Enforcement Regulations," *Fed. Reg.* 38:27017, 1973.

231. U.S. Consumer Product Safety Commission, "Labeling Asbestos-Containing Household Products; Enforcement Policy," *Fed. Reg.* 51:33910, 1986.

232. U.S. Department of Health and Human Services, Committee to Coordinate Environmental and Related Programs (CCERP), Executive Committee, "Risk Assessment and Risk Management of Toxic Substances, A Report to the Secretary, April 1985," *Determining Risks to Health*, U.S. Department of Health and Human Services (Dover MA: Auburn House Publishing Co., 1986).

233. U.S. Department of Health and Human Services, *Determining Risks to Health* (Dover, MA: Auburn House Publishing Co., 1986).

234. U.S. Department of Health and Human Services, Memorandum of Understanding Between the National Institute of Environmental Health Sciences and the National Toxicology Program, fiscal year 1986.

235. U.S. Department of Health and Human Services and U.S. Environmental Protection Agency, "Notice of the First Priority List of Hazardous Substances That Will Be the Subject of Toxicological Profiles," *Fed. Reg.* 52:12866, 1987.

236. U.S. Department of Health and Human Services, Food and Drug Administration, "General Principles for Evaluating the Safety of Compounds Used in Food-Producing Animals," typescript, July 1983.

237. U.S. Department of Health and Human Services, Food and Drug Administration, "Acrylonitrile Copolymers Used To Fabricate Beverage Containers; Final Decision," *Fed. Reg.* 42:48528, 1977.

238. U.S. Department of Health and Human Services, Food and Drug Administration, "Compounds Used in Food-Producing Animals," *Fed. Reg.* 38:19226, July 19, 1973.

239. U.S. Department of Health and Human Services, Food and Drug Administration, "Correction of Listing of D&C Orange No. 17 for Use in Externally Applied Drugs and Cosmetics," *Fed. Reg.* 52:8081, Feb. 19, 1987.

240. U.S. Department of Health and Human Services, Food and Drug Administration, "Correction of Listing of D&C Red No. 19 for Use in Externally Applied Drugs and Cosmetics," *Fed. Reg.* 52:5083, Feb. 19, 1987.

241. U.S. Department of Health and Human Services, Food and Drug Administration, "D&C Green No. 6; Listing as a Color Additive in Externally

Applied Drugs and Cosmetics," *Fed. Reg.* 47: 14138, Apr. 2, 1982.

242. U.S. Department of Health and Human Services, Food and Drug Administration, "Listing of D&C Orange No. 17 for Use in Externally Applied Drugs and Cosmetics," *Fed. Reg.* 51:28331, Aug. 7, 1986.

243. U.S. Department of Health and Human Services, Food and Drug Administration, "Listing of D&C Red No. 19 for Use in Externally Applied Drugs and Cosmetics," *Fed. Reg.* 51:28346, Aug. 7, 1986.

244. U.S. Department of Health and Human Services, Food and Drug Administration, "Listing of D&C Red No. 8 and D&C Red No. 9 for Use in Ingested Drug and Cosmetic Lip Products and Externally Applied Drugs and Cosmetics," *Fed. Reg.* 51:43677, 43899, Dec. 5, 1986.

245. U.S. Department of Health and Human Services, Food and Drug Administration, "Policy for Regulating Carcinogenic Chemicals in Food and Color Additives; Advance Notice of Proposed Rulemaking," *Fed. Reg.* 47:14464, Apr. 2, 1982.

246. U.S. Department of Health and Human Services, Food and Drug Administration, "Sponsored Compounds in Food-Producing Animals; Proposed Rule and Notice," *Fed. Reg.* 50:45530, Oct. 31, 1985.

247. U.S. Department of Health and Human Services, Food and Drug Administration, Advisory Committee on Protocols for Safety Evaluation, "Panel on Carcinogenesis Report on Cancer Testing in the Safety Evaluation of Food Additives and Pesticides," *Toxicol. Appl. Pharmacol.* 20:419-438, 1971.

248. U.S. Department of Health and Human Services, Food and Drug Administration, Bureau of Foods, *Toxicologic Principles for the Safety Assessment of Direct Food Additives and Color Additives Used in Food* (Washington, DC: 1982).

249. U.S. Department of Health and Human Services, Food and Drug Administration, comments on unpublished OTA draft report, "Identifying and Regulating Carcinogens" (1987).

250. U.S. Department of Health and Human Services, National Cancer Institute, Ad Hoc Committee on the Evaluation of Low Levels of Environmental Chemical Carcinogens, "Evaluation of Environmental Carcinogens" (Report to the Surgeon General, Apr. 22, 1970), *Chemicals and the Future of Man*, Hearings before Subcommittee on Executive Reorganization and Government Research, Committee on Government Operations, U.S. Senate, Apr. 6-7, 1971, pp. 180-190

(Washington, DC: U.S. Government Printing Office, 1970).

251. U.S. Department of Health and Human Services, National Cancer Institute, *Guidelines for Carcinogen Bioassay in Small Rodents*, NCI-CG-TR-1, DHEW Pub. No. (NIH) 76-801 (Washington, DC: U.S. Government Printing Office, February 1976).

252. U.S. Department of Health and Human Services, National Institute of Occupational Safety and Health, Division of Standards Development and Technology Transfer, "Evaluation of Epidemiologic Studies Examining the Lung Cancer Mortality of Underground Uranium Miners," report prepared for U.S. Mine Safety and Health Administration, May 9, 1985.

253. U.S. Department of Health and Human Services, National Toxicology Program, *NTP Tech. Bull.* 1(3):1-2, 1980.

254. U.S. Department of Health and Human Services, National Toxicology Program, "Announcement of Completed Short-term Toxicology Studies on Three Chemicals: Request for Comments," *Fed. Reg.* 51:31376, 1986.

255. U.S. Department of Health and Human Services, National Toxicology Program, "Explanation of Levels of Evidence of Carcinogenic Activity," March 1986.

256. U.S. Department of Health and Human Services, National Toxicology Program, "General Statement of Work for the Conduct of Acute, 14-Day Repeated Dose, 90-day Subchronic, and 2-Year Chronic Studies in Laboratory Animals," typescript, July 1984.

257. U.S. Department of Health and Human Services, National Toxicology Program, *Annual Plan, FY 1986*, DHHS Pub. No. NTP-86-086 (Research Triangle Park, NC, May 1986).

258. U.S. Department of Health and Human Services, National Toxicology Program, Board of Scientific Counselors, *Report of the NTP Ad Hoc Panel on Chemical Carcinogenesis Testing and Evaluation* (Research Triangle Park, NC, Aug. 17, 1984).

259. U.S. Department of Health and Human Services, National Toxicology Program, Board of Scientific Counselors, summary minutes, meeting of Apr. 30 and May 1, 1985, attachment 4, p. 2.

260. U.S. Department of Health and Human Services, National Toxicology Program, Board of Scientific Counselors, summary minutes, meeting of Nov. 25, 1986.

261. U.S. Department of Health and Human Services, National Toxicology Program, Chemical Evalu-

ation Committee, minutes, meeting of Sept. 16, 1986.

262. U.S. Department of Health and Human Services, National Toxicology Program, *First Annual Report on Carcinogens* (Research Triangle Park, NC, July 1980).

263. U.S. Department of Health and Human Services, National Toxicology Program, *Fourth Annual Report on Carcinogens*, Pub. No. NTP 85-001 (Research Triangle Park, NC, 1985).

264. U.S. Department of Health and Human Services, National Toxicology Program, *Review of Current DHHS, DOE, and EPA Research Related to Toxicology, Fiscal Year 1986*, Pub. No. NTP-86-087 (Research Triangle Park, NC, 1987).

265. U.S. Department of Health and Human Services, National Toxicology Program, *Second Annual Report on Carcinogens* (Research Triangle Park, NC, December 1981).

266. U.S. Department of Health and Human Services, National Toxicology Program, *Third Annual Report on Carcinogens* (Research Triangle Park, NC, September 1983).

267. U.S. Department of Health, Education, and Welfare, Food and Drug Administration, Division of Pharmacology, *Appraisal of the Safety of Chemicals in Foods, Drugs, and Cosmetics* (Topeka, KS: Association of Food and Drug Officials of the United States, 1959).

268. U.S. Department of Health, Education, and Welfare, Public Health Service, "Establishment of a National Toxicology Program," *Fed. Reg.* 43:53060, 1978.

269. U.S. Department of Labor, Mine Safety and Health Administration, "Ionizing Radiation Standards for Underground Metal and Nonmetal Mines; Proposed Rule," *Fed. Reg.* 51:45678, 1986.

270. U.S. Department of Labor, Mine Safety and Health Administration, "Recodification of Safety and Health Standards for Metal and Nonmetal Mines, Final Rule," *Fed. Reg.* 50:4048, 1985.

271. U.S. Department of Labor, Occupational Safety and Health Administration, "Access to Employee Exposure and Medical Records," 29 *Code of Federal Regulations* 1910.20.

272. U.S. Department of Labor, Occupational Safety and Health Administration, "Final Standard on Certain Carcinogens," *Fed. Reg.* 39:3756, 1974.

273. U.S. Department of Labor, Occupational Safety and Health Administration, "Hazard Communication," 29 *Code of Federal Regulations* 1910.1200.

274. U.S. Department of Labor, Occupational Safety and Health Administration, "Identification, Classification and Regulation of Potential Occupa-

tional Carcinogens; Conforming Deletion," *Fed. Reg.* 46:4889, Jan. 19, 1981.

275. U.S. Department of Labor, Occupational Safety and Health Administration, "Identification, Classification and Regulation of Potential Occupational Carcinogens," *Fed. Reg.* 47:187, Jan. 5, 1982.

276. U.S. Department of Labor, Occupational Safety and Health Administration, "Identification, Classification and Regulation of Potential Occupational Carcinogens," *Fed. Reg.* 45:5002, Jan. 22, 1980.

277. U.S. Department of Labor, Occupational Safety and Health Administration, "List of Substances Which May Be Candidates for Further Scientific Review and Possible Identification, Classification, and Regulation as Potential Occupational Carcinogens," *Fed. Reg.* 45:53672, Aug. 12, 1980.

278. U.S. Department of Labor, Occupational Safety and Health Administration, "Occupational Exposure to Coal Tar Pitch Volatiles; Modification of Interpretation, Final Interpretation," *Fed. Reg.* 48:2764, 1983.

279. U.S. Environmental Protection Agency, "Additional U.S. Environmental Protection Agency Guidance for the Health Assessment of Suspect Carcinogens With Specific Reference to Water Quality Criteria" (draft, June 21, 1982), *PCBs and Dioxin Cases*, Hearing before Subcommittee on Oversight and Investigations, Committee on Energy and Commerce, U.S. House of Representatives, Nov. 19, 1982, pp. 105-124 (Washington, DC: U.S. Government Printing Office, 1982).

280. U.S. Environmental Protection Agency, "Asbestos; Proposed Mining and Import Restrictions and Proposed Manufacturing Importation and Processing Prohibitions," *Fed. Reg.* 51:3738, Jan. 29, 1986.

281. U.S. Environmental Protection Agency, "Data Requirements for Pesticide Registration; Final Rule," *Fed. Reg.* 49:42856, Oct. 24, 1984e.

282. U.S. Environmental Protection Agency, "Development of Water Quality Based Permit Limitations for Toxic Pollutants; National Policy," *Fed. Reg.* 49:9017, 1984.

283. U.S. Environmental Protection Agency, "Formaldehyde; Determination of Significant Risk; Advance Notice of Proposed Rulemaking," *Fed. Reg.* 49:21870-98, May 23, 1984.

284. U.S. Environmental Protection Agency, "Guidelines for Carcinogen Risk Assessment," *Fed. Reg.* 51:33992, 1986.

285. U.S. Environmental Protection Agency, "Guidelines for Exposure Assessment," *Fed. Reg.* 51: 34042, 1986.

286. U.S. Environmental Protection Agency, "Guidelines for Health Risk Assessment of Suspect Developmental Toxicants," *Fed. Reg.* 51:34028, 1986.

287. U.S. Environmental Protection Agency, "Guidelines for Mutagenicity Risk Assessment," *Fed. Reg.* 51:34006, 1986.

288. U.S. Environmental Protection Agency, "Hazardous Waste Management System: Identification and Listing of Hazardous Waste, Final Rule," *Fed. Reg.* 45:33084, 1980.

289. U.S. Environmental Protection Agency, "Hazardous Waste Management System: Indentification and Listing of Hazardous Waste," *Fed. Reg.* 45:33084, 1980.

290. U.S. Environmental Protection Agency, "Hazardous Waste Guidelines and Regulations," *Fed. Reg.* 43:58946, 1978.

291. U.S. Environmental Protection Agency, "Hazardous Waste Management System; Dioxin-Containing Waste," *Fed. Reg.* 50:1978, 1985.

292. U.S. Environmental Protection Agency, "Health and Safety Data Reporting; Submission of Lists and Copies of Health and Safety Studies," *Fed. Reg.* 51:27562, 1986.

293. U.S. Environmental Protection Agency, "Health Risk and Economic Impact Assessments of Suspected Carcinogens: Interim Procedures and Guidelines," *Fed. Reg.* 41:402, 1976.

294. U.S. Environmental Protection Agency, "Inert Ingredients in Pesticide Products; Policy Statement," *Fed. Reg.* 52:13305, 1987.

295. U.S. Environmental Protection Agency, "Inert Ingredients in Pesticide Products; Policy Statement," *Fed. Reg.* 52:13305, 1987.

296. U.S. Environmental Protection Agency, "Intent To List Chloroform as a Hazardous Air Pollutant," *Fed. Reg.* 50:39626, 1985.

297. U.S. Environmental Protection Agency, "List of Office of Toxic Substances Documents Submitted to the Office of Technology Assessment in Response to the Request of May 16, 1986," letter to OTA.

298. U.S. Environmental Protection Agency, "National Emission Standards for Hazardous Air Pollutants; Policy and Procedures for Identifying, Assessing, and Regulating Airborne Substances Posing a Risk of Cancer," *Fed. Reg.* 44:58642, 1979.

299. U.S. Environmental Protection Agency, "National Primary Drinking Water Regulations; Synthetic Organic Chemicals, Inorganic Chemicals and Microorganisms," *Fed. Reg.* 50:46936, 1985.

300. U.S. Environmental Protection Agency, "National Primary Drinking Water Regulations; Volatile Synthetic Organic Chemicals, Proposed Rulemaking," *Fed. Reg.* 49:24330, 1984.

301. U.S. Environmental Protection Agency, "National Primary Drinking Water Act Regulations; Volatile Synthetic Organic Chemicals," *Fed. Reg.* 50:46880, 1985.

302. U.S. Environmental Protection Agency, "National Primary Drinking Water Regulations; Synthetic Organic Chemicals, Inorganic Chemicals and Microorganisms, Proposed Rulemaking," *Fed. Reg.* 50:46936, 1985.

303. U.S. Environmental Protection Agency, "National Revised Primary Drinking Water Regulations, Volatile Synthetic Organic Chemicals in Drinking Water, Advance Notice of Proposed Rulemaking," *Fed. Reg.* 47:9350, 1982.

304. U.S. Environmental Protection Agency, "National Revised Primary Drinking Water Regulations," *Fed. Reg.* 48:45502, 1983.

305. U.S. Environmental Protection Agency, "National Revised Primary Drinking Water Regulations, Advanced Notice of Proposed Rulemaking," *Fed. Reg.* 48:45502, 1983.

306. U.S. Environmental Protection Agency, "Notice of the First Priority List of Hazardous Substances That Will Be the Subject of Toxicological Profiles," *Fed. Reg.* 52:12866, Apr. 17, 1987.

307. U.S. Environmental Protection Agency, "Notification Requirements, Reportable Quantity Adjustments," *Fed. Reg.* 50:13456, 1985.

308. U.S. Environmental Protection Agency, "Proposed Comprehensive Assessment Information Rule; Proposed Rule," *Fed. Reg.* 51:35762, Oct. 7, 1986.

309. U.S. Environmental Protection Agency, "Proposed Guidelines for Carcinogen Risk Assessment; Request for Comments," *Fed. Reg.* 49: 46294, Nov. 23, 1984.

310. U.S. Environmental Protection Agency, "Proposed Guidelines for Exposure Assessment; Request for Comments," *Fed. Reg.* 49:46304, Nov. 23, 1984.

311. U.S. Environmental Protection Agency, "Proposed Guidelines for Health Risk Assessment of Suspect Developmental Toxicants and Request for Comments," *Fed. Reg.* 49:46324, Nov. 23, 1984.

312. U.S. Environmental Protection Agency, "Proposed Guidelines for Health Risk Assessment of Chemical Mixtures and Request for Comments; Notice," *Fed. Reg.* 50:1170, Jan. 9, 1985.

313. U.S. Environmental Protection Agency, "Proposed Guidelines for Mutagenicity Risk Assessment; Request for Comments," *Fed. Reg.* 49: 46314, Nov. 23, 1984.

314. U.S. Environmental Protection Agency, "Reportable Quantity Adjustments," *Fed. Reg.* 52:8140, 1987.

315. U.S. Environmental Protection Agency, "Respondent's Brief in Support of Proposed Findings, Conclusions and Order, In Re: Stevens Industries, Inc., et. al. [consolidated DDT hearing]" I.F. & R. Docket Nos. 63 et seq., Apr. 5, 1972.

316. U.S. Environmental Protection Agency, "Superfund Programs; Reportable Quantity Adjustments," *Fed. Reg.* 51:34534, 1986.

317. U.S. Environmental Protection Agency, "Toxic Substances Control Act Interagency Testing Committee (ITC)—Background," typescript, December 1986.

318. U.S. Environmental Protection Agency, "Toxic Substances Control Act Test Guidelines; Final Rules," *Fed. Reg.* 50:39252, Sept. 27, 1985.

319. U.S. Environmental Protection Agency, "Toxic Substances: Glycidol and Its Derivatives; Response to the Interagency Testing Committee," *Fed. Reg.* 48:57562, 1983.

320. U.S. Environmental Protection Agency, "Toxic Substances; Submission of Lists and Copies of Health and Safety Studies on Certain Substances Subject to the 1984 RCRA Amendments," *Fed. Reg.* 51:2890, Jan. 22, 1986.

321. U.S. Environmental Protection Agency, "Unsubstituted Phenylenediamines, Proposed Test Rule," *Fed. Reg.* 51:472, 1986.

322. U.S. Environmental Protection Agency, "Water Pollution Control; National Primary Drinking Water Regulations," *Fed. Reg.* 52:20672, 1987.

323. U.S. Environmental Protection Agency, "Water Quality Criteria Documents; Availability," *Fed. Reg.* 45:79318, Nov. 28, 1980.

324. U.S. Environmental Protection Agency, Office of Air and Radiation, "National Emission Standards for Hazardous Air Pollutants; Policy and Procedures for Identifying, Assessing, and Regulating Airborne Substances Posing a Risk of Cancer," *Fed. Reg.* 44:58642, 1979.

325. U.S. Environmental Protection Agency, Office of Air and Radiation, "National Emission Standards for Hazardous Air Pollutants; Vinyl Chloride," *Fed. Reg.* 50:1183, Jan. 9, 1985.

326. U.S. Environmental Protection Agency, Office of Drinking Water, *Health Advisory for Aldicarb* (Washington, D.C. September 1985).

327. U.S. Environmental Protection Agency, Office of Emergency and Remedial Response and Office of Waste Programs Enforcement, "Guidance on Feasibility Studies Under CERCLA," EPA/540/G-85/003, June 1985.

328. U.S. Environmental Protection Agency, Office of Pesticide Programs, Hazard Evaluation Division, personal communication, June 1986.

329. U.S. Environmental Protection Agency, Office of Pesticides Programs, Hazard Evaluation Division, Standard Evaluation Procedure, *Oncogenicity Potential: Guidance for Analysis and Evaluation of Long-Term Rodent Studies*, prepared by O.E. Paynter, EPA-540/9-85-019 (Washington, DC, June 1985).

330. U.S. Environmental Protection Agency, Office of Pesticide Programs, Hazard Evaluation Group, personal communication, June 1986.

331. U.S. Environmental Protection Agency, Office of Pesticide Programs, "Report on the Status of Chemicals in the Special Review Program, Registration Standards Program, and Data Call-in Program," typescript, March 1986.

332. U.S. Environmental Protection Agency, Office of Pesticides and Toxic Substances, *Pesticide Assessment Guidelines, Subdivision F, Hazard Evaluation: Human and Domestic Animals*, Series 83: Chronic and Long-term Studies, EPA-540/9-82-025 (Washington, DC, October 1982).

333. U.S. Environmental Protection Agency, Office of Research and Development, Environmental Research Laboratory, *Frequency of Organic Compounds Identified in Water*, EPA-600/4-76-062, December 1976.

334. U.S. Environmental Protection Agency, Office of Solid Waste and Emergency Response, *The New RCRA: A Fact Book*, Environmental Protection Agency, (Washington, DC, October 1985).

335. U.S. Environmental Protection Agency, Office of Toxic Substances, "Chemical Selection Process—1986," typescript, June 11, 1986.

336. U.S. Environmental Protection Agency, Office of Toxic Substances, "CSB Existing Chemical Assessment Tracking System: CHIP Report," Mar. 19, 1987, printout.

337. U.S. Environmental Protection Agency, Office of Toxic Substances, "CSB Existing Chemical Assessment Tracking System: CHIP Report," Mar. 19, 1987, provided to OTA.

338. U.S. Environmental Protection Agency, Office of Toxic Substances, "Procedures Governing Testing Consent Agreements and Test Rules Under the Toxic Substances Control Act; Interim Final Rule," *Fed. Reg.* 51:23706, June 30, 1986.

339. U.S. Environmental Protection Agency, Office of Toxic Substances, "TSCA Report to Congress for Fiscal Year 1986," typescript, June 1987.

340. U.S. Environmental Protection Agency, Office of Toxic Substances, "TSCA Section 8(e): Overview, Outreach, and Benefits of Implementation," typescript, May 1, 1986.

341. U.S. Environmental Protection Agency, Office of Toxic Substances, comments on unpublished OTA draft report, "Identifying and Regulating Carcinogens" (1987).

342. U.S. Environmental Protection Agency, Office of Toxic Substances, Exposure Evaluation Division, *Methods for Assessing Exposure to Chemical Substances*, vol. 1: Introduction, EPA-560/5-85-001 (Washington, DC, July 1985).

343. U.S. Environmental Protection Agency, Office of Toxic Substances, personal communication, 1986.

344. U.S. Environmental Protection Agency, Office of Toxic Substances, "Health and Safety Data Reporting; Submission of Lists and Copies of Health and Safety Studies on Certain Substances, Final Rule," *Fed. Reg.* 52:16022, 1987.

345. U.S. Environmental Protection Agency, Office of Water Regulations and Standards, "Numeric Criteria for Toxic Pollutants in State Water Quality Standards," typescript, Apr. 28, 1986.

346. U.S. Environmental Protection Agency, response to OTA request for information, July 1986.

347. U.S. Interagency Regulatory Liaison Group, "Scientific Bases for Identification of Potential Carcinogens and Estimation of Risks," *J. Nat. Cancer Inst.* 63(1):241-268, July 1979, and *Fed. Reg.* 44:39858, July 6, 1979.

348. U.S. National Cancer Advisory Board, Subcommittee on Environmental Carcinogenesis, "General Criteria for Assessing the Evidence for Carcinogenicity of Chemical Substances," *J. Nat. Cancer Inst.* 58:461, February 1977.

349. U.S. Office of Management and Budget, Comments on OSHA's Proposed Rulemaking: Occupational Exposure to Formaldehyde, March 1986.

350. U.S. Office of Management and Budget, *Regulatory Program of the United States Government: Apr. 1, 1986-Mar. 31, 1987* (Washington, DC: U.S. Government Printing Office, 1986).

351. U.S. Office of Science and Technology Policy, "Chemical Carcinogens: A Review of the Science and Its Associated Principles, February 1985," *Fed. Reg.* 50:10372, Mar. 14, 1985; reprinted in *Environ. Health Perspectives* 67:201, 1986.

352. U.S. Regulatory Council, "Statement on Regulation of Chemical Carcinogens; Policy and Request for Public Comments," *Fed. Reg.* 44:60038-49, Oct. 17, 1979.

353. Ulsamer, A.G., White, P.D., and Preuss, P.W., "Evaluation of Carcinogens: Perspective of the Consumer Product Safety Commission," *Handbook of Carcinogen Testing*, H.A. Milman and E.K. Weisburger (eds.) (Park Ridge, NJ: Noyes Publications, 1985).

354. *United States* v. *GAF Corp.*, Federal Supplement 389 (1975), p. 1379.

355. Warnick, H., Office of Pesticide Programs, U.S. Environmental Protection Agency, personal communication, July 1986.

356. Weinstein, I.B., letter, *Science* 219:794-796, 1983.

357. Weisburger, E.K., "History of the Bioassay Program of the National Cancer Institute," *Prog. Exp. Tumor Res.* 26:187-201, 1983.

358. Weisburger, J.H., and Williams, G.M., "Chemical Carcinogens," *Toxicology: The Basic Science of Poisons*, J. Doull, C.D. Klassen, and M.O. Amdur (eds.) 2d ed. (New York: MacMillan, 1980).

359. Weisburger, J.H., and Williams, G.M., "Carcinogen Testing: Current Problems and New Approaches," *Science* 214:401, 1981.

360. Wilbourn, J., Haroun, L., Hesletine, E., et al., "Response of Experimental Animals to Human Carcinogens: An Analysis Based Upon the IARC Monographs Program," *Carcinogenesis* 7:1853, 1986.

361. Wilk, A., Food and Drug Administration, National Toxicology Program, Chemical Evaluation Committee, comments, minutes, meeting of Sept. 16, 1986.

362. Wines, M., "Scandals at EPA May Have Done In Reagan's Move To Ease Cancer Controls," *National Journal* 15(25):1264, June 18, 1983.

363. Wolfe, S., statement presented at public meeting on National Toxicology Program *Annual Report*, Washington, DC, Apr. 21, 1987.

364. Wolfe, S.M., "Standards for Carcinogens: Science Affronted by Politics," *Origins of Human Cancer*, H.H. Hiatt, J.D. Watson, and J.A. Winsten (eds.) (Cold Spring Harbor, NY: Cold Spring Harbor Laboratory, 1977).

365. Wrenn, G.C., Director of Health Standards Programs, U.S. Occupational Safety and Health Administration, quoted in "Cracking Down on the Causes of Cancer," T.B. Clark, *National Journal* 10:2056, Dec. 30, 1978.

366. Young, F., Commissioner, Food and Drug Administration, "Testimony before U.S. House of Representatives Committee on Agriculture, Subcommittee on Department Operations, Research, and Foreign Agriculture," November 1986.

367. Zeiger, E., Tainer, B., Haworth, S., et al., "Salmonella Mutagenicity Tests, AIII: Results From the Testing of 255 Chemicals," *Environ. Mutagen.* (in press).

Printed and bound by CPI Group (UK) Ltd, Croydon, CR0 4YY

22/10/2024

01777600-0020